I0484281

National Management Measures for the Control of Nonpoint Pollution from Agriculture

U.S. Environmental Protection Agency
Office of Water (4503T)
1200 Pennsylvania Avenue, NW
Washington, D.C. 20460

EPA-841-B-03-004

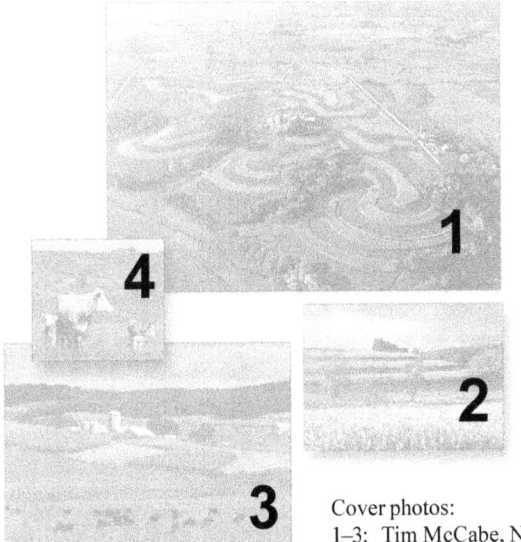

Cover photos:
1–3: Tim McCabe, Natural Resources Conservation Service
4: Lynn Betts, Natural Resources Conservation Service

Disclaimer

This document provides guidance to States, Territories, authorized Tribes, and the public regarding management measures that may be used to reduce nonpoint source pollution from agricultural activities. This document refers to statutory and regulatory provisions which contain legally binding requirements. This document does not substitute for those provisions or regulations, nor is it a regulation itself. Thus, it does not impose legally-binding requirements on EPA, States, Territories, authorized Tribes, or the public and may not apply to a particular situation based upon the circumstances. EPA, State, Territory, and authorized Tribe decision makers retain the discretion to adopt approaches on a case-by-case basis that differ from this guidance where appropriate. Interested parties are free to raise questions and objections about the appropriateness of the application of the guidance to a particular situation, and EPA will consider whether or not the recommendations in this guidance are appropriate in that situation. EPA may change this guidance in the future.

Acknowledgments

Steven A. Dressing, formerly of the Nonpoint Source Control Branch, Office of Water, U.S. Environmental Protection Agency, Washington, DC, was the primary author of this guidance document.
Many individuals assisted in this effort, including the following:

John Kosco, formerly of the Municipal Support Division, Office of Water, U.S. EPA, Washington, DC
Thomas Davenport, Region 5, U.S. EPA, Chicago, IL
David Rathke, Region 8, U.S. EPA, Denver, CO
Don Meals, consultant, Burlington, VT
Tommy C. Daniel, Department of Agronomy, University of Arkansas, Fayetteville, AR
Brent Hallock, Soil Science, Department, California Polytechnic State University, San Luis, Obispo, CA
Ray Knighton, USDA-CRES, Washington, D.C., formerly of the Soil Science Department, North Dakota State University, Fargo, ND
Jerry Hatfield, USDA-ARS, Washington, DC
Robert Goo, Nonpoint Source Control Branch, Office of Water, U.S. EPA, Washington, DC
Roger Dean, Region 8, U.S. EPA, Denver, CO
Amy Sosin, Department of Justice, Washington, DC
Chris Laabs, Watershed Branch, Office of Water, U.S. EPA, Washington, DC
Joan Warren, Watershed Branch, Office of Water, U.S. EPA, Washington, DC
Kristen Martin Dors, formerly of Region 6, U.S. EPA, Dallas, TX
Steven W. Coffey, Division of Soil and Water Conservation, NC Department of Environment and Natural Resources, Raleigh, NC
Judith A. Gale, Galeforce Consulting, Raleigh, NC
Richard E. Phillips, retired, Biological and Agricultural Engineering Department, North Carolina State University, Raleigh, NC
Ron Marlow, USDA-NRCS, Washington, DC
Alan Dixon, Registration Support Branch, Office of Pesticide Programs, Washington, DC
Sharon Buck, formerly of the Nonpoint Source Control Branch, Office of Water, U.S. EPA, Washington, DC
Stuart Lehman, Nonpoint Source Control Branch, Office of Water, U.S. EPA, Washington, DC
Katie Flahive, Nonpoint Source Control Branch, Office of Water, U.S. EPA, Washington, DC

The following team from North Carolina State University, Raleigh, NC, contributed as subcontractors to Tetra Tech, Inc., providing much of the writing and editing:

Laura Lombardo, Daniel E. Line, Garry L. Grabow, Jean Spooner, Terry W. Pollard, Janet M. Young and Catherine Scache, of the NCSU Water Quality Group, Deanna L. Osmond and Rich McLaughlin of the Soil Science Department, and Frank J. Humenik, Animal Waste Management Programs, College of Agriculture and Life Sciences. In addition, George Townsend and Leslie Shoemaker from Tetra Tech, Inc. provided valuable contributions.

Public comment was solicited in the Federal Register, October 17, 2000, on the draft version of the guidance. Comments were received from approximately 50 individuals. These comments were valuable in making this a better document and EPA appreciates the efforts of these individuals.

Table of Contents

List of Figures and Tables

Introduction

The nation's aquatic resources are among its most valuable assets. While environmental protection programs in the United States have successfully improved water quality during the past 25 years, many challenges still remain. Although significant strides have been made in reducing the impacts of discrete pollutant sources, aquatic ecosystems remain impaired, primarily due to complex pollution problems caused by nonpoint source (NPS) pollution.

The most recent national water quality inventory shows that, as of 2000, 39% of assessed stream miles, 45% of assessed lake acres, and 51% of assessed estuary acres are impaired. The leading causes of impairment are nutrients, siltation, metals, and pathogens. State inventories indicate that agriculture, including crop production, animal operations, pastures, and rangeland, impacts 18% of the total river and stream miles assessed, or 48% of the river and streams identified as impaired (EPA, 2002).

> Agriculture is listed as a source of pollution for 48% of the impaired river miles reported in the United States.

The Purpose and Scope of this Guidance

This guidance document is intended to provide technical information to state program managers and others on the best available, economically achievable means of reducing NPS pollution of surface and ground water from agriculture. The guidance provides background information about agricultural NPS pollution, where it comes from and how it enters the nation's waters, discusses the broad concept of assessing and addressing water quality problems on a watershed level, and presents up-to-date technical information about how to reduce agricultural NPS pollution. This document is not intended to be a "how to" technical guide for natural resource assessment, planning, design, and implementation.

The causes of agricultural NPS pollution, specific pollutants of concern, and general approaches to reducing the impact of such pollutants on aquatic resources are discussed in the Overview (Chapter 2). A general discussion of best management practices (BMPs) and the use of combinations of individual practices (BMP systems) to protect surface and ground water is given in Chapter 3. Management measures for nutrient management; pesticide management; erosion and sediment control; managing facility wastewater, manure and runoff from animal feeding operations; grazing management; and irrigation water management are described in Chapter 4. Also in Chapter 4 are discussions of BMPs that can be used to achieve the management measures, including cost and effectiveness information. Chapter 5 summarizes watershed planning principles, and Chapters 6 and 7 give overviews of nonpoint source monitoring and pollutant load estimation, respectively.

> This guidance is designed to provide current information to state program managers on controlling agricultural nonpoint source pollution.

While the scope of this guidance is broad, covering diverse agricultural NPS pollutants from a range of sources, there are a number of issues that are not covered. Such issues include nutrient transfer over long distances (e.g., the

This document does not impose legally-binding requirements on EPA, the states, or the public.

This guidance does NOT replace the 1993 *Guidance Specifying Management Measures for Sources of Nonpoint Pollution in Coastal Waters*.

shipping of feed from one state to another in which the resulting animal waste is then applied to fields), animal nutrition (e.g., changing the nutrient mix fed to livestock as an approach to managing nutrients in animal waste), alternatives for manure (such as composting or regional distribution of manure from farms that do not need it to farms that can use it), odor control, and methane production. Furthermore, because it is national in scope, this document cannot address all practices or techniques specific to local or regional soils, climate, or agronomic conditions. In addition, new BMPs are being developed as a result of ongoing agricultural research. Readers should consult with state or local agencies including the United States Department of Agriculture (USDA)–Natural Resources Conservation Service (NRCS), Cooperative Extension, land grant universities, conservation districts, and agricultural organizations for additional information on agricultural nonpoint source pollution controls applicable to their local area.

This document provides guidance to states, territories, authorized tribes, and the public regarding management measures that may be used to reduce nonpoint source pollution from agricultural activities. This document refers to statutory and regulatory provisions which contain legally binding requirements. This document does not substitute for those provisions or regulations, nor is it a regulation itself. Thus, it does not impose legally-binding requirements on EPA, states, territories, authorized tribes, or the public and may not apply to a particular situation based upon the circumstances. EPA, state, territory, and authorized tribe decision makers retain the discretion to adopt approaches on a case-by-case basis that differ from this guidance where appropriate. EPA may change this guidance in the future.

Readers should note that this guidance is entirely consistent with the *Guidance Specifying Management Measures for Sources of Nonpoint Pollution in Coastal Waters* (EPA, 1993a) published under Section 6217 of the Coastal Zone Act Reauthorization Amendments of 1990 (CZARA). This guidance, however, does not supplant or replace the 1993 coastal management measures guidance for the purpose of implementing programs under Section 6217.

Under CZARA, states that participate in the Coastal Zone Management Program under the Coastal Zone Management Act are required to develop coastal nonpoint pollution control programs that ensure the implementation of EPA's management measures in their coastal management area. The 1993 guidance continues to apply to that program.

This document modifies and expands upon supplementary technical information contained in the Coastal Management Measures Guidance both to reflect circumstances relevant to differing inland conditions and to provide current technical information. It does not set new or additional standards for either CZARA Section 6217 Coastal Nonpoint Pollution Control Programs or Clean Water Act Section 319 Nonpoint Source Management Programs. It does, however, provide information that can be used by government agencies, private sector groups, and individuals to understand and apply measures and practices to address agricultural sources of nonpoint source pollution.

What is Nonpoint Source Pollution?

Nonpoint source pollution generally results from precipitation, land runoff, infiltration, drainage, seepage, hydrologic modification, or atmospheric deposition. As runoff from rainfall or snowmelt moves, it picks up and transports natural pollutants and pollutants resulting from human activity, ultimately depositing them into rivers, lakes, wetlands, coastal waters, and ground water. Technically, the term *nonpoint source* is defined to mean any source of water pollution that does not meet the legal definition of *point source* in Section 502(14) of the Clean Water Act of 1987:

> The term **point source** means any discernible, confined and discrete conveyance, including but not limited to any pipe, ditch, channel, tunnel, conduit, well, discrete fissure, container, rolling stock, concentrated animal feeding operation, or vessel or other floating craft from which pollutants are or may be discharged. This term does not include agricultural stormwater discharges and return flows from irrigated agriculture.

Although diffuse runoff is generally treated as nonpoint source pollution, runoff that enters and is discharged from conveyances such as those described above is treated as a point source discharge and hence is subject to the permit requirements of the Clean Water Act. In contrast, nonpoint sources are not subject to federal permit requirements. Point sources generally enter receiving water bodies at some identifiable site(s) and carry pollutants whose generation is controlled by some internal process or activity, rather than weather. Point source discharges such as municipal and industrial waste waters, runoff or leachate from solid waste disposal sites and concentrated animal feeding operations, and storm sewer outfalls from large urban centers are regulated and permitted under the Clean Water Act.

While it is imperative that water program managers understand and manage in accordance with legal definitions and requirements, the non-legal community often characterizes nonpoint sources in the following ways:

- Nonpoint source discharges enter surface and/or ground waters in a diffuse manner at intermittent intervals related mostly to meteorological events.

- Pollutant generation arises over an extensive land area and moves overland before it reaches surface waters or infiltrates into ground waters.

- The extent of NPS pollution is related to uncontrollable climatic events and to geographic and geologic conditions and varies greatly from place to place and from year to year.

- The extent of NPS pollution is often more difficult or expensive to monitor at the point(s) of origin, as compared to monitoring of point sources.

- ❏ Abatement of nonpoint sources is focused on land and runoff management practices, rather than on effluent treatment.

- ❏ Nonpoint source pollutants may be transported and/or deposited as airborne contaminants.

Nonpoint source pollutants that cause the greatest impacts are sediments, nutrients, toxic compounds, organic matter, and pathogens. Hydrologic modification can also cause adverse effects on the biological, physical, and chemical integrity of surface and ground waters.

Section 319 requires states to assess NPS pollution and implement management programs.

National Efforts to Address Nonpoint Source Pollution

Nonpoint Source Program — Section 319 of the Clean Water Act

During the first 15 years of the national program to abate and control water pollution (1972–1987), EPA and the states focused most of their water pollution control activities on traditional point sources. These point sources are regulated by EPA and the states through the National Pollutant Discharge Elimination System (NPDES) permit program established by Section 402 of the 1972 Federal Water Pollution Control Act (Clean Water Act). Discharges of dredged and fill materials into wetlands have also been regulated by the U.S. Army Corps of Engineers and EPA under Section 404 of the Clean Water Act.

As a result of the above activities, the nation has greatly reduced pollutant loads from point source discharges and has made considerable progress in restoring and maintaining water quality. However, the gains in controlling point sources have not solved all of the nation's water quality problems. Recent studies and surveys by EPA and by states, tribes, territories, and other entities, indicate that the majority of the remaining water quality impairments in our nation's rivers, streams, lakes, estuaries, coastal waters, and wetlands result from NPS pollution and other nontraditional sources, such as urban storm water discharges and combined sewer overflows.

Section 319 authorizes EPA to provide grants to assist state NPS pollution control programs.

In 1987, in view of the progress achieved in controlling point sources and the growing national awareness of the increasingly dominant influence of NPS pollution on water quality, Congress amended the Clean Water Act to provide a national framework to address nonpoint source pollution. Under this amended version, referred to as the 1987 Water Quality Act, Congress revised Section 101, "Declaration of Goals and Policy," to add the following fundamental principle:

> It is the national policy that programs for the control of nonpoint sources of pollution be developed and implemented in an expeditious manner so as to enable the goals of this Act to be met through the control of both point and nonpoint sources of pollution.

More importantly, Congress enacted Section 319 of the 1987 Water Quality Act, which established a national program to address nonpoint sources of water pollution. Under Section 319, states address NPS pollution by assessing NPS pollution problems and causes within the state and implementing management programs to control the NPS pollution. Section 319 authorizes EPA to issue grants to states to assist them in implementing management programs or portions of management programs which have been approved by EPA. For additional information and a list of state contacts, see www.epa.gov/owow/nps.

National Estuary Program

EPA also administers the National Estuary Program under Section 320 of the Clean Water Act. This program focuses on point and NPS pollution in geographically targeted, high-priority estuarine waters. In this program, EPA assists state, regional, and local governments in developing and implementing comprehensive conservation and management plans that recommend priority corrective actions to restore estuarine water quality, fish populations, and other designated uses of the waters.

Pesticides Program

Another program administered by EPA that controls some forms of NPS pollution is the pesticides program under the Federal Insecticide, Fungicide, and Rodenticide Act (FIFRA). Among other provisions, this program authorizes EPA to control pesticides that may threaten ground and surface water. FIFRA provides for the registration of pesticides and enforceable label requirements, which may include maximum rates of application, restrictions on use practices, and classification of pesticides as "restricted use" pesticides (which restricts use to certified applicators trained to handle toxic chemicals).

Coastal Nonpoint Pollution Control Program

In November 1990, Congress enacted the Coastal Zone Act Reauthorization Amendments (CZARA). These amendments were intended to address several concerns, including the impact of NPS pollution on coastal waters.

The Federal Coastal Nonpoint Pollution Control Program (6217) is designed to enhance state and local efforts to manage land use activities that degrade coastal habitats and waters.

To more specifically address the impacts of NPS pollution on coastal water quality, Congress enacted Section 6217, *Protecting Coastal Waters* (codified as 16 U.S.C. Section 1455b). Section 6217 provides that each state with an approved Coastal Zone Management Program must develop and submit to EPA and the National Oceanic and Atmospheric Administration (NOAA) for approval a Coastal Nonpoint Pollution Control Program. The purpose of the program "shall be to develop and implement management measures for nonpoint source pollution to restore and protect coastal waters, working in close conjunction with other state and local authorities."

Coastal Nonpoint Pollution Control Programs are not intended to supplant existing coastal zone management programs and NPS management programs. Rather, they are intended to serve as an update and expansion of existing NPS management programs and are to be coordinated closely with the coastal zone management programs that states and territories are already implementing pursuant to the Coastal Zone Management Act of 1972. The legislative history indicates that the central purpose of Section 6217 is to strengthen the links between federal and state Coastal Zone Management and Water Quality Programs and to enhance state and local efforts to manage land use activities that degrade coastal waters and habitats.

Section 6217(g) of CZARA requires EPA to publish, in consultation with NOAA, the U.S. Fish and Wildlife Service, and other federal agencies, "guidance for specifying management measures for sources of nonpoint pollution in coastal waters." Management measures are defined in Section 6217(g)(5) as:

> economically achievable measures for the control of the addition of pollutants from existing and new categories and classes of nonpoint sources of pollution, which reflect the greatest degree of pollutant reduction achievable through the application of the best available nonpoint source control practices, technologies, processes, siting criteria, operating methods, or other alternatives.

EPA published *Guidance Specifying Management Measures for Sources of Nonpoint Source Pollution in Coastal Waters* (EPA, 1993a). In EPA's (1993a) document, management measures for urban areas; agricultural sources; forestry; marinas and recreational boating; hydromodification (channelization and channel modification, dams, and streambank and shoreline erosion); and wetlands, riparian areas, and vegetated treatment systems were defined and described. The management measures for controlling agricultural NPS pollution discussed in Chapter 4 of this document are based on those outlined by EPA (1993a).

Source Water Protection Program

The 1996 Amendments to the Safe Drinking Water Act provided for source water assessment and protection programs to prevent drinking water contamination. States are required to develop comprehensive Source Water Assessment Programs (SWAPs) that will: identify the areas that supply public tap water; inventory contaminants and assess water system susceptibility to contamination and inform the public of the results. EPA is responsible for the review and approval of state SWAPs. Several programs specifically address ground water protection.

In selected watersheds, the RCWP showed that implementation of agricultural BMPs improved water quality.

Rural Clean Water Program (RCWP)

The Rural Clean Water Program (RCWP), an NPS pollution control program implemented by USDA and EPA, was conducted from 1980 to 1990 as an experimental effort to address agricultural NPS pollution in watersheds across the country.

The objectives of the RCWP were to:

☐ Achieve improved water quality in the approved project area in the most cost-effective manner possible while providing food, fiber, and a quality environment;

❏ Assist agricultural landowners and farm operators in reducing agricultural NPS water pollutants and improving water quality in rural areas to meet water quality standards or goals; and

❏ Develop and test programs, policies, and procedures for the control of agricultural NPS pollution.

Twenty-one experimental projects were funded across the United States. Each project included implementation of BMPs to reduce NPS pollution and water quality monitoring to evaluate the effects of BMPs. The BMPs were targeted to critical areas in each project — sources of NPS pollutants identified as having significant impacts on the impaired water resource. Landowner participation was voluntary, with cost-sharing and technical assistance offered as incentives for implementing BMPs.

The linkage of water quality monitoring to land treatment efforts in the RCWP helped improve targeting of BMPs to sources most in need of treatment. Water quality findings from the RCWP projects were also used to adjust and refine agricultural NPS programs and BMPs. Additional details are available in the project evaluation report (EPA, 1993c).

2002 Farm Bill Conservation Provisions

Technical and financial assistance for landowners seeking to conserve, improve, and sustain our soil and other natural resources is authorized by the federal government under provisions of the Food Security and Rural Investment Act (Farm Bill). The following sections summarize provisions in the 2002 Act relating directly to installation and maintenance of BMPs. For additional information, see the U.S. Department of Agriculture's website at www.usda.gov.

Environmental Quality Incentives Program (EQIP) — The EQIP was established by the 1996 Farm Bill to provide a voluntary conservation program for farmers and ranchers who face serious threats to soil, water, and related natural resources. Funding increases are authorized from $200 million to $1.1 billion between 2002 and 2007. EQIP offers financial, technical, and educational help to install or implement structural, vegetative, and management practices designed to conserve soil and other natural resources. The law dictates that 60% of the available monies be directed to livestock-related concerns. Cost-sharing generally pays up to 75% of the costs for certain conservation practices. Incentive payments may be made to encourage producers to perform land management practices such as nutrient management, manure management, integrated pest management, irrigation water management, and wildlife habitat management. Cost-share for construction of animal waste management facilities is now allowed for livestock operations over 1,000 animal units.

> Many Farm Bill programs provide funds for land treatment. Please contact your state or local USDA office for details.

Conservation Reserve Program (CRP) — First authorized by the Food Security Act of 1985 (Farm Bill), this is a voluntary program that offers annual rental payments, incentive payments, and cost-share assistance for establishing long-term, resource-conserving cover crops on highly erodible land. Conservation Reserve Program contracts are issued for a duration of 10 to 15 years for up to 39.2 million acres of cropland and marginal pasture. Land can be accepted into the CRP through a competitive bidding process where all offers are ranked using an environmental benefits index, or through continuous sign-up for eligible lands where certain special conservation practices (e.g. filter strips and riparian buffers) will be implemented.

Conservation Security Program — This 2002 Farm Bill program provides incentive payments to producers who adopt or maintain existing conservation practices. Producers may receive up to 20,000, 35,000, or 45,000 dollars per year for practice falling into 3 tiers. The higher payments go to the more comprehensive sets of practices. The program contracts are for 5 to 10 years.

The Conservation Reserve Enhancement Program (CREP) is a 1996 initiative continued in the 2002 Farm Bill. CREP is a joint, state-federal program designed to meet specific conservation objectives. CREP targets state and federal funds to achieve shared environmental goals of national and state significance. The program uses financial incentives to encourage farmers and ranchers to voluntarily protect soil, water, and wildlife resources.

Wetlands Reserve Program (WRP) — The WRP is a voluntary program to restore and protect wetlands and associated lands. Participants may sell a permanent or 30-year conservation easement or enter into a 10-year cost-share agreement with USDA to restore and protect wetlands. The landowner voluntarily limits future use of the land, yet retains private ownership. The NRCS provides technical assistance in developing a plan for restoration and maintenance of the land. The landowner retains the right to control access to the land and may lease the land for hunting, fishing, and other undeveloped recreational activities. The acreage is expanded by 1.2 million acres to 2.275 million acres in 2002.

Wildlife Habitat Incentives Program (WHIP) — This program is designed for people who want to develop and improve wildlife habitat on private lands. Plans are developed in consultation with the NRCS and local Conservation District. USDA will provide technical assistance and cost-share up to 75% of the cost of installing the wildlife practices. Participants may get bonus payments for agreements over 15 years.

Forest Land Enhancement Program (FLEP) — Authorized in the 2002 Farm Bill, the FLEP creates a new title for Forestry. It replaces and expands the Stewardship Incentive program and Forestry program. The new Forest Land Enhancement program will provide up to $100 million over six years to private, non-industrial Forest owners. The new title also provides $210 million to help fight fire on private land and address prevention.

Grazing Reserve Program (GRP) — This 2002 provision will use 30 year easements and rental agreements to improve management of up to 2 million acres of private grazing land. 500,000 acres are to be reserved for protected tracts of 40 acres or less as native grasslands. Restoration costs may go as high as 75%.

Funding Sources

For information on sources of funding to address nonpoint source pollution, see EPA's Nonpoint Source website at www.epa.gov/owow/nps/funding.html.

Overview

2

Agricultural Nonpoint Source Pollution

State water quality assessments continue to show that nonpoint source pollution is the leading cause of impairments in surface waters of the U.S. According to these assessments, agriculture is the most wide-spread source of pollution for assessed rivers and lakes. Agriculture impacts 18% of assessed river miles and 14% of assessed lake acres. The state reports also indicate that agriculture impacts 48% of impaired river miles and 41% of impaired lake acres (EPA, 2002).

The primary agricultural NPS pollutants are nutrients, sediment, animal wastes, salts, and pesticides. Agricultural activities also have the potential to directly impact the habitat of aquatic species through physical disturbances caused by livestock or equipment. Although agricultural NPS pollution is a serious problem nationally, a great deal has been accomplished over the past several decades in terms of sediment and nutrient reduction from privately-owned agricultural lands. Much has been learned in the recent past about more effective ways to prevent and reduce NPS pollution from agricultural activities.

The purpose of this chapter is to describe the general causes of agricultural NPS pollution, the specific pollutants and problems of concern, and the general approaches that have been found most effective in reducing the impact of such pollutants and problems on aquatic resources.

Nutrients

Nitrogen (N) and phosphorus (P) are the two major nutrients from agricultural land that degrade water quality. Nutrients are applied to agricultural land in several different forms and come from various sources, including:

❐ Commercial fertilizer in a dry or fluid form, containing nitrogen, phosphorus, potassium (K), secondary nutrients, and micronutrients;

❐ Manure from animal production facilities including bedding and other wastes added to the manure, containing N, P, K, secondary nutrients, micronutrients, salts, some metals, and organics;

❐ Municipal and industrial treatment plant sludge, containing N, P, K, secondary nutrients, micronutrients, salts, metals, and organic solids;

❐ Municipal and industrial treatment plant effluent, containing N, P, K, secondary nutrients, micronutrients, salts, metals, and organics;

❐ Legumes and crop residues containing N, P, K, secondary nutrients, and micronutrients;

❐ Irrigation water;

❐ Wildlife; and

❐ Atmospheric deposition of nutrients such as nitrogen, phosphorus, and sulphur.

Commercial fertilizers and manure are the primary sources of crop nutrients for agriculture.

In addition, decomposition of organic matter and crop residue may be a source of mobile forms of nitrogen, phosphorus, and other essential crop nutrients.

Surface water runoff from agricultural lands may transport the following pollutants:

❑ Particulate-bound nutrients, chemicals, and metals, such as phosphorus, organic nitrogen, and metals applied with some organic wastes;

❑ Soluble nutrients and chemicals, such as nitrogen, phosphorus, metals, and many other major and minor nutrients;

❑ Particulate organic solids, oxygen-demanding material, and bacteria, viruses, and other microorganisms applied with some organic waste; and

❑ Salts.

Ground water infiltration from agricultural lands to which nutrients have been applied may transport the following pollutants:

❑ Soluble nutrients and chemicals, such as nitrogen, phosphorus, metals;

❑ Other major and minor nutrients;

❑ Salts; and

❑ Bacteria and other pathogens applied with some organic waste.

> Overloading with nitrogen and phosphorus causes eutrophication which reduces the suitability of waterways for beneficial uses.

All plants require nutrients for growth. Nitrogen and phosphorus generally are present in aquatic environments at background or natural levels below 0.3 and 0.01 mg/L, respectively. When these nutrients are introduced into a stream, lake, or estuary at higher rates, aquatic plant productivity may increase dramatically. This process, referred to as cultural eutrophication, may adversely affect the suitability of the water for other uses.

Excessive aquatic plant productivity results in the addition to the system of more organic material, which eventually dies and decays. Bacteria decomposing this organic matter produce unpleasant odors and deplete the oxygen supply available to other aquatic organisms. Depleted oxygen levels, especially in colder bottom waters where dead organic matter tends to accumulate, can reduce the quality of fish habitat and encourage the propagation of fish that are adapted to less oxygen or to warmer surface waters. Anaerobic conditions can also cause the release of additional nutrients from bottom sediments.

Highly enriched waters will stimulate algae production, consequently increasing turbidity and color. In addition, certain algae can produce severe taste and odor problems that impair the quality of drinking water sources (EPA, 1999a). For example, the City of Tulsa, OK spends an additional $100,000 a year to correct taste and odor problems, resulting from extreme algae growth in the city's drinking water source (Lassek, 1997). Excess algae growth may also interfere with recreational activities such as swimming and boating. Algae growth is also believed to be harmful to coral reefs (e.g., Florida coast). Furthermore, the increased turbidity results in less sunlight penetration and availability to submerged aquatic vegetation (SAV). Since SAV provides habitat for small or juvenile fish, the loss of SAV has severe consequences for the food chain. Tampa Bay is an example in which nutrients are believed to have contributed to SAV loss.

Nitrogen

All forms of transported nitrogen are potential contributors to eutrophication in lakes, estuaries, and some coastal waters. In general, though not in all cases, nitrogen availability is the limiting factor for plant growth in marine ecosystems. Thus, the addition of nitrogen can have a significant effect on the natural functioning of marine ecosystems.

Eutrophication in coastal waters has been linked to increased nutrient loads from rivers, as evidenced by increasing incidence of noxious algal blooms and hypoxia in bottom waters (Justic et al., 1995.) The Gulf of Mexico has experienced midsummer hypoxia (low dissolved oxygen) since the early 1970s. From 1993 through 1999, the extent of bottom-water hypoxia ranged from about 6,200 to 7,700 square miles (16,000 to 20,000 km2), greater than twice the surface area of the Chesapeake Bay (Rabalais et al., 1999). The hypoxia is thought to be due to eutrophication resulting from high nutrient loading to the Gulf. Recent analysis has shown that about 89 percent of the annual total nitrogen flux to the Gulf (1.57 million metric tons) was from nonpoint sources, and the remaining 11 percent was from municipal and industrial point sources (Goolsby and Battaglin, 2000).

The toxic dinoflagellate *Pfiesteria piscicida*, implicated in causing about 50% of the major fish kills in North Carolina's estuaries and coastal waters from 1991 to 1993, has been linked to conditions of over-enrichment of nutrients such as nitrogen and phosphorus (Burkholder, 1996). More research is needed to determine the specific physical, chemical, and biological factors that promote outbreaks of *Pfiesteria piscicida.* Pfiesteria-like species have also been tracked to eutrophic sudden-death fish kill sites in estuaries, coastal waters, and aquaculture facilities from the mid-Atlantic through the Gulf Coast (Burkholder et al., 1995).

In addition to eutrophication, excessive nitrogen causes other water quality problems. Dissolved ammonia at concentrations above 0.2 mg/L may be toxic to fish, especially trout. Also, nitrates in drinking water are potentially dangerous to newborn infants. Nitrate is converted to nitrite in the digestive tract, which reduces the oxygen-carrying capacity of the blood (methemoglobinemia), resulting in brain damage or even death. The U.S. Environmental Protection Agency has set a limit of 10 mg/L nitrate-nitrogen in water used for human consumption (EPA, 1989a).

> Excessive ammonia can be toxic to fish.

Nitrogen is naturally present in soils but must be added to meet crop production needs. Nitrogen is added to the soil primarily by applying commercial fertilizers and manure, but also by growing legumes (biological nitrogen fixation) and incorporating crop residues. Not all nitrogen that is present in or on the soil is available for plant use at any one time. Applied nitrogen may be stored in the soil as organic material, soil organic matter (humus), or adsorbed to soil particles. For example, in the eastern Corn Belt, it is normally assumed that about 50% of applied nitrogen is assimilated by crops during the year of application (Nelson, 1985). Organic nitrogen normally constitutes the majority of the soil nitrogen. It is slowly converted (2 to 3% per year) to the more readily plant-available inorganic ammonium or nitrate. Nitrogen conversions are governed by carbon to nitrogen rations of crop residue and environmental conditions (e.g., temperature, moisture).

The chemical form of nitrogen affects its impact on water quality. The most biologically important inorganic forms of nitrogen are ammonium (NH4-N), nitrate (NO3-N), and nitrite (NO2-N). Organic nitrogen occurs as particulate matter, in living organisms, and as detritus. It occurs in dissolved form in compounds such as amino acids, amines, purines, and urea.

> Nitrate-nitrogen can readily leach below the root zone into shallow ground water and can threaten water supplies if it exceeds water quality standards.

Nitrate-nitrogen is highly mobile and can move readily below the crop root zone, especially in sandy soils. It can also be transported in surface runoff. Ammonium, on the other hand, becomes adsorbed to the soil and is lost primarily with eroding sediment. Even if nitrogen is not in a readily available form as it leaves the field, it can be converted to an available form either during transport or after delivery to water bodies.

Data collected in the U.S. Geological Survey NAWQA program sites showed that nitrate concentrations in ground water were highest in samples from wells in agricultural areas, with concentrations exceeding the drinking water standard of 10 mg/L in about 12% of domestic wells (Mueller and Helsel, 1996). Over the period 1986 – 1992, annual flow-weighted mean nitrate concentrations in ground water in the highly agricultural Big Spring basin of Iowa ranged from 5.7 mg/L in the very dry water year 1989 to 12.5 mg/L in the very wet water year 1991 (Rowden et al.,1995).

Across the U.S., nitrate levels in ground water are associated with source availability (i.e., population density, nitrogen inputs in fertilizer, manure, and atmospheric sources) and regional environmental factors (i.e., soil drainage characteristics, precipitation, cropland acres) (Spalding and Exner, 1993; Nolan et al., 1997). In Iowa's Big Spring basin, for example, the proportion of land in corn directly affected nitrogen concentrations and loads to surface and ground water because the greatest nitrogen inputs were fertilizers applied to corn (Rowden et al.,1995). In general, areas with high nitrogen input, well-drained soils, and high cropland areas have the highest potential for ground water contamination by nitrate (Nolan et al., 1997). Large areas of ground water where nitrate concentrations exceed the 10 mg/L limit occur in regions of irrigated cropland on well-drained soils; most of these areas are west of the Mississippi River where irrigation is necessary (Spalding and Exner, 1993). In the eastern U.S., localized nitrate-nitrogen contamination occurs beneath cropped, well-drained soils that receive excessive applications of fertilizer and manure, notably in the middle Atlantic states and the Delmarva Peninsula.

Soil drainage has reduced ground water nitrate problems in the Corn Belt states, because extensive tiling and ditching intercept soil water and carry it to surface water. High nitrogen inputs in such areas are more likely to affect surface water than ground water (Nolan et al., 1997). Studies in Walnut Creek, Iowa, showed that nitrate levels in the stream ranged from 10 to 20 mg/L (Hatfield et al., 1995). Walnut Creek, like many Midwestern streams, is fed by subsurface drainage, and high nitrate levels originated from the bottom of the root zone (1 – 1.2 m) in corn-soybean cropland in the watershed.

Phosphorus

Phosphorus can also contribute to the eutrophication of both freshwater and estuarine systems. Studies on the Cannonsville Reservoir, New York, showed that eutrophication was accelerated by phosphorus loading (Brown et al., 1986).

The low dissolved oxygen levels associated with eutrophication impacted fish populations, and use of the lake for recreational fishing was much less than at nearby Pepacton Reservoir. Moreover, the accelerated phosphorus loadings also contributed to the impairment of the drinking water supply for New York City because both reservoirs serve as major drinking water sources for the New York City water supply system. Also, nutrients are the major cause of use impairment in Lake Champlain, Vermont, with phosphorus the main culprit (Vermont Agency of Natural Resources, 1996). It is estimated that 55 – 66% of the NPS phosphorus load to Lake Champlain is derived from agricultural activities (Meals and Budd, 1998; Hegman et al., 1999).

While phosphorus typically plays the controlling role in freshwater systems, in some estuarine systems both nitrogen and phosphorus can limit plant growth. Algae consume dissolved inorganic phosphorus and convert it to the organic form. Phosphorus is rarely found in concentrations high enough to be toxic to higher-level organisms.

Phosphorus can be found in the soil in dissolved, colloidal, or particulate forms. Although the phosphorus content of most soils in their natural condition is low (between 0.01 and 0.2% by weight), soil test data indicate that decades of P application to agricultural land in excess of crop removal have resulted in widespread increases in soil P levels in the U.S. and elsewhere (Sims, 1993; Sharpley et al., 1993; Sims et al., 2000). Long-term trends in soil test values show that soil P in many areas of the world is excessive, relative to crop require-ments; the greatest concern occurs with animal-based agriculture, where farm and watershed-scale P surpluses and over-application of P to soils are common (Sims et al., 2000). Manures are normally applied at rates needed to meet crop nitrogen needs, yet the ratio of nitrogen to phosphorus in most manures results in over-application of phosphorus (Sharpley et al., 1996).

> Most often, phosphorus is sediment-attached. Phosphorus may also be dissolved. Either form can contribute to eutrophication.

The main forces controlling P movement from land to water are transport (runoff, infiltration, and erosion) and source factors (surface soil P and manage-ment of fertilizer/manure applications) (Sharpley et al., 1993; Daniel et al., 1998). Erosion processes control particulate P movement, while runoff processes drive dissolved P movement. Particulate P movement is a complex function of rainfall, irrigation, runoff, and soil management factors affecting erosion. Movement of dissolved P is a function of sorption/desorption, dissolution, and extraction of P from soil and plant material by water. Whereas surface runoff is typically the dominant pathway of P loss from agricultural land, there is increas-ing evidence that leaching of P from some soil types, especially on tile-drained fields, can present a threat to water quality (Beauchemin et al. 1998; Schoumans and Groenendijk, 2000; Simard et al., 2000).

Farm practices, such as manure or fertilizer applications and tillage, largely determine the quantity of P available in the soil to be moved by transport factors. Accumulation of P near the soil surface (0 – 2 inches) has been widely observed to influence the concentration and loss of P in runoff. Significant linear relation-ships have been demonstrated on a variety of soils and cropping systems be-tween the amount of soil test P in surface soil and dissolved P concentrations in surface runoff (Sharpley et al., 1993; Sharpley, 1995b; Pote et al., 1996; Pote et al., 1999; Sims et al., 2000; Sharpley et al., 2000; Sims, 2000). Soil P saturation status, rather than simply soil test P value, is thought to be a better predictor of runoff P loss, especially as the theoretical basis to establish environmental soil test P limits, because it integrates the effect of soil type (Sharpley, 1995b; Sims et al., 2000).

While there is little doubt that increased P concentrations at the soil surface contribute to higher P concentrations in runoff, the value of using soil test P as the sole predictor of transportable P is questionable (Coale, 2000). Consideration of hydrology is critical to understanding P export from a watershed. (Daniel et al., 1998). Chemical soil tests quantify concentration of soluble, biologically available, and potentially desorbable P in soils, but they provide no information on transport processes and management practices that influence movement of P from soil to water. They also do not characterize direct release of P from fertilizers, animal manure, and biosolids applied to soils (Sims et al., 2000).

Although soil P content is clearly important in determining the concentration of P in agricultural runoff, surface runoff and erosion potential, as well as mismanagement of fresh P inputs will often override soil P levels in determining P export. Use of a single threshold value for soil test P is too limited in its prediction of surface runoff P to be the only criterion to guide P management (Sharpley, 2000). Data from soil P testing must be integrated with understanding of transport processes and information on P management to predict P loss to water.

Lemunyon and Gilbert (1993) described an index for identifying soils, landforms, and management practices that could cause phosphorus problems in water bodies. The index uses soil erosion rates, runoff, soil test values of available phosphorus, and fertilizer and organic phosphorus application rates to assess the potential for phosphorus movement from the site. Sharpley (1995a) applied the Lemunyon and Gilbert phosphorus index to 30 watersheds in the Southern Plains, and concluded that the index is a valuable tool for identifying sources where phosphorus management is most needed. Several recommendations were made for improving the accuracy and utility of the index.

Gburek et al. (2000a and 2000b) have stressed that management of watershed phosphorus export should focus not just on areas of high soil P or P saturation but on critical source areas (CSA) that represent the intersection of surface runoff source areas (i.e., areas of actual or potential transport mechanisms) with areas of high soil P and high fertilizer/manure application. It is suggested that management of phosphorus loss from agricultural watersheds must focus on identifying, targeting, and remediating these spatially variable areas.

Runoff and erosion can carry some phosphorus to nearby water bodies. Dissolved inorganic phosphorus (orthophosphate phosphorus) is probably the only form directly available to algae, but eutrophication can be stimulated by the bioavailable phosphorus derived from the upper 5 cm of agricultural soils (Sharpley, 1985). Bioavailable phosphorus consists of dissolved phosphorus and a portion of particulate phosphorus that varies from site to site. Sharpley (1993) developed a method using iron-oxide impregnated paper to estimate the amount of phosphorus in soil that is available for algal growth. This method covers both dissolved and adsorbed phosphorus. Particulate and organic phosphorus delivered to water bodies may later be released as dissolved phosphorus and made available to algae when the bottom sediment of a stream becomes anaerobic, causing water quality problems.

Sediment

Sediment is the result of erosion. It is the solid material, both mineral and organic, that is in suspension, is being transported, or has been moved from its site of origin by wind, water, gravity, or ice. The types of erosion associated with agriculture that produce sediment are (1) sheet and rill erosion, (2) ephemeral and classic gully erosion, (3) wind erosion, and (4) streambank erosion. Soil erosion can be characterized as the transport of particles that are detached by rainfall, flowing water, or wind. Eroded soil is either redeposited on the same field or transported from the field in runoff or by wind.

Soil loss reduces nutrients and deteriorates soil structure, causing a decrease in the productive capacity of the land from which it is eroded. Wind erosion may cause abrasion of crops and structures by flying soil particles, air pollution by particles in suspension, transport of sediment-attached nutrients and pesticides, and burial of structures and crops by drifting soil.

> Sediment threatens water supplies and recreation, and causes harm to plant and fish communities.

Sediment affects the use of water in many ways. Suspended solids reduce the amount of sunlight available to aquatic plants, cover fish spawning areas and food supplies, smother coral reefs, clog the filtering capacity of filter feeders, and clog and harm the gills of fish. Turbidity interferes with the feeding habits of certain species of fish. These effects combine to reduce fish, shellfish, coral, and plant populations and decrease the overall productivity of lakes, streams, estuaries, and coastal waters. Recreation is limited because of the decreased fish population and the water's unappealing, turbid appearance. Turbidity also reduces visibility, making swimming less safe.

Deposited sediment reduces the transport capacity of roadside ditches, streams, rivers, and navigation channels. Decreases in capacity can result in more frequent flooding. Sediment can also reduce the storage capabilities of reservoirs and lakes and necessitate more frequent dredging.

The use of Highland Silver Lake, Illinois, as a public water supply was impaired by high turbidity levels and sedimentation (EPA, 1990b). Similarly, sediment surveys revealed that Lake Pittsfield, also in Illinois, was losing storage capacity at a rate of 1.08%, which would cause the lake to fill in with sediment in 92 years if no efforts had been made to control erosion (Davenport and Clarke, 1984). Due to erosion control efforts the rate of storage capacity loss has been reduced from 15% over 13 years to 10% over the subsequent 18 years (EPA, 1996). In addition, a water supply intake on Long Creek, North Carolina, was clogged due to erosion from surrounding lands, necessitating annual dredging of the water supply intake pool (EPA, 1996).

At current rates of sedimentation, Morro Bay, California, could be lost as an open water estuary within 300 years unless erosion control efforts are stepped up (EPA, 1996). Sedimentation has been associated with the lack of ocean-run trout in tributary streams, as well as significant economic losses to the oyster industry in the bay. Also, a trout fishery in Long Pine Creek, Nebraska, was impaired by high sediment loadings from streambank erosion and irrigation discharge (Hermsmeyer, 1991). Irrigation return flows with high sediment loads and streambank erosion caused negative impacts to salmonid spawning and recreational uses of Rock Creek, Idaho (Yankey et al., 1991).

Chemicals such as some pesticides, phosphorus, and ammonium are transported with sediment in an adsorbed state. Changes in the aquatic environment, such as decreased oxygen concentrations in the overlying waters or the development of anaerobic conditions in the bottom sediments, can cause these chemicals to be released from the sediment. Adsorbed phosphorus transported by the sediment may not be immediately available for aquatic plant growth but does serve as a long-term contributor to eutrophication.

Sediments from different sources vary in the kinds and amounts of pollutants that are adsorbed to the particles. For example, sheet, rill, ephemeral gully, and wind erosion mainly move soil particles from the surface or plow layer of the soil. Sediment that originates from surface soil has a higher pollution potential than that from subsurface soils. The topsoil of a field is usually richer in nutrients and other chemicals because of past fertilizer and pesticide applications, as well as nutrient cycling and biological activity. Topsoil is also more likely to have a greater percentage of organic matter. Sediment from gullies and streambanks usually carries less adsorbed pollutants than sediment from surface soils.

> Sediment from topsoil, often containing higher levels of nutrients and pesticides, can be a greater threat to water quality compared to subsoil sediment.

Soil eroded and delivered from cropland as sediment usually contains a higher percentage of finer and less dense particles than the parent soil on the cropland. This change in composition of eroded soil is due to the selective nature of the erosion process. For example, larger particles are more readily detached from the soil surface because they are less cohesive, but they also settle out of suspension more quickly because of their size. Organic matter is not easily detached because of its cohesive properties, but once detached it is easily transported because of its low density. Clay particles and organic residues will remain suspended for longer periods and at slower flow velocities than will larger or more dense particles. This selective erosion can increase overall pollutant delivery per ton of sediment delivered because small particles have a much greater adsorption capacity than larger particles. As a result, eroding sediments generally contain higher concentrations of phosphorus, nitrogen, and pesticides than the parent soil (i.e., they are enriched).

Animal Wastes

Animal waste (manure) includes the fecal and urinary wastes of livestock and poultry; process water (such as from a milking parlor); and the feed, bedding, litter, and soil with which they become intermixed. The following pollutants may be contained in manure and associated bedding materials and could be transported by runoff water and process wastewater from confined animal facilities:

> Runoff containing animal waste that reaches surface water can result in oxygen depletion and fish kills.

❏ Oxygen-demanding substances;

❏ Nitrogen, phosphorus, and many other major and minor nutrients;

❏ Organic solids;

❏ Salts;

❏ Bacteria, viruses, and other microorganisms;

❏ Metals; and

❏ Sediments.

When such runoff, process wastewater or manure enters surface waters, excess nutrients and organic materials are added. Increased nutrient levels can cause excessive growth of aquatic plants and algae. The decomposition of aquatic plants depletes the oxygen supply in the water, creating anoxic or anaerobic conditions which can lead to fish kills. Amines and sulfides are produced in anaerobic waters, causing the water to acquire an unpleasant odor, taste, and appearance. Methane, a greenhouse gas, can also be produced in anaerobic waters. Such waters can be unsuitable for drinking, fishing, and other recreational uses. Investigations in Illinois have demonstrated the impacts of animal waste on water quality, including fish kills associated with a hog facility, a cattle feeding operation, and surface application of liquid waste on frozen or snow-covered ground (Ackerman and Taylor, 1995). In addition, North Carolina experienced six spills from animal waste lagoons in the summer of 1995, totaling almost 30 million gallons. This total included a spill of 22 million gallons of swine waste into the New River, which killed fish along a 19-mile downstream area (EPA Office of Inspector General, 1997).

A study of Herrings Marsh Run in the coastal plain of North Carolina showed that nitrate levels in stream and ground water were highest in areas with the greatest concentration of swine and poultry production (Hunt et al., 1995). Orthophosphate levels were affected only slightly by animal waste applications since most of the phosphorus was bound by the soil. In addition, runoff from feedlots has long been associated with severe stream pollution. Feedlots, which are devoid of vegetation and subjected to severe hoof action, generate runoff containing large amounts of bacteria, which may cause violations of water quality standards (Baxter-Potter and Gilliland, 1988).

Diseases can be transmitted to humans through contact with animal or human feces. Runoff from fields receiving manure will contain extremely high numbers of microorganisms if the manure has not been incorporated or the microorganisms have not been subject to stress. Shellfishing and beach closures can result from high fecal coliform counts. Although not the only source of pathogens, animal waste has been responsible for shellfish contamination in some coastal waters.

The pathogen *Cryptosporidium*, a protozoan parasite, is common in surface waters, especially those containing high amounts of sewage contamination or animal waste. Without advanced filtration technology, *Cryptosporidium* may pass through water treatment filtration and disinfection processes in sufficient numbers to cause health problems, such as the gastrointestinal disease cryptosporidiosis. The most serious consequences of cryptosporidiosis tend to be focused on people with severely weakened immune systems. In 1993, Milwaukee, Wisconsin, which draws its water from Lake Michigan, experienced an outbreak of cryptosporidiosis, affecting 400,000 people, with more than 4,000 hospitalized and over 50 deaths attributed to the disease (EPA, 1997c). While the source of contamination is uncertain, the problem was linked to suboptimal performance of the water treatment plant, together with unusually heavy rainfall and runoff. The watersheds of two rivers which discharge into Lake Michigan contain slaughterhouses, human sewage discharges, and cattle grazing ranges (Lisle and Rose, 1995).

Giardia is another commonly identified pathogen in surface waters. *Giardia* is the intestinal parasite that causes the disease giardiasis. Giardiasis is sometimes referred to as "backpacker's disease" since the disease frequently occurs in hikers and nature lovers who unwittingly drink water from contaminated springs or streams. However, several community-wide outbreaks of giardiasis have been linked to contaminated municipal drinking water (CDC, n.d.). The commonly associated symptoms of giardiasis are persistent diarrhea, weight loss, abdominal cramps, nausea, and dehydration. With proper treatment and a healthy immune system, giardiasis is not deadly, but it can be life threatening to AIDS patients, small children, the elderly, or someone recovering from major surgery. The best strategy to protect a drinking water supply from *Giardia* contamination is the physical removal of the organism. This can be accomplished by controlling land use within a watershed to prevent degradation of the source water and by utilizing a properly designed and operated water filtration plant.

Viruses in animal waste also pose a potential health threat to humans. Enteric viruses are the most significant virus group affecting water quality and human health (EPA, 2001). There are over 100 different types of enteric viruses, all considered pathogenic to man (EPA, 1984). When ingested, enteric viruses may attack the gastrointestinal track or the respiratory system, sometimes, fatally. More typically, infection causes sore throat, diarrhea, fever and nausea. Enteric viruses may be found in livestock excrement from barnyards, pastures, rangelands, feedlots, and uncontrolled manure storage areas; and areas of land application of manure and sewage sludge (NCSU, 2001). When animal waste is applied to agricultural land for irrigation or fertilization purposes, enteric viruses can survive in soil for periods of weeks or even months (EPA, 1984). Enteric viruses in land applied manure or sewage sludge can leach into ground water and/or eventually be transported by overland flow into surface water bodies, thus creating a potential for the contamination of water resources. Management measures should be instituted in all situations in which sludge is used for irrigation or fertilization, to prevent the contamination of vegetables and drinking water sources by enteric viruses (EPA, 1984).

Since pathogenic organisms present in polluted waters are generally difficult to identify and isolate, scientists typically choose to monitor indicator organisms. Indicator organisms are usually nonpathogenic bacteria assumed to be associated with pathogens transmitted by fecal contamination but are more easily sampled and measured. Fecal indicators are used to develop water quality criteria to support designated uses, such as primary contact recreation and drinking water supply. For example, studies conducted by USEPA have demonstrated that the risk to swimmers of contracting gastrointestinal illness seems to be predicted better by enterococci than by fecal coliform bacteria since the die off rate of fecal coliform bacteria is much greater than the enterococci die off rate (EPA, 2001). Moreover, a comparison of various fecal indicators of potential pathogens with disease incidence revealed that elevated levels of enterococci bacteria were most strongly correlated with gastroenteritis in both fresh and marine recreational waters (EPA, 1986). The USEPA believes that enterococci is best suited as an indicator organism for predicting the presence of gastrointestinal illness-causing pathogens in fresh water and marine waters and recommends that people do not swim in fresh waters that contain 33 or more enterococci per 100 milliliters (mL) or marine waters with 35 or more enterococci per 100 mL (EPA, 2000b).

Animal wastes contain large numbers of bacteria and other microorganisms. Although many of these organisms tend to die rapidly outside the animal, some can survive under favorable conditions. Microorganisms can survive for extended periods in fecal deposits on pasture, in soils, and in aquatic sediments (Thelin and Gifford, 1983; Kress and Gifford, 1984; Sherer et al., 1992). Conditions that promote die-off of microorganisms after land application include low soil moisture, low pH, high temperatures, direct solar radiation, and predation by protozoa. Manure storage generally promotes die-off, although pathogens can remain dormant at certain temperatures. Composting the wastes can be quite effective in decreasing the number of pathogens.

In a review of literature regarding the impacts of long-term animal waste applications on soil characteristics, it was concluded that positive impacts include buildup of soil organic matter, increased soil fertility, and improvement of soil physical properties (Wood and Hattey, 1995). Negative impacts include nitrate pollution of ground water, phosphorus contamination of surface water, and potential toxicity to crops from elevated concentrations of metals or other trace elements. For example, copper and zinc concentrations can build up where poultry litter and hog manure are applied.

The method, timing, and rate of manure application are significant factors in determining the likelihood that water quality contamination will result. Manure is generally more likely to be transported in runoff when applied to the soil surface than when incorporated into the soil. Spreading manure on frozen ground or snow can result in high concentrations of nutrients being transported from the field during rainfall or snowmelt, especially when the snowmelt or rainfall events occur soon after spreading (Robillard and Walter, 1986). Binding of phophorus with soil particles also increases as soil temperature increases. Winter spreading of manure onto corn fields in Vermont increased phosphorus export by up to 1500%, with up to 15% of the applied phosphorus lost in runoff (Meals, 1996). Soil type, crops, anticipated yields, and crop nutrient uptake are other factors that should be considered when determining the likelihood of manure contaminated runoff.

When application rates of manure for crop production are based on N, the P and K rates applied normally exceed plant requirements (Westerman et al., 1985). The soil generally has the capacity to adsorb much of the phosphorus from manure applied on land, but this capacity is not unlimited. As previously mentioned, however, nitrates are easily leached through soil into ground water or to return flows, and phosphorus can be transported by eroded soil.

Salts

Salts are a product of the natural weathering process of soil and geologic material. They are present in varying degrees in all soils and in fresh water, coastal waters, estuarine waters, and ground waters.

In soils that have poor subsurface drainage, high salt concentrations are created within the root zone where most water extraction occurs. The accumulation of soluble and exchangeable sodium leads to soil dispersion, structure breakdown, decreased infiltration, and possible toxicity; thus, salts often become a serious problem on irrigated land, both for continued agricultural production and for water quality considerations. High salt concentrations in streams can harm

Accumulation of excess sodium reduces agricultural production, and runoff of saline water harms aquatic ecosystems.

freshwater aquatic plants just as excess soil salinity damages agricultural crops. While salts are generally a more significant pollutant for freshwater ecosystems than for saline ecosystems, they may also adversely affect anadromous fish. Although they live in coastal and estuarine waters most of their lives, anadromous fish depend on freshwater systems near the coast for crucial portions of their life cycles.

The movement and deposition of salts depend on the amount and distribution of rainfall and irrigation, the soil and underlying strata, evapotranspiration rates, and other environmental factors. In humid areas, dissolved mineral salts have been naturally leached from the soil and substrata by rainfall. In arid and semi-arid regions, salts have not been removed by natural leaching and are concentrated in the soil. Soluble salts in saline and sodic soils consist of calcium, magnesium, sodium, potassium, carbonate, bicarbonate, sulfate, and chloride ions. They are fairly easily leached from the soil. Sparingly soluble gypsum and lime also occur in amounts ranging from traces to more than 50% of the soil mass.

Irrigation water, whether from ground or surface water sources, has a natural base load of dissolved mineral salts. As the water is consumed by plants or lost to the atmosphere by evaporation, the salts remain and become concentrated in the soil. This is referred to as the "concentrating effect."

The total salt load carried by irrigation return flow is the sum of the salt remaining in the applied water plus any salt picked up from the irrigated land. Irrigation return flows provide the means for conveying the salts to the receiving streams or ground water reservoirs. If the amount of salt in the return flow is low in comparison to the total stream flow, water quality may not be degraded to the extent that use is impaired. However, if the process of water diversion for irrigation and the return of saline drainage water is repeated many times along a stream or river, water quality will be progressively degraded for downstream irrigation use as well as for other uses.

Another related issue is selenium toxicity. Selenium is a natural element in soil, found in a variety of geologic formations, including Cretaceous sediments in the western U.S. Selenium is essential to human and animal health in very small amounts, but is toxic to some organisms when ingested in excessive quantities (Letey et al., 1986). The major threat posed by selenium is the leaching of its soluble, oxidized form (selenate) from seleniferous soils and movement of leachate to shallow ground water and ultimately surface waters. It is in the aquatic environment where selenium enters the food chain through plants, which then become the food base for higher organisms such as insects, fish or birds. Accumulation and concentration of selenium as it moves up the food chain can become toxic (Letey et al., 1986).

In the western U.S., irrigation of soils from seleniferous parent materials can accelerate the natural leaching process. In the early 1980's, irrigation drainage water laden with high concentrations of selenium caused congenital deformities and mortality of waterfowl at Kesterson Reservoir, a National Wildlife Refuge in central California (Long et al., 1990). Concern over this incident prompted the U.S. Department of Interior to establish the National Irrigation Water Quality Program in 1985, to evaluate the potential for toxic effects of selenium in other irrigated areas of the west (Nolan and Clark, 1997).

Pesticides

The term *pesticide* includes any substance or mixture of substances intended for preventing, destroying, repelling, or mitigating any pest or intended for use as a plant regulator, defoliant, or desiccant. The principal pesticidal pollutants that may be detected in surface water and in ground water are the active and inert ingredients and any persistent degradation products. Pesticides and their degradation products may enter ground and surface water in solution, in emulsion, or bound to soil colloids. A study of 303 wells from across the Midwest showed that pesticide metabolites were found more frequently than the parent compounds (Kolpin et al., 1996). For example, the metabolite alachlor ethanesulfonic acid was detected nearly 10 times more frequently than alachlor in the 153 wells where both chemicals were analyzed. For simplicity, the term *pesticides* will be used to represent "pesticides and their degradation products" in the following sections.

Despite the documented benefits of using pesticides (insecticides, herbicides, fungicides, miticides, nematicides, etc.) to control plant pests and enhance production, these chemicals may, in some instances, cause impairments to the uses of surface water and ground water. Some types of pesticides are resistant to degradation and may persist and accumulate in aquatic ecosystems.

Many studies have evaluated pesticides in runoff and in streams, generally finding that the concentration can be relatively high near the application site soon after application with significant reductions further downstream and with time. Seasonal pulses of some of the most widely used pesticides can exceed lifetime maximum contaminant levels (MCL) established by the U.S. EPA, however the annual means on which those regulations are based are rarely exceeded (Larson et al., 1997).

Monitoring of seven Lake Erie tributaries from 1983 to 1993 detected maximum atrazine concentrations of 6.80 to 68.40 ug/L, and maximum concentrations of alachlor, metolachlor, metribuzin, cyanazine, and linuron ranging of 1.16 to 64.94, 5.39 to 96.92, 1.49 to 25.15, 1.36 to 24.77, and 1.92 to 15.5 ug/L, respectively (Baker, 1993). The long-term time-weighted mean concentrations in these cases, however, were all below EPA's maximum contaminant levels and lifetime health advisory levels for drinking water. In a related study, it was determined that alachlor and atrazine were the most frequently detected pesticides in drinking water supplies in Ohio (Baker and Richards, 1991). Although chronic health standards were not exceeded, public water supplies derived from rivers or reservoirs draining agricultural watersheds were more likely to have detectable residues of pesticides than other water supplies.

Pesticides have a wide range for potential harm to the environment due to the large variations in both chemical makeup and application schedule. Generally speaking, pesticides with higher levels of toxicity and persistence are more likely to create problems. Toxicity can be defined in terms of short-term (acute) and longer-term (chronic) effects. Acute effects usually occur soon after spraying, as in the case of a fish kill from drift or runoff. Chronic effects can occur when a pesticide is present in an environment over months or years at concentrations high enough to trigger a response by one or more organisms. Some of the pesticides banned years ago, such as DDT, had these effects on many birds and other organisms. Most pesticides currently in use have few reported chronic effects at levels commonly found in the environment.

Persistence is a measure of how long the chemical remains in the environment, which can be from days to years. A more persistent pesticide could present more of a risk for environmental contamination. The use of highly persistent pesticides is generally limited to situations where repeated applications would be undesirable, such as in termite control around buildings or vegetation control along right-of-ways.

The threat to water quality is often dependent upon the combination of application location and method. The highest risk occurs when aerial insecticide spraying is located near open water. This poses such a high risk because the chance for drift is greatest in aerial spraying compared to other application methods and insecticides are more likely to affect aquatic organisms than other types of pesticides. However, pesticide residues in runoff and ground water also pose a risk to water quality. Herbicides, compared to other pesticides, are more likely to travel by means of surface runoff or ground water as they are more widely used and are persistent enough to be detected many weeks after application. Concentrations of pesticides in ground water are generally low because soil retains most of the infiltrated pesticide residue. In areas where pesticides are widely applied, surface water has an annual cycle of higher residues during the growing season and much lower residues during the rest of the year.

The primary routes of pesticide transport to aquatic systems are through (Maas, 1984):

- ☐ Direct application;
- ☐ Runoff;
- ☐ Aerial drift;
- ☐ Leaching;
- ☐ Volatilization and subsequent atmospheric deposition; and
- ☐ Uptake by biota and subsequent movement in the food web.

The amount of field-applied pesticide that leaves a field in the runoff (either dissolved or adsorbed) and enters a stream primarily depends on:

- ☐ The intensity and duration of rainfall or irrigation;
- ☐ The length of time between pesticide application and rainfall occurrence;
- ☐ The amount of pesticide applied and its soil/water partition coefficient;
- ☐ The length and degree of slope and soil composition;
- ☐ The extent of exposure to bare (vs. residue or crop-covered) soil;
- ☐ Proximity to streams;
- ☐ Soil loss/erosion rate;
- ☐ Soil organic carbon content;
- ☐ The method of application; and
- ☐ The extent to which runoff and erosion are controlled with agronomic and structural practices.

Pesticide losses are generally greatest when rainfall is intense and occurs shortly after pesticide application, a condition for which water runoff and erosion losses are also greatest.

A study of herbicides and nutrients in storm runoff from nine stream basins in the Midwestern states from 1990-1992 showed sharp increases in triazine herbicides (e.g., atrazine) in the post-planting period (Scribner et al., 1994). Atrazine levels increased from 1.0 ug/L to peaks of 10-75 ug/L. EPA's maximum contaminant level (MCL) for atrazine in public water supplies is 3.0 ug/L. In this and many other studies, EPA MCLs are utilized as reference points for assessing water quality. It should be noted that an exceedance of the MCL in these surface or ground water quality monitoring studies does not necessarily indicate violation of a water quality standard.

In the Scribner et. al study (1994), it was concluded that transport of herbicides to streams was seasonal, with peaks from early May to early July. In a related study of 76 Midwestern reservoirs from April 1992 through September 1993, atrazine was the most frequently detected and persistent herbicide, followed by alachlor ethane sulfonic acid, deethylatrazine, deisopropylatrazine, metolachlor, cyanazine amide, and cyanazine (Scribner et al., 1996). Eight reservoirs had concentrations of one or more herbicides exceeding EPA's maximum contaminant levels or health advisory levels for drinking water during late April through mid-May, 1992, while 16 reservoirs had these high contaminant levels in late June through July, 1992. The annual average concentrations on which the MCLs are based are usually not exceeded, however, because residues drop to low or undetectable levels at other times of the year.

Research at the 5,600-ha Walnut Creek watershed in Iowa also showed that atrazine levels in runoff increased to above the MCL with heavy rains after chemical application. The total loss of atrazine and metolachlor in stream flow was about 1% of the amount applied each year. Herbicide concentrations in tile drains were often near the detection limit of 0.2 ug/L, while only atrazine and metolachlor exceeded 3.0 ug/L once in more than 1,700 ground water samples. Water balance studies indicated that the predominant flow path in the prairie-pothole watershed is from the bottom of the root zone into the stream through tile drains (Hatfield et al., 1995).

Concentrations of atrazine, alachlor, cyanazine, and metolachlor in Midwestern streams and reservoirs increased suddenly during rainstorms following herbicide applications (Goolsby et al., 1995). Atrazine levels less than 0.2 ug/L also persist year-round in Midwestern streams, partly due to the discharge of contaminated waters from surface and ground water reservoirs.

Elevated monthly average pesticide concentrations in Lake Erie tributaries usually occur in May to August, and smaller tributaries had higher maximum concentrations, more frequent concentrations below the detection limit, and fewer intermediate concentrations than larger tributaries (Richards and Baker, 1993).

From calculations combining estimated pesticide use data with measured load data, it was estimated that less than 2% of applied pesticides reached surface waters in the Mississippi River basin (Larson et al., 1995). Since the relative percentages of specific pesticides reaching the rivers were often not in agreement with projected runoff potentials, it was concluded that soil characteristics, weather, and agricultural management practices are more important than chemical properties in the delivery of pesticides to surface waters. Richards and Baker (1993) concluded that average pesticide concentrations in Lake Erie tributaries are correlated with amount applied, but are also affected by chemical properties and modes of application of the pesticides.

The rate of pesticide movement through the soil profile to ground water is inversely proportional to the pesticide adsorption partition coefficient or Kd (a measure of the degree to which a pesticide is adsorbed by the soil versus dissolved in the water). The larger the Kd, the slower the movement and the greater the quantity of water required to leach the pesticide to a given depth. Other factors affecting pesticide movement include pesticide solubility as well as soil pH and temperature.

Pesticides can be transported to receiving waters either in dissolved form or attached to sediment. Dissolved pesticides may be leached to ground water supplies. Both the degradation and adsorption characteristics of pesticides are highly variable.

Pesticides have been widely detected in ground water, with concentrations usually much lower than in surface water but with greater longevity (Barbash and Resek, 1996). The most common detected are corn and soybean herbicides, which were reported to occur in up to 30% of samples in a national water quality assessment (Barbash et al., 2001). Of those with detections, 98% were below 1.0 ug/L and only exceeded the MCL in 2 of 2,227 sites. In another study, herbicides, including atrazine, prometron, metolachlor, and alachlor were detected in 24 percent of shallow aquifers in the Midwest sampled by USGS (Burkhart and Kolpin, 1993). Reported concentrations for all compounds were less than 0.5 ug/l. In Walnut Creek, Iowa, herbicides were not generally found in concentrations above 0.2 ug/l in shallow ground water (Hatfield et al. 1993). In the Mid-Atlantic region, pesticide compounds, including atrazine and its metabolites, metolachlor, prometron, and simazine, have been detected in about half of ground water samples analyzed, but rarely at concentrations exceeding established MCLs (Ator and Ferrari, 1997). The occurrence of pesticides in ground water of the Mid-Atlantic region was related to land cover and rock type: agricultural and urban land use practices are likely sources of pesticides, and rock type affects the movement of these compounds into and through the ground water system. Recently, Kolpin et al. (2000) found that one or more pesticides were detected at nearly half of 2500 USGS NAWQA ground water sites sampled across the United States. Observed pesticide concentrations were generally low. Pesticides were commonly detected beneath both agricultural and urban areas.

Habitat Impacts

The functioning condition of riparian-wetland areas is a result of interaction among geology, soil, water, and vegetation. Riparian-wetland areas are functioning properly when adequate vegetation is present to

- ❏ Dissipate stream energy associated with high water flows, thereby reducing erosion and improving water quality;
- ❏ Filter sediment and aid floodplain development;

❏ Support denitrification of nitrate-contaminated ground water as it is discharged into streams;

❏ Improve floodwater retention and ground water recharge;

❏ Develop root masses that stabilize banks against fluvial erosion (scouring) and gravitational bank collapse (slumping);

❏ Develop diverse ponding and channel characteristics to provide the habitat and the water depth, duration, and temperature necessary for fish production, waterfowl breeding, and other uses; and

❏ Support biodiversity.

Numerous land uses, such as silviculture, agriculture, and urbanization, have the potential to degrade riparian habitats. Improper livestock grazing affects all four components of the water-riparian system: banks and shores, water column, channel morphology, and aquatic and bordering vegetation (Platts, 1990). The potential effects of improper grazing management or improper use of grazing lands include:

Shore/banks

❏ Shear or sloughing of streambank soils by hoof or head action.

❏ Water, ice, and wind erosion of exposed streambank and channel soils because of loss of vegetative cover.

❏ Elimination or loss of streambank vegetation.

❏ Reduction of the quality and quantity of streambank undercuts.

❏ Increasing streambank angle (laying back of streambanks), which increases water width, decreases stream depth, and alters or eliminates fish habitat.

Water Column

❏ Excessive withdrawal from streams to irrigate grazing lands.

❏ Drainage of wet meadows or lowering of the ground water table to facilitate grazing access.

❏ Pollutants (e.g., sediments) in return water from grazed lands, which are detrimental to the designated uses such as fisheries.

❏ Changes in magnitude and timing of organic and inorganic energy (i.e., solar radiation, debris, nutrients) inputs to the stream.

❏ Increase in fecal contamination.

❏ Changes in stream morphology, such as increases in stream width and decreases in stream depth, including reduction of stream shore water depth.

❏ Changes in timing and magnitude of stream flow events from changes in watershed vegetative cover.

❏ Increase in stream temperature.

Channel

❏ Changes in channel morphology.

❏ Altered sediment transport processes.

> Riparian-wetland vegetation is essential for stable aquatic ecosystems.

Riparian Vegetation

- ☐ Changes in plant species composition (e.g., shrubs to grass to forbs).

- ☐ Reduction of floodplain and streambank vegetation including vegetation hanging over or entering into the water column.

- ☐ Decrease in plant vigor.

- ☐ Changes in timing and amounts of organic energy leaving the riparian zone.

- ☐ Elimination of riparian plant communities (i.e., lowering of the water table allowing xeric plants to replace riparian plants).

Water temperature plays a key role in the life of fish and other aquatic organisms by influencing their distribution, growth rate, and survival (Barthalow, 1989; Holmes and Regier, 1990; Armour 1991), as well as migration patterns, egg maturation, incubation success, competitive ability, and resistance to parasites, diseases, and pollutants (Armour 1991). Increases in water temperature can also cause shifts in algal communities from cold-water diatoms to warm-water green and blue-green species which can cause other water quality problems (Horner et. al, 1994). In addition, water temperature affects the rates of in-stream chemical reactions, the self-purification capacity of streams, and their aesthetic and sanitary qualities (Feller 1981). Changes in channel morphology leading to an increased stream width and decreased depth, as well as loss of riparian vegetation, have the potential to alter stream temperature. A wider and shallower stream has a greater surface area and a greater air-water interface, where most energy exchanges occur; hence, the surface area of the stream is directly related to water temperature changes. Also, losses in riparian vegetation expose the stream to greater temperature fluctuations, resulting in potentially higher temperatures during the day and cooler temperatures at night. Riparian vegetation acts to moderate stream temperatures by absorbing short-wave radiation during the day and insulating the stream from loss of long-wave radiation at night.

Improperly managed livestock grazing can significantly contribute to streambank erosion and riparian habitat degradation. In a study of 60 streams in the Intermountain West, it was found that grazed stream habitats were substantially degraded with poor riparian conditions (Robinson and Minshall, 1995). Problems associated with improper grazing management included reduced riparian cover, exposed streambanks, high sediment levels, elevated water temperatures, higher nutrient levels, and a shifting to more stress-tolerant invertebrates.

Soil erosion, primarily from poor grazing management and poorly maintained riparian areas, is causing excessive sedimentation to the Missouri River in South Dakota (Osmond et al., 1997). This sedimentation has impaired recreational uses and hydropower generation, and has increased flooding in the cities of Pierre and Ft. Pierre. Improper livestock grazing management has also contributed to declines in anadromous fish populations in the Upper Grande Ronde Basin in Oregon (Osmond et al., 1997). Increased stream water temperature and loss of habitat, caused largely by the loss in riparian vegetation, are key factors in the decline (Hafele, 1996). Improper grazing management in the Morro Bay, California, watershed has stripped riparian areas of their vegetation and decreased streambank stability, contributing to the excessive erosion in the watershed

---

Improper livestock grazing can have devastating impacts on streambanks, hydrology, water quality, and aquatic habitat.

(Osmond et al., 1997). Sedimentation has caused negative impacts to both the oyster industry and anadromous fish species. Streambank erosion in Peacheater Creek, Oklahoma, has impaired aquatic habitat (Osmond et al., 1997).

Mechanisms to Control Agricultural Nonpoint Source Pollution

There exists a considerable amount of jargon associated with the mechanisms to control nonpoint source pollution. Terms include *best management practices (BMPs), management practices, accepted agricultural practices, management measures, BMP systems, management practice systems, resource management systems (RMSs), total resource management systems*, and the like. Some of these terms are based in legislation or regulations such as the management measures specified by EPA for the section 6217 coastal nonpoint pollution control program (EPA, 1993a) and Vermont's accepted agricultural practices (Vermont Department of Agriculture, 1995), while other terms are found in technical manuals, journal articles, and informational materials.

The meanings of the terms also vary. Most practitioners consider *BMPs* to be individual practices or groups of practices that serve specific functions such as excluding livestock or routing water safely away from eroding or contaminated areas. *Management measures* are generally groups of affordable management practices that are used together in a system to achieve more comprehensive goals such as minimizing the delivery of sediment from a farm to receiving waters or maximizing the efficiency with which nutrients are applied to croplands to achieve reasonable yields. RMSs generally go beyond management measures in that they may contain practices that address natural resource concerns other than water quality, and must meet criteria for soil, water, air, and related plant, animal, and human resources. Since the focus of this guidance is water quality issues, the full complement of issues addressed in a typical RMS is not addressed. For example, water quality performance expectations are contained in the management measures, but criteria for animal resources are absent. Resource management *planning concepts* are discussed briefly in this chapter, however.

Because definitions of terms overlap, there is no clear hierarchy or levels of control that can be adopted for this guidance and agreed upon by all readers, but the following statements apply:

❏ Complete RMSs are not presented in this guidance, but resource management planning concepts are discussed. The water quality aspects and some of the soil, air, and plant criteria of an RMS are addressed through the management measures.

❏ Individual management practices are the building blocks for management practice systems and management measures.

❏ Implementation of all six management measures, as appropriate, will result in a comprehensive, technology-based water quality protection plan on most[1] farms.

---

[1] In some cases, additional control practices may be needed to address problems that are not anticipated by the management measures.

Management Measures

Management measures are defined under section 6217 of CZARA as:

> economically achievable measures for the control of the addition of pollutants from existing and new categories and classes of nonpoint sources of pollution, which reflect the greatest degree of pollutant reduction achievable through the application of the best available nonpoint source control practices, technologies, processes, siting criteria, operating methods, or other alternatives.

The management measures specified by EPA for section 6217 contain performance expectations and, in many cases, specific actions that are to be taken to prevent or minimize nonpoint source pollution (EPA, 1993a). For example, the performance expectations for erosion and sediment control for agriculture are "to minimize the delivery of sediment from agricultural lands to surface waters" or "to settle the settleable solids and associated pollutants in runoff delivered from the contributing area for storms up to and including a 10-year, 24-hour frequency." Individual management practices or specific actions needed to achieve these performance expectations are not included in the management measure statement. The management measure for pesticides, however, includes both performance expectations ("reduce contamination of surface water and ground water from pesticides") and specific practices and actions such as anti-backflow devices on hoses, and calibration of pesticide spray equipment. Thus, in most cases, there is considerable flexibility to determine *how* to best achieve the performance expectations for EPA's section 6217 management measures.

EPA's six management measures for agriculture are described in Chapter 4.

Management Practices

"Best" management practices, BMPs, are designed to reduce the quantities of pollutants that are generated at and/or delivered from a source to a receiving water body. In EPA's guidance for section 6217, the term *management practice* is used in lieu of *BMPs* since "best" can be a highly subjective and site-specific label. For example, the BMP manuals used by States to implement the Clean Water Act section 319 program are not identical although much consistency exists across States. Even within States, a practice may be considered best in one area (e.g., coastal plain) but inappropriate in another area (e.g., mountains). Criteria for determining what is best may include extent of pollution prevention or pollutant removal, ease of implementation, ease of maintenance and operation, durability, attractiveness to landowner (e.g., how willing will farmers be to implement the practice in a voluntary program?), cost, and cost-effectiveness. The relative importance assigned these and other criteria in judging what is best varies across States, within States, and among landowners, often for very good reasons (e.g., irrigation water management considerations are very different in western States with low rainfall and water rights laws, versus midwestern States with diminishing ground-water reserves, versus eastern States with plentiful rainfall and surface waters). For these reasons, this guidance is consistent with the section 6217 management measures guidance in its use of the term "management practice" rather than "BMP."

Management practices can be structural (e.g., waste treatment lagoons, terraces, or sediment basins) or managerial (e.g., rotational grazing, nutrient management, pesticide management, or conservation tillage). Management practices generally do not stand alone in solving water quality problems, but are used in combinations to build *management practice systems*. For example, soil testing is a good practice for nutrient management, but without estimates of realistic yield; good water management; appropriate planting techniques and timing; and proper nutrient selection, rates, and placement; the performance expectations for nutrient management cannot be achieved.

Each practice, in turn, must be selected, designed, implemented, and maintained in accordance with site-specific considerations to ensure that the practices function together to achieve the overall management goals. For example, a grassed waterway must be designed to handle all of the water that will be conveyed to it from upland areas, including all water re-routed with diversions and drainage pipes. Design standards and specifications must be compatible for practices to work together as effective systems.

A summary of agricultural management practices and how they function in systems is given in Chapter 3. Management practices that can be used to achieve each of the six agricultural management measures are described in Chapter 4.

Resource Management Planning Concepts

Resource management planning, also known as conservation planning, for agricultural operations is a natural resource problem solving and management process. The process integrates economic, social (including cultural resources), and ecological considerations to meet goals and objectives. It involves setting of personal, environmental, economic, and production goals for the farm or ranch. The challenge in resource management planning is to balance the short-term demands for production of food, fiber, wood, and other agricultural products, with long-term sustainability of a quality environment.

Resource management systems are combinations of conservation practices and resource management, identified by land or water uses, for the treatment of all natural resource concerns for soil, water, air, plants, and animals that meets or exceeds the quality criteria for resource sustainability. The quality criteria are described in the USDA Natural Resources Conservation Service (NRCS) Field Office Technical Guide (FOTG). See Appendix B for additional information on the FOTG.

Resource management planning is preferred by land managers who have a negative reaction to "single purpose plans" that address individual economic or natural resource issues. Essential goals for a farm or ranch resource management plan include:

❒ Improving or ensuring profitability by finding solutions that save money, increase sales, improve product quality, or simplify/reduce the work;

❒ Reducing water pollution through application of appropriate systems of management practices;

A resource management plan for the farm serves to maintain quality of life while achieving goals for profitability and water quality.

❏ Coordinating regulatory input so that implementation of the resource management plan will assure compliance with all applicable regulations impacting the agricultural operation; and

❏ Incorporating the farm or ranch family's personal goals for quality of life.

NRCS and its cooperating conservation partners use a three-phase, nine-step planning process. This process is very dynamic, frequently requiring planners to cycle back to previous steps in order to fully achieve the goals set for the plan. Many states are developing their own resource management planning protocols. An example of one of these efforts is the Idaho One Plan. The Idaho program was developed to reduce diverse agency requirements and to produce a user-friendly product that allows farmers and ranchers to develop resource management plans unique to their operations.

Individuals interested in resource management planning should contact their local NRCS office, soil and water conservation district, cooperative extension service, land grant university, state department of agriculture, or other appropriate agency to learn more about locally available information.

Management Practices

Management practices are implemented on agricultural lands for a variety of purposes, including protecting water resources, protecting terrestrial or aquatic wildlife habitat, and protecting the land resource from degradation by wind, salt, and toxic levels of metals. The primary focus of this guidance is on agricultural management practices that control the generation and delivery of pollutants into water resources or remediate or intercept pollutants before they enter water resources.

NRCS maintains a National Handbook of Conservation Practices (USDA–NRCS, 1977), updated continuously, which details nationally accepted management practices. These practices can be viewed at the USDA-NRCS web site at *www.nrcs.usda.gov/technical/efotg/*. In addition to the NRCS standards, many States use locally determined management practices that are not reflected in the NRCS handbook. Readers interested in obtaining information on management practices used in their area should contact their local Soil and Water Conservation District or local USDA office. Two very helpful handbooks for farmers in the Midwest are *60 Ways Farmers Can Protect Surface Water* (Hirschi et al., 1997), and *50 Ways Farmers Can Protect their Ground Water* (Hirschi et al., 1993).

How Management Practices Work to Prevent Nonpoint Source Pollution

Management practices control the delivery of nonpoint source (NPS) pollutants to receiving water resources by

- ❐ minimizing pollutants available (source reduction);

- ❐ retarding the transport and/or delivery of pollutants, either by reducing water transported, and thus the amount of the pollutant transported, or through deposition of the pollutant; or

- ❐ remediating or intercepting the pollutant before or after it is delivered to the water resource through chemical or biological transformation.

Management practices are generally designed to control a particular pollutant type from specific land uses. For example, conservation tillage is used to control erosion from irrigated or non-irrigated cropland. Management practices may also provide secondary benefits by controlling other pollutants, depending on how the pollutants are generated or transported. For example, practices which reduce erosion and sediment delivery often reduce phosphorus losses since phosphorus is strongly adsorbed to silt and clay particles. Thus, conservation tillage not only reduces erosion, but also reduces transport of particulate phosphorus.

In some cases, a management practice may provide environmental benefits beyond those linked to water quality. For example, riparian buffers, which reduce

Management practices can minimize the delivery and transport of agriculturally derived pollutants to surface and ground waters. Although a wide variety of BMPs are available, all require regular inspection and maintenance.

phosphorus and sediment delivery to water bodies, also serve as habitat for many species of birds and plants.

Sometimes, however, management practices used to control one pollutant may inadvertently increase the generation, transport, or delivery of another pollutant. Conservation tillage, because it creates increased soil porosity (i.e., large pore spaces), may increase nitrate leaching through the soil, particularly when the amount and timing of nitrogen application is not part of the management plan. Tile drains, used to reduce runoff and increase soil drainage, can also have the undesirable effect of concentrating and delivering nitrogen directly to streams (Hirschi et al., 1997). In order to reduce the nitrogen pollution caused by tile drains, other management practices, such as nutrient management for source reduction and biofilters that are attached to the outflow of the tile drains for interception, may be needed. On the other hand, practices which reduce runoff may contribute to reduced in-stream flows, which have the potential to adversely impact habitat. Therefore, management practices should only be chosen after a thorough evaluation of their potential impacts and side-effects.

Control of surface transport may increase leaching of pollutants.

Water Quality Effects of USDA-NRCS Practices

USDA-NRCS conservation practices can be structural (e.g., Waste Treatment Lagoons; Terraces; Sediment Basins; or Fences) or agronomic (e.g., Prescribed Grazing; Nutrient Management; Pest Management; Residue Management; or Conservation Cover.) Not all USDA-NRCS conservation practices are applicable in all areas of the United States. When and where applicable, their effects on water quality may vary based on many factors. Some of these factors include climate, soils, topography, geology, existing cultural and management activities, as well as modifications made to the practice standards that govern how the practices are to be applied in local settings.

Guidance identifying expected effects of USDA-NRCS conservation practices has been prepared and is being kept up to date by discipline and resource specialist in each state. Technical guidance for water quality effects is found in the Conservation Practice Physical Effects (CPPE) documents in Section V of the NRCS Field Office Technical Guide (FOTG). Table 3-1 is a simplified table developed from the CPPE in the Oregon FOTG Section V. This table shows the kind of information available at the local level that can be used to help evaluate the effects of specific conservation practices. For example, in the area for which this guidance was prepared it has been determined that Contour Buffer Strips (NRCS Practice Code 332) can be expected to have beneficial effects on surface water quality, but because the practice increases infiltration it can be expected to have detrimental effects on ground water quality.

Table 3-1. NRCS conservation practices, pollutants potentially controlled, and sources of pollutants (USDA-NRCS, 1977).

NRCS Code	CONSERVATION PRACTICES	Ground Water					Surface Water								
		Pesticides	Nutrients & Organic	Salinity	Heavy Metals	Pathogens	Pesticides	Nutrients & Organic	Salinity	Heavy Metals	Pathogens	Temperature	Low Dissolved Oxygen	Suspended Sediments & Turbidity	Aquatic Habitat Suitability
322	Channel Vegetation	D			B	B	B	B	B	B	B	B		B	B
327	Conservation Cover		B	B	B	B	B	B		B	B	B	B	B	B
656	Constructed Wetland						B	B	B	B	B		B		B
332	Contour Buffer Strips	D	D	D	D	D	B	B	B	B	B		B	B	B
342	Critical Area Planting	D	B	B	B	B	B	B	B	B	B	B	B	B	B
400	Floodwater Diversion	B	B				B	B	B	B	B	B	B	B	
490	Forest Site Preparation							D					D	D	D
412	Grassed Waterway						B	B					B	B	B
561	Heavy Use Area Protection						B	D						B	
422	Hedgerow Planting						B	B			B			B	
441	Irrigation System - Micro	B	B		B	B	B	B	B	B	B	B	B	B	B
442	Irrigation System - Sprinkler	D	D	D	D	D	B	B	B	B	B	B	B	B	B
634	Manure Transfer		B	B				B	B	B	B	B	B	B	
484	Mulching	D	D	D	D	D	B	B	B	B	B	B	B	B	
590	Nutrient Management		B	B	B	B		B	B	B	B		B		B
528A	Prescribed Grazing		B				B	B			B	B		B	B
344	Residue Management, Seasonal	D	D	D	D	D	B	B	B	B	B	B	B	B	B
391	Riparian Forest Buffer	B	B	B	B	B	B	B	B	B	B	B		B	B
350	Sediment Basin	D	D	B	D	D	B		B	B	B			B	B
351	Well Decommissioning	B	B	B	B	B									
657	Wetland Restoration						B	B			B		B	B	B

B - Beneficial effects expected
D - Detrimental effects expected
Blank - Not Rated

Management Practice Systems

If multiple sources of a pollutant exist, more than one management practice system will be needed to provide effective control.

Water quality problems cannot usually be solved with one management practice because single practices do not typically provide the full range and extent of control needed at a site. Multiple practices are combined to build *management practice systems* that address treatment needs associated with pollutant generation from one or more sources, transport, and remediation. Management practice systems are generally more effective in controlling the pollutant since they can be used at two or more points in the pollutant delivery process. For example, the objective of many agricultural NPS pollution projects is to reduce the delivery of soil from cropland to water bodies. A system of management practices can be designed to reduce soil detachment, erosion potential, and off-site transport of eroded soil. Such a system could include conservation tillage to reduce soil detachment and cropland erosion. Grassed waterways could be included to carry concentrated flows from the fields in a non-erosive manner, while filter strips might be used to filter sediment from water leaving the field in shallow, uniform flow (Hirschi et al., 1997). Sediment retention basins could be added to trap sediment and runoff from the farm if other practices failed to provide the level of control needed.

Similarly, if nitrogen is the pollutant of concern, nutrient management can be used to minimize the availability of nitrogen for transport from cropland. This can be achieved by matching the application rate with crop needs, based upon soil testing, analysis of nutrient sources, and realistic yield expectations. Proper timing of nutrient application will also reduce nitrogen availability since the time frame over which the applied nitrogen is available but not used by the crop is minimized. Conservation tillage can help reduce overland transport of nitrogen by reducing erosion and runoff, and nutrient management will minimize subsurface losses due to the resulting increased infiltration. Filter strips can be used to decrease nitrogen transport by increasing infiltration, and through uptake of available nitrogen by the field border crop. Nitrogen not controlled by nutrient management, conservation tillage, and filter strips can be intercepted and remediated through denitrification in riparian buffers.

A set of practices does not constitute an effective management practice system unless the practices are selected and designed to function together to achieve water quality goals reliably and efficiently. In the Oregon RCWP project (see Chapter 1 for a discussion of RCWP), dairy farmers installed animal waste management systems to reduce fecal coliform runoff into an important shellfish-producing estuary. Although 12 practices (waste storage, guttering, dike, drains, etc.) initially comprised the animal waste management systems, these systems were not as effective as needed because the practices addressed manure storage but not land application of the manure. Utilization of manure was added as a practice which enabled implementation of complete management practice systems that successfully addressed the need for managing land application to achieve water quality goals (Gale et al., 1993).

Types of Management Practice Systems

Management practice systems can be separated into three categories:

☐ repetitive treatment,

❑ necessary diversification, or

❑ a combination of the first two.

Systems that combine individual management practices to treat a pollutant at different points in the pollutant delivery process achieve management objectives through repetitive treatment. The above examples for sediment and nitrogen control both employ repetitive treatment. Conservation tillage, grassed water-ways, field borders, and sediment retention basins control soil particles and runoff at various stages in the pollutant delivery process. Nutrient management, conservation tillage, field borders, and riparian buffers provide similar repetitive treatment to control nitrogen losses in the second example.

In some cases a management practice cannot be used without an accompanying practice. For example, if it is necessary to install fence to keep cows from a stream, watering devices may be needed to provide drinking water for the cows. This is an example of necessary diversification.

Some management practice systems include both treatment redundancy and necessary diversification. An example of such a system is an animal waste management system in which some components are included to help others function. For example, diversions and subsurface drains may be necessary to convey runoff and wastes to a waste treatment lagoon for treatment. While the diversions and subsurface drains may not provide any measurable pollution control of their own, they are essential to the overall performance of the animal waste management system. Other components, such as lagoons and waste utilization plans, are added to provide repetitive treatment.

Site-Specific Design of Management Practice Systems

There is no single, *ideal* management practice system for controlling a particular pollutant in all situations. Rather, the system should be designed based on the type of pollutant; the source of the pollutant; the cause of the pollution at the source; the agricultural, climatic, and environmental conditions; the pollution reduction goals; the economic situation of the farm operator; the experience of the system designers; and the willingness and ability of the producer to implement and maintain the practices. The relative importance of these and other factors will vary depending upon other considerations such as whether the implementation is voluntary (e.g., State cost-sharing program) or mandatory (e.g., discharge permits).

An example of site-specific design of management practice systems can be found in the Rural Clean Water Program (RCWP) which was discussed in Chapter 1. A similar water quality problem existed in RCWP projects conducted in Utah and Florida (Gale et al., 1993). In both projects, eutrophication was caused partly by excess phosphorus contained in dairy runoff. Animal waste management systems were installed in both projects. In the Florida project, seven individual management practices (referred to as "BMPs" in the RCWP) were needed to control the animal manure in barnyard areas, whereas only five BMPs were needed in Utah (Table 3-2). Some BMPs were used in both projects, while other BMPs were used in one but not both projects. Differences existed because the regions in which the two projects were located have significantly different climatic, ecological, and soil characteristics, requiring different approaches to mitigate animal waste problems. In Florida, annual rainfall is approximately 50 inches per year, whereas annual

Table 3-2. Animal waste management BMP systems used in two agricultural pollution control projects (Utah and Florida).

NRCS Code	Individual Animal Waste Management BMPs	UT	FL
312	Waste Management System	**	**
313	Waste Storage Structure	**	**
356	Dike		**
362	Diversion	**	**
425	Waste Stirage Pond	**	**
428	Concrete Lining		**
633	Waste Utilization	**	

NRCS = Natural Resources Conservation Service, U.S. Department of Agriculture
Source for NRCS codes: USDA—NRCS, 1977

precipitation in Utah is approximately 16 inches per year. Surface water is largely derived from snowmelt in Utah. Dikes were used in the Florida project to prevent runoff and phosphorus from entering the drainage ditches. These dikes were not needed in Utah due to the lower rainfall producing less runoff.

Practices Must Fit Together for Systems to Perform Effectively

Each practice in a management practice system must be selected, designed, implemented, and maintained in accordance with site-specific considerations to ensure that the practices function together to achieve the overall management goals. If, for example, nutrient management, conservation tillage, filter strips, and riparian buffers are used to address a nitrogen problem, then planting and nutrient applications need to be conducted in a manner consistent with conservation tillage goals and practices (e.g., injecting rather than broadcasting and incorporating fertilizer). In addition, runoff from the fields must be conveyed evenly to the filter strips which, in turn, must be capable of delivering the runoff to the riparian buffers in accordance with design standards and specifications.

Management Measures

This guidance document is intended to provide technical information to state program managers and others on the best available, economically achievable means of reducing NPS pollution of surface and ground water from agriculture. The guidance provides background information about agricultural NPS pollution, where it comes from and how it enters the nation's waters, discusses the broad concept of assessing and addressing water quality problems on a watershed level, and presents up-to-date technical information about how to reduce agricultural NPS pollution.

Management measures for nutrient management, pesticide management, erosion and sediment control, facility wastewater and runoff from confined animal facilities, grazing management, and irrigation water management are described in Chapter 4. Also in Chapter 4 are discussions of BMPs that can be used to achieve the management measures, including cost and effectiveness information.

4A: Nutrient Management

Management Measure for Nutrients

Develop, implement, and periodically update a nutrient management plan to: (1) apply nutrients at rates necessary to achieve realistic crop yields, (2) improve the timing of nutrient application, and (3) use agronomic crop production technology to increase nutrient use efficiency. When the source of the nutrients is other than commercial fertilizer, determine the nutrient value and the rate of availability of the nutrients. Determine and credit the nitrogen contribution of any legume crop. Soil and plant tissue testing should be used routinely. Nutrient management plans contain the following core components:

1. Farm and field maps showing acreage, crops, soils, and waterbodies. The current and/or planned plant production sequence or crop rotation should be described.

2. Realistic yield expectations for the crop(s) to be grown, based primarily on the producer's actual yield history, State Land Grant University yield expectations for the soil series, or local NRCS information for the soil series.

> To reduce water pollution caused by nitrogen and phosphorus, develop and implement a broad-based nutrient management plan.

3. A summary of the nutrient resources available to the producer, which at a minimum include:

    ☐ Soil test results for pH, phosphorus, nitrogen, and potassium;

    ☐ Nutrient analysis of manure, sludge, mortality compost (birds, pigs, etc.), or effluent (if applicable);

    ☐ Nitrogen contribution to the soil from legumes grown in the rotation
    (if applicable); and

❑ Other significant nutrient sources (e.g., irrigation water, atmospheric deposition).

4. An evaluation of field features based on environmental hazards or concerns, such as:

    ❑ Sinkholes, shallow soils over fractured bedrock, and soils with high leaching potential;

    ❑ Subsurface drains (e.g., tile drains);

    ❑ Lands near surface water;

    ❑ Highly erodible soils;

    ❑ Shallow aquifers;

    ❑ Combinations of excessively well drained soils and high rainfall seasons, resulting in very high potential for surface runoff and leaching; and

    ❑ Submarine seeps, where nutrient-laden ground water from upland areas can directly enter the ocean through tidal pumping (e.g. along the coastline of Maui, Hawaii).

5. Use of the limiting nutrient concept to establish the mix of nutrient sources and requirements for the crop based on a realistic yield expectation.

6. Identification of timing and application methods for nutrients to provide nutrients at rates necessary to achieve realistic crop yields, reduce losses to the environment, and avoid applications as much as possible to frozen soil and during periods of leaching or runoff.

7. Provisions for the proper calibration and operation of nutrient application equipment.

Management Measure for Nutrients: Description

The goal of this management measure is to minimize nutrient losses from agricultural lands occurring by edge-of-field runoff and by leaching from the root zone. Once nitrogen, phosphorus, or other nutrients are applied to the soil, their movement is largely controlled by the movement of soil and water and must therefore be managed through other control systems such as erosion control and irrigation water management. Effective nutrient management abates nutrient movement by minimizing the quantity of nutrients available for loss (source reduction). This is usually achieved by developing a nutrient budget for the crop, applying nutrients at the proper time with proper methods, applying only the types and amounts of nutrients necessary to produce a crop, and considering the environmental hazards of the site. In cases where manure is used as a nutrient source, manure holding areas may be needed to provide capability to apply manure at optimal times.

The focus of nutrient management is to increase the efficiency with which applied nutrients are used by crops, thereby reducing the amount available to be transported to both surface and ground waters. In many instances, nutrient management results in the use of less commercial fertilizer and, therefore, a reduction in production costs. However, where there has not been a balanced use of nutrients in the past, the application of this management measure may result in more nutrients being applied.

While the nutrient management plan may have many components, the principle is simple: minimize total losses.

The best approach to *minimizing nutrient transport* to surface and ground waters depends upon whether the nutrient is in the dissolved phase or is attached to soil particles. For dissolved nutrients, effective management includes source reduction and reduction of water runoff or leaching. Erosion and sediment transport controls are necessary to reduce transport of nutrients attached to soil particles. Practices that focus on controlling the transport of smaller soil particle sizes (e.g., clays and silts) are most effective because these are the soil fractions that transport the greatest share of adsorbed nutrients.

Sources of Nutrients

Nitrogen (N), phosphorus (P), and potassium (K) are the primary nutrients applied in most agricultural operations. Nutrient management plans typically focus mainly on N and P, the nutrients of greatest concern for water quality.

The major sources of nutrients include:

- ☐ Commercial fertilizers

- ☐ Manures, sludges, and other organic materials

- ☐ Crop residues and legumes in rotation

- ☐ Irrigation water

- ☐ Soil reserves

> Nutrient management planning is enhanced by knowledge of the nitrogen and phosphorus cycles.

Because these two elements behave very differently, basic understanding of how N and P are cycled in the soil-crop system is an important foundation for effective nutrient management.

Nutrient Cycles

Nitrogen is continually cycled among plants, soil organisms, soil organic matter, water, and the atmosphere (Figure 4a-1) in a complex series of biochemical

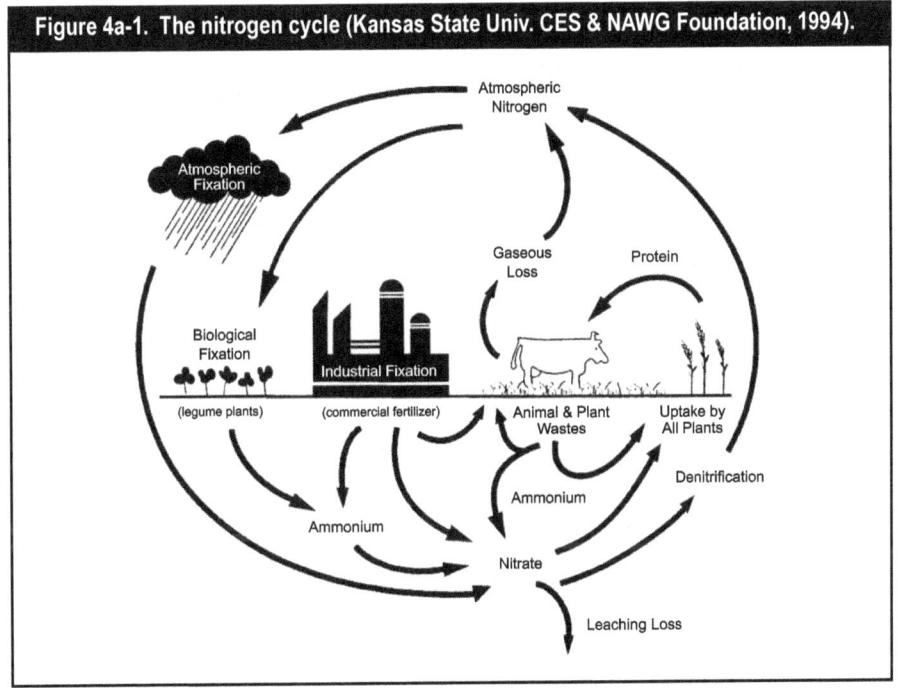

Figure 4a-1. The nitrogen cycle (Kansas State Univ. CES & NAWG Foundation, 1994).

transformations. Some N forms are highly mobile, while others are not. At any given time, most of the N in the soil is held in soil organic matter (decaying plant and animal tissue) and the soil humus. *Mineralization* processes slowly transform the N in soil organic matter by microbial decomposition to ammonium ions (NH4+), releasing them into the soil where they can be strongly adsorbed and relatively immobile. Plants can use the ammonium, however, and it may be moved with sediment or suspended matter. *Nitrification* by soil microorganisms transforms ammonium ions (either mineralized from soil organic matter or added in fertilizer) to nitrite (NO2-) and then quickly to nitrate (NO3-), which is easily taken up by plant roots. Nitrate, the form of N most often associated with water quality problems, is soluble and mobile in water. *Immobilization* includes processes by which ammonium and nitrate ions are converted to organic-N, through uptake by plants or microorganisms, and bound in the soil. *Denitrification* converts nitrate (NO3) into nitrite (NO2) and then to nitrous oxide (N2O) and gaseous nitrogen (N) through microbial action in an anaerobic environment.

A nitrogen molecule may pass through this cycle many times in the same field. The processes in the nitrogen cycle can occur simultaneously and are controlled by soil organisms, temperature, and availability of oxygen and carbon in the soil. The balance among these processes determines how much N is available for plant growth and how much may be lost to ground water, surface water, or the atmosphere.

Phosphorus lacks an atmospheric connection (although it can be transported via airborne soil particles) and is much less subject to biological transformation, rendering the P cycle considerably simpler (Figure 4a-2). Most of the P in soil occurs as a mixture of mineral and organic materials. A large amount of P (50–75%) is held in soil organic matter which is slowly broken down by soil microorganisms. Some of the organic P is released into soil solution as phosphate ions that are immediately available to plants. The phosphate ions released by decomposition or added in fertilizers are strongly adsorbed to soil particles and are rapidly immobilized in forms that are unavailable to plants. The equilibrium

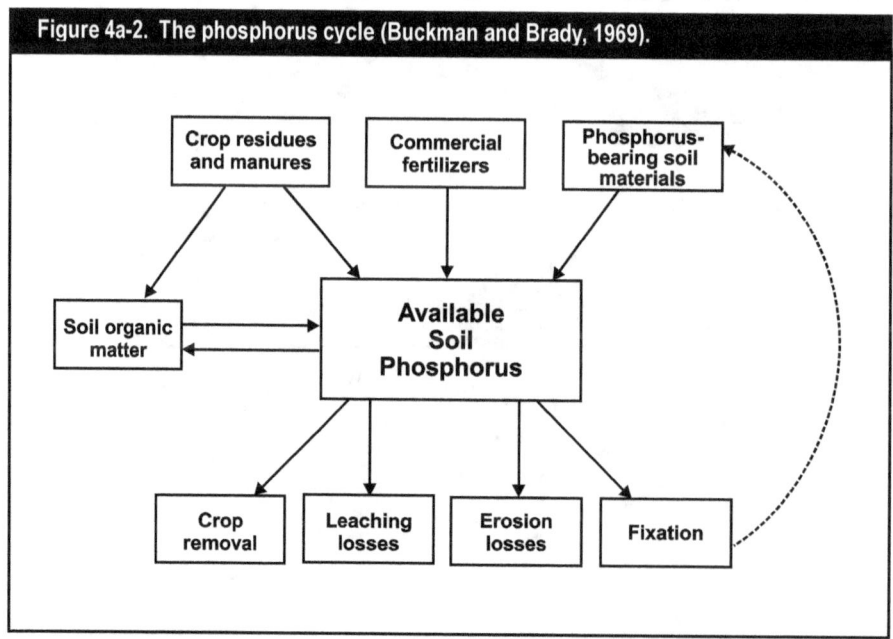

Figure 4a-2. The phosphorus cycle (Buckman and Brady, 1969).

level of dissolved P in the soil solution is controlled by the chemical environment of the soil (e.g. pH, oxidation-reduction, iron concentration) and by the P content of the soil.

Commercial Fertilizers

Fertilizers represent the largest single source of N, P, and K applied to most cropland in the U.S. Major commercial fertilizer N sources include anhydrous ammonia, urea, ammonium nitrate, and ammonium sulfate. Major P fertilizer sources include monoammonium phosphate, diammonium phosphate, triple superphosphate, ammonium phosphate sulfate, and liquids. The predominant source of potassium (K) fertilizer is potassium chloride. Descriptions of common fertilizer materials are given in Table 4a-1. The use of any particular material or blend is governed by the characteristics of the formulation (such as volatilization potential and availability rate), suitability for the particular crop, crop needs, existing soil test levels, economics, application timing and equipment, and handling preferences of the producer. An example of general fertilizer

Table 4a-1. Common fertilizer minerals.

Common Name	Chemical Formula	Analysis (%)		
		N	P_2O_5	K_2O
Nitrogen materials				
Ammonium nitrate	NH_4NO_3	34	0	0
Ammonium sulfate	$(NH_4)_2SO_4$	21	0	0
Ammonium nitrate-urea	$NH_4NO_3+(NH_2)_2CO$	32	0	0
Anhydrous ammonia	NH_3	82	0	0
Aqua ammonia	NH_4OH	20	0	0
Urea	$(NH_2)_2CO$	46	0	0
Phosphate materials				
Superphosphate	$Ca(H_2PO_4)_2$	0	20-46	0
Ammoniated superphosphate	$Ca(NH_4H_2PO_4)_2$	5	40	0
Monoammonium phosphate	$NH_4H_2PO_4$	13	52	0
Diammonium phosphate	$(NH_4)_2HPO_4$	18	46	0
Urea-ammonium phosphate	$(NH_2)_2CO+(NH_4)_2HPO_4$	28	28	0
Potassium materials				
Muriate of potash	KCl	0	0	60
Monopotassium phosphate	KH_2PO_4	0	50	40
Potassium hydroxide	KOH	0	0	70
Potassium nitrate	KNO_3	13	0	45
Potassium sulfate	K_2SO_4	0	0	50

Source: Pennsylvania State University. 1997. *The Penn State Agronomy Guide, 1997-1998,* University Park, PA. Cornell Cooperative Extension. 1997. *1997 Cornell Recommendations for Integrated Field Crop Management.* Resource Center, Cornell University, Ithaca, NY.

Precision Farming
A New Era of Production

The Precisely Tailored Practice

Precision farming, also known as site-specific management, is a fairly new practice that has been attracting increasing attention both within and outside the agricultural industry over the past few years. It is a practice concerned with making more educated and well-informed agricultural decisions. Precision farming provides tools for tailoring production inputs to specific plots (or sections) within a field. The size of the plots typically range from one to three acres, depending on variability within the field and the farmer's preference. By treating each plot as much or as little as needed, farmers can potentially reduce the costs of seed, water, and chemicals; increase overall crop yields; and reduce environmental impacts by better matching inputs to specific crop needs. Rather than applying fertilizer or pesticides to an entire field at a single rate of application, farmers first test the soil and crop yields of specific plots and then apply the appropriate amount of fertilizer, water, and/or chemicals needed to alleviate the problems in those sections of the field. Precision farming requires certain technology, which is an added cost, as well as increased management demands.

Precision farming is changing the way farmers think about their land. They are increasingly concerned not with the average needs of the entire field, but with the actual needs of specific plots, which can fluctuate from one square meter to the next. The practice of precision farming acknowledges the fact that conditions for agricultural production vary across space and over time. With this in mind, precision farmers are now making management decisions more specific to time and place rather than regularly scheduled and uniform applications.

The Computer-Aided Approach

The approach of precision farming involves using a wide range of computer-related information technologies, many just recently introduced to production agriculture, to precisely match crops and cultivation to the various growing conditions. The key to successfully using the new technologies available to the precision farmer to maximize possible benefits associated with this approach is *information*. Data collection efforts begin before crop production and continue until after the harvest. Information-gathering technologies needed prior to crop production include grid soil sampling, past yield monitoring, remote sensing, and crop scouting. These data collection efforts are even further enhanced by obtaining precise location coordinates of plot boundaries, roads, wetlands, etc., using a global positioning system (GPS).

Other data collection takes place during production through "local" sensing instruments mounted directly on farm machinery. Variable rate technology (VRT) uses computerized controllers to change rates of inputs such as seed, pesticides, and nutrients through planters, sprayers, or irrigation equipment. For example, soil probes mounted on the front of fertilizer spreaders can continuously monitor electrical conductivity, soil moisture, and other variables to predict soil nutrient concentrations and accordingly adjust fertilizer application "on-the-fly" at the rear of the spreader. Other direct sensors available include yield monitors, grain quality sensors, salinity meter sleds, weather monitors, and spectroscopy devices. Optical scanners can be used to detect soil organic matter, to recognize weeds, and to instantaneously alter the amount or application of herbicides applied.

The precision farmer can then take the information gathered in the field and analyze it on a personal computer. The personal computer can help today's farmer organize and manage the information collected more effectively. Computer programs, including spreadsheets, databases, geographic information systems (GIS), and other types of application software, are readily available. By tying specific location coordinates obtained from the GPS in with the other field data obtained, the farmer can use the GIS capability to create overlays and draw analytical relationships for site-specific patterns of soils, crop yields, input applications, drainage patterns, and other variables of interest over a particular distance or time period.

GIS can also be integrated with other decision support systems (DSS), such as process models and artificial intelligence systems, to simulate anything from crop growth and financial expectations to the generation and movement of nutrients and pesticides through the environment. Today's precision farmer can also use expert systems, information systems based on input from human experts, to retrieve advice on when to spray for specific pests, when to till, and so forth. These systems are continuously modified for the farmer's field based on past, current, and expected conditions represented by soil, weather, pest level, and other data input from the GIS.

The Technology-Driven Future

Further technological advances will make the coming years decisive for the precision farming industry. There's no saying what the future holds for this new era of agricultural production. Listed below are just a few of the technological advances projected to hit this industry in the years to come.

- ☐ Onboard grain quality analyzers will check both physical and chemical attributes (including smell).

- ☐ High-precision soil testing will move from the lab to the field, with fiberoptic spectrometers attached to real-time onboard computers.

- ☐ Micro-ecology will be tested along with water runoff and air samples.

- ☐ Immunochemical assays will measure chemical residues on leaf surfaces or monitor plant health and productivity.

- ☐ A wide range of sensors, monitors, and controllers such as shaft monitors, pressure transducers, and servo motors will be used to collect accurate data.

- ☐ Weather monitors will be mounted on sprayers, or "talk" directly to local weather station networks as they simultaneously change droplet size or spray patterns, as well as rates and products, on the go.

- ☐ Remote imaging technologies will be used to assess crop health and management practice implementation.

- ☐ Guidance on control systems will guarantee straight rows, control depth, and optimize inputs.

- ☐ Crop models will optimize economic and environmental variables. Farmers will buy insurance directly from the underwriter, who will also rely on remote sensing and risk modeling.

- ☐ Wearable computers with voice recognition and head-mounted displays will guide farmers through equipment maintenance and crop scouting.

Although precision farming has not yet been widely adopted to date, this practice continues to attract increasing attention both on and off the farm. Much of the off-the-farm enthusiasm for precision farming can be attributed to the eminent good sense of matching input application to plant needs. Precision farming is simply a more finely tuned version of the kinds of BMPs already recommended at the field level. Because this technology is still somewhat new to the industry, there is much more to learn about the potential overall impact of precision farming on water and air quality relative to conventional techniques. But one thing is certain: precision farming has the potential to enhance economic return (by cutting costs and raising yields) and to reduce environmental risk (by reducing the impacts of fertilizers, pesticides, and erosion).

Table 4a-2. Fertilizer recommendations for corn in New York State (Cornell Cooperative Extension, 1997).

Soil Management Group	Years Following Sod	NITROGEN (N)[5],[6],[7] — Type of Plowed Sod						PHOSPHORUS (P_2O_5) — Soil Test Phosphorus Levels[8]					POTASSIUM (K_2O) — Soil Test Potassium Levels[8]				
		Grass No Manure	Grass Manure	Less than 50% Legume No Manure	Less than 50% Legume Manure	Greater than 50% Legume No Manure	Greater than 50% Legume Manure	Very Low	Low	Medium	High	Very High	Very Low	Low	Medium	High	Very High
Soil group I–Clayey soils, fine-textured soils in northern New York, near lakes and along the Hudson River. Examples: Vergennes, Kingsbury, Hudson, Rhinebeck, Schoharie, Odessa.	1	10–30	10–30	10–30	10–30	10–30	10–30	70	60	40	20	0	50	40	30	20	0
	2	50–100	10–40	30–80	10–20	20–70	10–30										
	3	70–110	10–50	60–100	10–40	60–100	10–40										
	4 or more	80–120	20–60	80–120	20–60	80–120	20–60										
Soil group II–Silty soils, medium- to moderately fine-textured soils of the central region. Examples: Cazenovia, Hilton, Honeoye, Lima, Ontario, Lansing, Mohawk, Chagrin, Teel.	1	10–30	10–30	10–30	10–30	10–30	10–30	70	60	40	20	0	60	60	40	20	0
	2	60–100	10–40	50–90	10–30	40–80	10–30										
	3	80–120	10–60	70–110	10–50	70–110	10–50										
	4 or more	90–130	30–70	90–130	30–70	90–130	30–70										
Soil group III–Silt loam soils, moderately coarse-textured acid soils of the Southern Tier, glacial outwash. Examples: Barbour, Chenango, Palmyra, Tioga, Mardin, Langfor, Tunkhannock.	1	10–30	10–30	10–30	10–30	10–30	10–30	70	60	40	20	0	80	70	50	25	0
	2	60–100	10–40	40–90	10–30	30–80	10–30										
	3	80–120	10–60	70–110	10–50	70–110	10–50										
	4 or more	90–130	30–70	90–130	30–70	90–130	30–70										
Soil group IV–Loamy soils, coarse- to medium-textured soils of northern New York and the Hudson Valley. Examples: Bombay, Broadalbin, Copake, Empeyville, Madrid, Sodus, Worth.	1	10–30	10–30	10–30	10–30	10–30	10–30	70	60	40	20	0	120	80	50	25	0
	2	60–110	10–50	50–90	10–30	40–90	10–30										
	3	80–120	10–60	70–120	10–60	70–110	10–50										
	4 or more	90–130	30–70	90–130	30–70	90–130	30–70										
Soil group V–Sandy soils, very coarse-textured soils on beach ridges, deltas, and sandy or gravelly outwash near mountains and the Hudson Valley. Examples: Alton, Colton, Windsor, Colonie, Elmwood, Junius, Suncook.	1	10–30	10–30	10–30	10–30	10–30	10–30	70	60	40	20	0	120	90	60	30	0
	2	40–100	10–40	20–80	10–20	20–70	10–30										
	3	60–110	10–50	50–100	10–40	50–100	10–40										
	4 or more	70–120	20–60	70–120	10–60	70–120	10–60										

[1]A more specific recommendation will be obtained from a complete soil test ...

recommendations for corn is shown in Table 4a-2. Commercial fertilizers offer the advantage of allowing exact formulation and delivery of nutrient quantities specifically tailored to the site, crop, and time of application in concentrated, readily available forms.

Organic Nutrient Sources

Organic nutrient sources, such as manure, sludge, and compost, can supply all or part of the N, P, and K needs for crop production. Organic nutrient sources offer additional advantages because they also contain secondary nutrients and micro-nutrients (e.g. iron, boron), add organic matter to the soil, provide nutrients to crops for several years after application, and provide a practical outlet to recycle manure and other farm organic materials. The use of manure is particularly important on livestock and poultry farms because nutrients can build up in the soil, be lost to the atmosphere, leach into ground water, or runoff to surface waters as more nutrients are brought onto the farm than leave in products sold. Table 4a-3 shows examples of estimated N and P mass balances for several New York dairy farms.

Table 4a-3. N and P mass balances on several New York dairy farms.					
	Nitrogen Size (# of cows)			**Phosphorus** Size (# of cows)		
	45	85	120	45	85	120
	—tons of N/yr— -			—tons of P/yr—-		
INPUT						
purchased fertilizer	1.0	2.2	4.6	1.2	0.9	1.3
purchased feed	3.8	9.7	21.4	1.0	1.7	5.4
legume N fixation	1.3	1.1	3.2	==	==	==
Total:	6.1	13.0	29.2	2.2	2.6	6.7
OUTPUT						
milk	2.0	3.8	6.3	0.4	0.7	1.1
meat	0.1	0.4	0.6	<0.1	0.1	0.2
crops sold	0.1	0.5	==	≤0.1	≤0.1	==
Total:	2.2	4.7	6.9	0.4	0.8	1.3
REMAINDER	**3.9**	**8.3**	**22.3**	**1.8**	**1.8**	**5.4**
remaining on farm	**64%**	**64%**	**76%**	**81%**	**69%**	**81%**

Source: Klausner, S. 1995. Nutrient Management: Crop Production and Water Quality. 95CUWFP1, Cornell University, Ithaca, NY.

The nutrient content of manure and other organic materials can vary greatly according to the type of animal, type of feed, storage and handling procedures, climate, and management. In order to use them efficiently, these materials must be analyzed for their nutrient content. Examples of average values for nutrient content of organic materials are shown in Table 4a-4; however, it is important to note that the nutrient content of manure even on neighboring farm operations may vary widely from the average.

A difficulty in using organic nutrient sources is that their nutrient content is rarely balanced for the specific soil and crop needs. For example, the ratio of N:P in applied manure is usually around 3 or less, while the ratio at which crops

Table 4a-4. Representative values for nutrients in manure, sludge, and whey, as applied.

		Total N	P_2O_5 [1]	K_2O [1]
SOLID MANURE				
Species	**% dry matter**	——lb/ton——		
Dairy cattle	18-22	6-17	4-9	2-15
Beef cattle	15-50	11-21	7-18	10-26
Swine	18	8-10	6-9	7-9
Poultry	22-76	20-68	16-64	12-45
Sheep	28	14-18	9-11	25-26
Horse	46	14	4	14
LIQUID MANURE				
Species	**% dry matter**	——lb/1000 gal——		
Dairy cattle	1-8	4-32	4-18	5-30
Beef cattle	1-11	4-40	9-27	5-34
Veal calf	3	24	25	51
Swine	1-4	4-36	2-27	4-22
Poultry	13	69-80	36-69	33-96
DIGESTED SLUDGE		——lb/1000 gal——		
		20	12	1
WHEY		——lb/1000 gal——		
		12	9	18

[1]Convert values for P2O5 and K2O to P and K by multiplying by 0.43 and 0.83, respectively.

Sources: Midwest Plan Service. 1985. *Livestock Waste Facilities Handbook.* Iowa State University, 1991a. Ames, IA. Klausner, S. 1995. *Nutrient Management: Crop Production and Water Quality.* 95CUWFP1, Cornell University, Ithaca, NY. University of Wisconsin-Extension and Wisconsin Dept. of Agriculture, Trade, and Consumer Protection. 1989. *Nutrient and Pesticide Best Management Practices for Wisconsin Farms.* WDATCP Technical Bulletin ARM-1, Madison, WI. University of Vermont. 1996. *Agricultural Testing Laboratory – Manure Analysis Averages, 1992-1996.* Dept. of Plant & Soil Science, University of Vermont, Burlington, VT.

use nutrients typically ranges from 5 to 7. Therefore, when manure is applied at rates based solely on N analysis and crop need for N, excess amounts of P are added. Because the amounts of P added in manure exceed the amounts removed by crops, continuous manure usage can result in accumulations of excess P in the soil, increasing the potential for P to be transported in runoff and erosion (Daniel et al., 1997).

Another difficulty in efficient use of manure nutrients involves nutrient availability. Not all nutrients in manure are immediately available for crop uptake. The organic N in manure, for example, must be mineralized before it can be used by plants, a process that may take 3 or more years to complete. Examples of average amounts of nutrients available for crop growth in the first year of application in Wisconsin are shown in Table 4a-5. Actual quantities of available nutrients at a specific site will depend on initial nutrient content of the manure, soil type, temperature, and soil moisture. Failure to account for this slow availability can result in under-supply of nutrients in a given year of manure application. Perhaps more critically, it must be recognized that when manure is applied to the same field over the years, each succeeding year requires the addition of

Credits for previous year manure applications and nitrogen-fixing crops should be considered in the plan for nitrogen management.

Table 4a-5. Nutrients available for crop use in the first year after spreading manure.

Animal	SOLID			LIQUID		
	N incorp.	N not incorp.	P₂O₅	N Incorp.	N not incorp.	P₂O₅
		——lbs/ton ——			——bs/1000 gal ——	
Dairy	4	3	3	10	8	8
Beef	4	4	5	12	10	14
Swine	5	4	3	15	12	6
Poultry	15	13	14	41	35	38

Source: University of Wisconsin-Extension and Wisconsin Dept. of Agriculture, Trade, and Consumer Protection. 1989. *Nutrient and Pesticide Best Management Practices for Wisonsin Farms.* WDATCP Technical Bulletin ARM-1, Madison, WI.

Table 4a-6. Quantity of livestock or poultry manure needed to supply 100 kg of Nitrogen over the cropping year with repeated applications of manure (Schepers and Fox, 1989).

Number of years applied	Quantity (metric tons) needed for manure with these percent N			
	0.25	1.0	2.0	4.0
1	154	22	7	1.4
2	79	16	6	1.4
3	54	13	5	1.4
4	41	11	5	1.3
5	33	10	4	1.3
10	17	7	3.7	1.3
15	12	6	3.3	1.2
20	9	5	3.0	1.2

less N to maintain an adequate supply of plant available N (Table 4a-6). Failure to consider this N carryover could lead to excessive application of N.

Since organic nutrient sources contain valuable nutrients and have soil-conditioning properties, application to land should never be considered disposal. In cases where organic nutrient sources are disposed of as waste with no regard given to their N and P content, excessive levels of available nutrients and losses to surface or ground waters are likely to occur.

Because of their ability to "fix" atmospheric nitrogen, legumes grown in rotation can represent a significant input of N into the soil of a crop field. Alfalfa has been reported to fix from 60 to 530 lb N/ac (pounds of nitrogen per acre); soybeans may fix from 13 to 275 lb N/ac. Some of this fixed N is removed in harvest, but some remains in crop residue or in the soil and is available for subsequent crops. Table 4a-7 shows representative values for residual N contributions from legume crops. Failure to account for such added N could result in excessive application of N from other sources.

Table 4a-7. Representative values for first-year nitrogen credits for previous legume crops.

Crop	Nitrogen Credit (lb N/ac)
Forages	
Alfalfa[a]	
>50%	80 – 120
25-50%	50 – 80
<25%	0 – 40
Red Clover and Trefoil[a]	
>50%	60 – 90
25-50%	40 – 60
<25%	0 – 30
Soybeans	1 lb N/ac for each bu/ac harvested up to 40 lb N/ac
Green Manure Crops (plowed down after growing season of seeding year)	
Sweet clover	80 - 120
Alfalfa 60 - 100	
Red clover	50 - 80
Vegetable Crops (residue not removed)	
Peas, snap beans, lima beans	10 - 20

[a] The percentage of stand of the particular crop.

Sources: Pennsylvania State University. 1997. The Penn State Agronomy Guide, 1997-1998, University Park, PA. University of Wisconsin-Extension and Wisconsin Dept. of Agriculture, Trade, and Consumer Protection. 1989. Nutrient and Pesticide Best Management Practices for Wisconsin Farms. WDATCP Technical Bulletin ARM-1, Madison, WI.

Irrigation Water

Irrigation water, if drawn from already nutrient-enriched sources, can supply significant amounts of N. In the Central Platte River Valley in Nebraska, ground water used to irrigate corn contributed an average of 41 lb N/ac, nearly one-third of the N fertilizer requirement (Schepers et al., 1986). Ground water used to irrigate potatoes in Wisconsin contributed an average of 51 lb N/ac, or 25% of the N added as fertilizer (Saffigna and Keeney, 1977). Table 4a-8 shows guidelines for calculating the N contribution from irrigation water.

Table 4a-8. Calculating N contributions from irrigation water.

N in water (mg/l)	Water Application Rate (acre-feet)			
	0.5	1.0	1.5	2
	lb N/ac			
2	3	5	8	11
4	5	11	16	22
6	8	16	24	32
8	11	13	32	43
10	13	27	40	54

Source: Kansas State University Cooperative Extension System and The National Association of Wheat Growers Foundation. 1994. Best Management Practices for Wheat. NAWG Foundation, Washington, D.C.

Soil Nutrients

The release of N, P, K, and micronutrients from soil reserves provides an additional source of plant-available nutrients. The amount of nutrient release depends on soil moisture, aeration, temperature, pH, and the amount of organic matter in the soil. The magnitude of this source can be assessed accurately only through soil testing.

Atmospheric Sources

Finally, atmospheric deposition can significantly contribute nutrients, especially N, to the soil. Because of the atmospheric linkages of the N cycle and industrial additions of N to the atmosphere, N loading from atmospheric deposition can be significant. From 1983-1994, average annual inorganic N deposition over the Chesapeake Basin ranged from 3.5 to 7.7 kg N/ha; average annual NO3+NH4 atmospheric deposition loading rates ranged from 6.7 to 7.8 kg N/ha (Wang et al., 1997). McMahon and Woodside (1997) cite wet NO3 and NH4 deposition rates of 9.8 kg N/ha/yr and 2.8 kg N/ha/yr, respectively, for the Albemarle-Pamlico Drainage Basin in North Carolina and Virginia. Examples of atmospheric deposition rates for various forms of N across the U.S. are given in Table 4a-9.

Table 4a-9. N loading in atmospheric deposition, NADP/NTN data, 1996.				
Location	Station	NH_4-N	NO_3-N	Inorganic N
		——— kg N/ha/yr ———		
Vermont	Mt. Mansfield (VT99)	1.78	2.95	4.73
North Carolina	Mt. Mitchell (NC45)	2.39	2.92	5.31
Florida	Quincy (FL14)	1.06	1.60	2.66
Wisconsin	Popple River (WI09)	1.93	2.16	4.10
Indiana	Purdue Ag Res Ctr (IN41)	3.29	3.64	6.94
Arkansas	Fayetteville (AR27)	2.55	2.24	4.80
Nebraska	North Platte Ag Exp Sta (NE99)	2.54	1.58	4.12
California	Davis (CA88)	2.18	0.82	3.00
Alaska	Poker Creek (AK01)	0.05	0.11	0.16
Hawaii[1]	Mauna Loa (HI00)	0.05	0.05	0.10

all data reported as N

[1] 1993

Source: National Atmospheric Deposition Program (NRSP-3)/National Trends Network (June 24, 1998). NADP/NTN Coord. Office, Illinois State Water Survey, 2204 Griffith Dr., Champaign, IL 61820.

Atmospheric deposition of P is generally very small. Ahl (1988) cited atmospheric deposition of 0.05–0.5 kg P/ha/yr in Canada. Annual P loading rates to the Chesapeake Basin have been estimated at 0.16 to 0.47 kg/ha (Wang et al., 1997). A similar P deposition rate of 0.16 kg/ha/yr has been measured in the Lake Champlain basin (VTDEC and NYS DEC, 1997). An estimated annual load of 0.66 kg P/ha by atmospheric deposition has been cited for the Albemarle-Pamlico Basin (McMahon and Woodside, 1997).

The most comprehensive collection of data on precipitation chemistry and atmospheric deposition is available from the National Atmospheric Deposition Program/National Trends Network (NADP/NTN) at: http://nadp.sws.uiuc.edu/. Data are available for precipitation chemistry, annual and seasonal wet deposition totals, isopleth maps of precipitation chemistry and wet deposition, and other variables for over 200 sites in the continental U.S., Alaska, Hawaii, Puerto Rico, and the Virgin Islands. While deposition data from the NADP network may not be exactly applicable to a specific site due to local factors such as elevation, air movement, or industrial emissions, NADP data can help provide an initial screening estimate of the possible significance of atmospheric nutrient sources. If atmospheric inputs are estimated to be significant, specific local data can be sought from university or agency research activities.

Nutrient Movement into Surface and Ground Water

Nutrients in harvested crops typically represent the largest <u>single</u> component of nutrient output from agricultural land. Table 4a-10 gives representative values for annual crop nutrient removal. However, crop uptake of added N and P is by no means complete. Overapplication of nutrients relative to crop need results in build-up of N and P surplus in agricultural soils. Nutrient surpluses have been documented at both the farm scale (Klausner, 1995) and the watershed scale (McMahon and Woodside, 1997; Cassell et al., 1998). Soil test values show that soil P in many areas is excessive, relative to crop requirements; the greatest concern occurs with animal-based agriculture, where farm and watershed-scale P surpluses and over-application of P to soils are common. (Breeuwsma et al., 1995; Lander et al., 1998; Sims et al., 2000). Accumulation of P in cropland soils may be especially high if the N requirement of the crop is met with animal waste, adding P in excess of crop P uptake (Figure 4a-3). The magnitude of potential loss of nutrients to surface and ground waters is directly related to accumulation of excessive nutrient levels in soils.

Some general principles govern nutrient movement. Site specific crop history, climate, soils, watershed, and farming characteristics result in specific local nutrient pathways and transformations.

Table 4a-10. Crop nutrient removal.			
Crop	**Yield** /ac	**N** —— lb/ac ——	**P**
Corn	125 bu	95	22
Corn silage	21 t	190	46
Grain sorghum	125 bu	65	33
Soybeans	40 bu	130	18
Wheat/rye	60 bu	90	26
Oats	80 bu	90	31
Barley	75 bu	105	20
Alfalfa	5 t	250	33
Orchardgrass	6 t	300	44
Tall fescue	3.5 t	135	29
Sugar beets	30 t	275	37

Sources: Pennsylvania State University. 1997. *The Penn State Agronomy Guide 1997-1998*, University Park, PA; Midwest Plan Service. 1985. *Livestock Waste Facilities Handbook*. Iowa State University, Ames, IA.

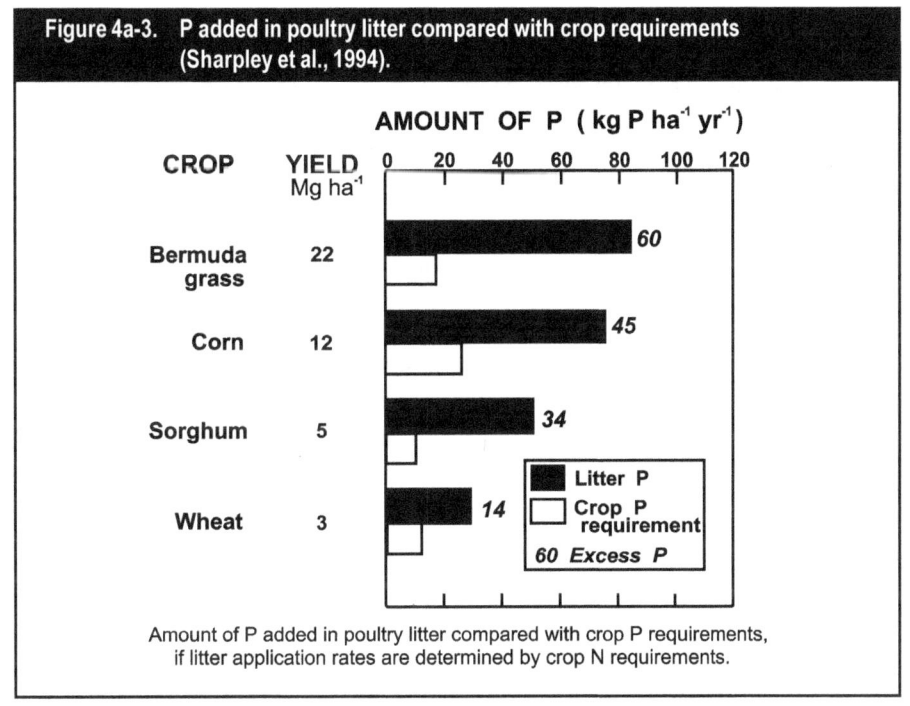

Figure 4a-3. P added in poultry litter compared with crop requirements (Sharpley et al., 1994).

Amount of P added in poultry litter compared with crop P requirements, if litter application rates are determined by crop N requirements.

N and P not removed in the harvested crop can become available for transport to surface and ground waters. The movement of applied nutrients is primarily driven by the movement of water and eroded soil, but the specific transport pathways are largely determined by the characteristics of the nutrient source, soil characteristics, and related environmental conditions (e.g., soil temperature). As noted in the earlier discussion of nutrient cycles, readily soluble nitrate moves easily in the liquid phase. Due to its strong affinity for soil particles, phosphorus usually moves primarily with eroding soil particles. Nitrogen can volatilize directly from fertilizers such as urea and ammonia and from surface-applied manure; N lost to the atmosphere in this way may be washed from the atmosphere by rain a great distance away. Nitrogen can also be lost to the atmosphere as harmless nitrogen gas through denitrification. Other factors influencing nutrient movement include topography, precipitation patterns, and, of course, land use and management.

Movement to Surface Waters

Transport of nutrients to surface waters depends on the availability of nutrients in the upper soil zone, how easily the nutrients and/or associated soil particles are detached, whether the chemical is transported in the dissolved form or attached to soil, and any deposition that may occur before delivery to a water-way. Nutrients are most susceptible to runoff loss while they are in a thin (<3 cm) layer at the soil surface where overland flow, chemicals, and soil intermix during runoff. Once nutrients are below this mixing zone, they are usually less vulnerable to ordinary runoff losses. Nitrate is an exception, as it can be readily leached through the soil.

Nitrogen can be delivered to surface waters through runoff, erosion, and subsur-face flow. Some N in the form of ammonium can be lost by erosion along with organic N attached to soil particles. Soluble N can be carried in surface runoff, but most soluble nitrate is lost via leaching through the soil. Leached nitrate may move into surface waters through shallow subsurface flow or be transported to deeper ground water. Drainage tiles may provide an important short circuit for delivery of N from shallow subsurface flow to surface waters. Concentrations of nitrate in tile drain flow are normally higher than levels found in surface runoff.

The majority of phosphorus lost from agricultural land is transported via surface runoff, mostly in particulate form attached to eroded soil particles. Because P is so strongly adsorbed to soil particles, the P level in the soil is a critical factor in determining loads of P delivered to surface waters (Daniel et al., 1997). Increased residual P levels in the surface soil can lead to increased P loadings to surface water, both attached to soil particles and in dissolved form. Soluble P losses from cropland can also be significant if runoff occurs very soon after heavy addition of phosphate fertilizer.

Runoff of Dissolved P

Phosphorus can be exported from agricultural land in particulate and dissolved forms. In most cases, the majority of P loss occurs in surface runoff in particulate form. However, dissolved P carried in surface runoff or subsurface flow may be a critical consideration because dissolved P tends to be immediately available to stimulate growth in receiving waters.

- Loss of dissolved P in runoff is often directly related to the P content of surface soils — linear relationships have been observed between dissolved P concentration in runoff and P content of surface soils in cropped and grassed watersheds (Daniel et al., 1997; Pote et al., 1999; Schoumans and Groenendijk, 2000).

- P losses from grassland may be high, particularly because fertilizers and animal waste are not usually incorporated into the soil. Significant phosphorus export has been measured in surface runoff and interflow from grazed grassland, with losses of over 0.5 kg P/ha during major storm events, especially when events closely followed inorganic fertilizer application (Haygarth and Jarvis, 1997).

- Soluble P losses may be greater from pasturelands than from croplands due to the presence of animal waste on the land surface, P release from plant decomposition, and low amounts of suspended sediment to sorb dissolved P (Baker et al., 1978; Sharpley and Menzel, 1987; Sharpley et al., 1992).

- In the Chesapeake Basin, dissolved P concentrations in storm runoff were higher from pastureland than from either cropland or forest (Correll et al., 1995).

Movement to Ground Water

The magnitude of nutrient loss to ground water, especially through leaching, depends on the availability of the chemical in the soil profile, the ease with which the nutrient form is detached from the soil, the rate and path of downward transport or percolation of water and chemicals, and any possible removal or deposition of the chemical before it reaches ground water. Nutrients may be introduced to ground water by direct routes such as abandoned wells, irrigation wells, sinkholes, or back-siphoning of nutrients when filling tanks. Such pathways are especially significant because transport through soil is bypassed, eliminating any opportunity for adsorption or uptake. While it is important to protect all ground water through the proper use of nutrients, in areas where ground water quality problems are known to exist, special emphasis should be placed on nutrient management planning and the careful use of nutrients.

Leaching of soluble nutrients to ground water can occur as chemicals are carried with precipitation or irrigation water moving downward past the root zone to the ground water table. Over-application of irrigation water can enhance leaching of nutrients to ground water by carrying dissolved nutrients quickly below the root zone. Ponded water in surface depressions due to large runoff events can be a significant source of nutrient transport to ground water, as ground water mounds underneath the depression (Zebarth and DeJong, 1989). Summer fallow may have a higher ground water contamination risk than continuous cropping because of the increased water storage in soil profiles that may increase deep percolation (Campbell et al., 1984; Bauder et al., 1993). Finally, idling of cropland either due to normal rotations or to commodity or conservation programs can in some cases initially increase nutrient leaching to ground water as nutrients are not taken up by growing plants and are available for leaching loss (Webster and Goulding, 1995).

> Increasing efficiency and reducing nutrient losses is founded upon the development of sound soil and water conservation principles.

Nitrogen in the form of nitrate is normally the nutrient most susceptible to leaching to groundwater. Nitrate not used by crops or denitrified by soil bacteria, is subject to leaching. Leaching potential is a function of soil type, crop, climate, tillage practices, fertilizer management, and irrigation and drainage management. Coarse textured soils pose a greater potential problem than fine textured soils, and crops with poor nitrogen use efficiencies present a greater hazard. In some studies, no-till systems have been shown to reduce nitrate leaching over conventional tillage, as well as proper crop rotation, especially those including a nitrogen-fixing crop (Meek et. al, 1995). However, other studies have shown that conservation tillage increases the infiltration rate of soils (Baker, 1993). Soil macroporosity and the proportion of rainfall moving through preferential flow paths often increase with the adoption of conservation tillage, potentially increasing the transmission of nitrates and other chemicals available in the upper soil to subsoils and shallow groundwater (Shipitalo et al., 2000). Over-irrigation, particularly on sandy soils, is a primary cause of nitrate leaching to groundwater.

Leaching of phosphorus to ground water is generally not a significant problem. However, organic soils and sandy soils, which lack the iron and aluminum oxides important for P adsorption, are exceptions; P losses in leaching from intensive cropping on such soils can be large. The degree of leaching will vary with soil structure, geologic conditions, climate, and management practices. Recent reports document phosphorus leaching in areas of intensive manure application to highly enriched soils over shallow water tables (Breeuwsma et al., 1995), or in areas of artificial drainage or preferential flow through soil macropores (Simard et al., 2000).

Nutrient Management Practices and Their Effectiveness

Nutrient Management Principles

There are several fundamental principles that should be applied to managing nutrients for both crop production and water quality protection. These principles focus on improving the efficiency of nutrient use and thereby reducing the potential for nutrient loss to surface or ground waters:

☐ Determine realistic yield goals, preferably on a field-by-field basis

☐ Account for available nutrients from all sources before making supplemental applications

☐ Synchronize nutrient applications with crop needs; N is needed most during active crop growth and N applied at other times may be lost

☐ Reduce excessive soil-P levels by balancing P inputs and outputs

Because of the complex cycling and multiple sources of N in the soil-crop system, careful accounting for all sources is often the most critical step in improving N management. Since the level of P in the soil is a major factor determining the amount of P lost from agricultural land, reducing soil P levels will ultimately reduce P delivery to surface and ground waters.

Additional practices may be needed to reduce detachment and transport of N and P and delivery to surface or ground waters. Erosion control practices are particularly critical to reduce losses of P and sediment-bound forms of N. Efficient water management can reduce leaching of soluble N from irrigated cropland, and improved irrigation practices can reduce water, sediment, and nutrient transport in tailwaters. Crop failure due to a lack of water leaves nutrients in the soil, rendering them vulnerable to leaching or runoff loss.

Nutrient Management Practices

Numerous practices are available to address the above principles. Many of these are specific to the cropping system, soils, climate, and management activities associated with particular crops and regions of the country. Readers are encouraged to contact their State Land Grant universities, NRCS, cooperative extension offices, State agriculture departments, or producer organizations for more site specific practices.

Soil and Water Conservation Districts, NRCS, or Extension offices can assist growers with the selection of nutrient management practices.

Following are practices, components, and sources of information that should be considered in the development of a nutrient management plan:

1. Use of soil surveys in determining soil productivity and identifying environmentally sensitive sites. Aerial photographs or maps and a soil map should be used. If the agricultural lands lie within a watershed that has been designated as having impaired surface or ground water quality associated with nutrients, then nutrient management plans should include an assessment of the potential for N or P from the agricultural lands to be contributing to the impairment.

2. Use of producer-documented yield history and other relevant information to determine realistic crop yield expectations. Appropriate methods include averaging the three highest yields in five consecutive crop years for the planning site or other methods based on criteria used

in developing the State Land Grant University's nutrient recommendations. Increased yields due to improved management and/or the use of new and improved varieties and hybrids should be considered when yield goals are set for a specific site.

3. Application of N and P at recommended rates for realistic yield goals. Through remote sensing and precision farming techniques, yield and fertilization can be optimized. Accurately located (e.g. via Global Positioning System, GPS) soil testing can help evaluate soil variability between and within fields, and use of on-the-go yield monitors and GPS-driven variable rate application can match inputs to soil and field variations and place nutrients where increased yield potential exists. Limit manure and sludge applications to phosphorus crop needs, supplying any additional nitrogen needs with nitrogen fertilizers or legumes.

   It may be necessary in some cases to route excess phosphorus in manures or sludge to fields that will be rotated into legumes, to other fields that will not receive manure applications the following year, or to sites with low runoff and low soil erosion potential.

   USDA has developed P application guidelines for situations where animal manure or other agricultural by-products are applied (see Table 4a-11). Producers unable to meet the P-based application rate requirement of the standard initially are encouraged to do so in a reasonable period of time using progressive planning approaches.

4. Soil testing for pH, phosphorus (Figure 4a-4), potassium, and nitrogen (Figure 4a-5). Preplant or midseason soil profile nitrate testing (e.g., a pre-sidedress nitrate test) should be used when appropriate. Sub-soil sampling for residual nitrate may be needed for irrigated croplands. Surface layer sampling (0-2 inches) for elevated soil P and soil acidity may be needed when there is permanent vegetation, non-inversion

> Soil, tissue, and manure testing provide useful information for nutrient management planning.

Table 4a-11. Allowable P Application Rates for Organic By-products (e.g., manure) A–NRCS, 1977, revised 1999).

The following guidelines are contained in USDA's Conservation Practice Standard 590 for Nutrient Management.

For phosphorus, one of the following options should be used to establish acceptable phosphorus application rates when manure or other organic by-products are applied:

- **Phosphorus Index (PI) Rating.** Nitrogen based manure application on Low or Medium Risk sites, phosphorus based or no manure application on High and Very High Risk Sites.**

- **Soil Phosphorus Threshold Values.** Nitrogen based manure application on sites on which the soil test phosphorus levels are below the threshold values. Phosphorus based or no manure application on sites on which soil phosphorus levels equal or exceed threshold values.**

- **Soil Test.** Nitrogen based manure application on sites on which there is a soil test recommendation to apply phosphorus. Phosphorus based or no manure application on sites on which there is no soil test recommendation to apply phosphorus.**

** Acceptable phosphorus based manure application rates shall be determined as a function of soil test recommendation or estimated phosphorus removal in harvested plant biomass. Guidance for developing these acceptable rates is found in the NRCS General Manual, Title 190, Part 402 (Ecological Sciences, Nutrient Management, Policy), and the National Agronomy Manual, Section 503).

Figure 4a-4. Example of soil test report (Pennsylvania State University, 1992a).

07/31/84	0004	700234	SOMERSET	25	NPBUU1	READINGTON
DATE	LAB NO.	SERIAL NO.	COUNTY	ACRES	FIELD	SOIL

THE PENNSYLVANIA STATE UNIVERSITY
COLLEGE OF AGRICULTURE
MERKLE LABORATORY - SOIL & FORAGE TESTING
UNIVERSITY PARK, PA 16802

SOIL TEST REPORT FOR:

COPY SENT TO:

P.A. PENN
RD1
ANYTOWN, PA   10000

ACME FERTILIZER CO.
MAIN STREET
ANYTOWN, PA   10000

SOIL NUTRIENT LEVELS:

			LOW	OPTIMUM	HIGH	EXCESSIVE
Soil pH	6.2		XXXXXXXXXXXXXX			
Phosphate	(P₂O₅)	114 lb/A	XXXXXXXXXXXX			
Potash	(K₂O)	178 lb/A	XXXXXXXXXXX			
Magnesium	(MgO)	230 lb/A	XXXXXXXXXXXXXX			

RECOMMENDATIONS FOR: *PLANTING CORN FOR GRAIN* (For other crops see ST 2 column 1)

YIELD GOAL   125.0 BUSHELS (PER ACRE)

LIMESTONE: 3400 lb/A Calcium Carbonate Equivalent

See Back
For Comments
1,2
3,4

PLANT NUTRIENT NEEDS:

NITROGEN (N)	PHOSPHATE (P₂O₅)	POTASH (K₂O)	MAGNESIUM (MgO)
130 lb/A	70 lb/A	90 lb/A	10 lb/A

8,11

MESSAGES:

* USE A STARTER FERTILIZER

* LIMESTONE RECOMMENDATION, IF ANY, IS TO BRING THE SOIL PH TO 6.0 - 6.5. MULTIPLY THE EXCHANGABLE ACIDITY BY 1000 TO ESTIMATE THE LIME REQUIREMENT FOR PH 6.5 - 7.0.

* RECOMMENDED LIMESTONE CONTAINING   .2% MGO WILL MEET THE MG REQUIREMENT.

* IF MANURE WILL BE APPLIED, SEE ST-10 "USE OF MANURE"

6,7

9

LABORATORY RESULTS

6.2	50	4.1	0.19	0.6	7.8	12.6	1.5	4.7	61.5
SOIL pH	P lb/A	ACIDITY	K	Mg	Ca	CEC	K	Mg	Ca
			EXCHANGEABLE CATIONS (meq/100 g)				% SATURATION		

OTHER TESTS: ORGANIC MATTER - 2.2 %

Figure 4a-5. Example of Penn State's soil quicktest form (Pennsylvania State University, 1992a).

PENNSTATE

PRE-SIDEDRESS SOIL NITROGEN TEST FOR CORN QUICKTEST EVALUATION PROJECT
- SOIL TEST INFORMATION AND REPORT FORM -

GROWER (PLEASE PRINT)

↑ NAME ↑

↑ STREET OR R. D. NO. ↑

↑ CITY, STATE, AND ZIP↑

↑ COUNTY ↑

DATE:

ANALYZED BY:

↑ AREA CODE ↑   TELEPHONE NO. ↑

Best time to call (8 am - 4:30 pm): _____

Please answer all of the following questions about this field:

1. What is the field ID (name or number)? _____   Corn Height _____ in.

2. What is the expected yield of the corn crop (bu/A or ton/A) in this field? _____

3. What was the previous crop? _____

   If this was a forage legume what was the % stand?

   (check one): ☐ 0-25%   ☐ 25-50 %   ☐ 50-100%

4. Was manure applied to this field? ☐ Yes ☐ No   If "yes" answer the following questions:

   When? ☐ Fall   ☐ Spring   ☐ Both   ☐ Daily

   Type? ☐ Cattle   ☐ Poultry   ☐ Swine   ☐ Horse   ☐ Sheep

   Estimate manure rate: _____ tons/acre - OR - _____ gallons/acre

   If incorporated how many days were there between spreading and incorporation? _____

5. What is the tillage program on this field? ☐ Conventional Tillage   ☐ Minimum Tillage   ☐ No-till

6. What would be your normal N fertilizer application rate for this field? _____ lbs. N/acre

Do not write below this line (to be completed by the analyst)

Quicktest Analysis Result & Recommendation

Individual Meter Readings	Average meter reading	Conversion factor	Average standard reading	Soil Nitrate-N (ppm)
	[ ] X [20] ÷ [ ] = [ ]			

Sidedress N Fertilizer Recommendation
(See table and guidelines on back of form) [ ] **lbs. N/acre**

If you have any questions about this test contact your Penn State Cooperative Extension Office

White copy- Grower
Yellow copy- Analyst
Pink copy- Agronomy Extension

5/91

tillage, or when animal manure or other organic by-products are broadcast or surface-applied.

5. Plant tissue testing, e.g. chlorophyll testing in corn.

6. Manure, sludge, mortality compost, and effluent testing.

7. Quantification of nutrient impacts from irrigation water, atmospheric deposition, and other important nutrient sources.

8. Use of proper timing, formulation, and application methods for nutrients that maximize plant utilization of nutrients and minimize the loss to the environment. This includes split applications and banding of the nutrients, use of nitrification inhibitors and slow-release fertilizers, and incorporation or injection of fertilizers, manures, and other organic sources. In addition, fall application of N fertilizer on coarse-textured soils should be avoided. Manure should be applied uniformly in accordance with crop needs, but surface application to no-till cropland should be avoided.

9. Coordination of irrigation water management with nutrient management. For example, in-field measurement of crop and soil N status during the growing season can be coupled with high-frequency irrigation to match N applications with crop needs and reduce N losses (Onken et al., 1995). Irrigation should also be managed to minimize leaching and runoff.

10. Use of small grain cover crops or deeply-rooted legumes to scavenge nutrients remaining in the soil after harvest of the principal crop, particularly on highly leachable soils. Consideration should be given to establishing a cover crop on land receiving sludge or animal waste if there is a high leaching potential. Sludge and animal waste should be incorporated or subsurface injected.

11. Use of buffer areas or intensive nutrient management practices to address concerns on fields where the risk of environmental contamination is high, such as:

   ❏ Karst topographic areas containing sinkholes and shallow soils over fractured bedrock,

   ❏ Subsurface drains (e.g., drain tile),

   ❏ Lands near surface water,

   ❏ High leaching index soils,

   ❏ Irrigated land in humid regions,

   ❏ Highly erodible soils,

   ❏ Lands prone to surface loss of nutrients, and

   ❏ Shallow aquifers and drinking water supplies.

   For example, nitrification inhibitors may be needed when conditions promote leaching, and banding or ridge application may render applied N or P less susceptible to leaching. Manure should not be applied to frozen or saturated soils, to shallow soils over fractured bedrock, or to excessively drained soils.

12. Use of soil erosion control practices to minimize runoff and soil loss.

13. Calibrate nutrient application equipment regularly.

14. A narrative accounting of the nutrient management plan that explains the plan and its use.

The best means for implementing and coordinating many of the above activities is through a comprehensive, site-specific nutrient management plan. Nutrient management plans should be reviewed annually to determine if modifications are needed for the next crop, and a thorough review of the plan should be done at least once every 5 years or once per crop rotation period. Application equipment should be calibrated and inspected for wear and damage periodically and repaired when necessary. Records of nutrient use and sources should be maintained along with other management records for each field. This information will be useful when it is necessary to update or modify the management plan.

A list of the required nutrient management plan elements for confined animal operations in the Pequea-Mill Creek (PA) National Monitoring Program project is shown Table 4a-12. Table 4a-13 shows a set of nutrient recommendations from a Vermont Crop Management Association. Table 4a-14 shows two summary tables from a sample plan.

Practice Effectiveness

Following is a summary of information regarding pollution reductions that can be expected from installation of nutrient management practices.

- ☐ The State of Maryland estimates that average reductions of 34 pounds of nitrogen and 41 pounds of P_2O_5 applied per acre can be achieved through the implementation of nutrient management plans (Maryland Department of Agriculture, 1990). These average reductions may be high because they apply mostly to farms that use animal wastes; average reductions for farms that use only commercial fertilizer may be lower.

- ☐ As of July 1990, the Chesapeake Bay drainage basin states of Pennsylvania, Maryland, and Virginia had reported that approximately 114,300 acres (1.4% of eligible cropland in the basin) had nutrient management plans in place (EPA, 1991a). The average nutrient reductions of TN and TP were 31.5 and 37.5 pounds per acre, respectively. The States initially focused nutrient management efforts on animal waste utilization. Because initial planning was focused on animal wastes (which have a relatively high total nitrogen and phosphorus loading factor), estimates of nutrient reductions attributed to nutrient management may decrease as more cropland using only commercial fertilizer is enrolled in the program.

- ☐ In Iowa, average corn yields remained constant while nitrogen use dropped from 145 pounds per acre in 1985 to less than 130 pounds per acre in 1989 and 1990 as a result of improved nutrient management. In addition, data supplied from nitrate soil tests indicated that at least 32% of the soils sampled did not need additional nitrogen for optimal yields (Iowa State University, 1991b).

- ☐ Data from the 66,640-acre Big Spring ground water basin in northeastern Iowa indicate that reduced application of nitrogen fertilizer associated with the 1983 payment-in-kind set-aside program resulted in reduced nitrate levels in ground water two years later (Hallberg et al., 1993). Based upon this analysis, it is postulated that water quality improvements at the watershed level will be definable over time in

Table 4a-12. Required nutrient management plan elements for confined animal operations in the Pequea-Mill Creek National Monitoring Program project, Pennsylvania.

A. **Farm Identification**

including location, receiving waters, size of operation, and farm maps of fields, soils, and slopes

B. **Summary of Plan**

Manure summary, including annual manure generation, use, and export

Nutrient application rates by field or crop

Summary of excess manure utilization procedures

Implementation schedule

Manure management and stormwater BMPs

C. **Nutrient Application**

Inventory of nutrient sources

Animal populations

Acreage and expected crop yields for each crop group

Nutrients necessary to meet expected crop yields

Nutrient content of manure

Nitrogen available from manure

Residual N from legumes and past manure applications

Planned manure application rate

Target spreading rates for manure application

Nitrogen balance calculation

Winter manure spreading procedures (if applicable)

D. **Alternative Manure Use**

Amount, destination, and use of manure exported to other landowners, brokers, markets, or used in other than agricultural application

E. **Barnyard Management**

F. **Storm Water Runoff Control**

Source: Penn State Cooperative Extension. 1997. *Pequea-Mill Creek Information Series.* Smoketown, PA.

Table 4a-13. Missisquoi Crop Management Association 1997 nutrient recommendations.

Crop	Field Name	Acres	Manure Applied In Fall	Recom. Manure Rate	Loads /Field 3375 gal	lb/A	N	P₂0₅	K₂0	Micronutrients	N	P₂0₅	K₂0	Mg	Need
						──── Recommended Fertilizer ────					After Manure & Fertilizer —Remaining Need—			Lime	
Corn	#7	9.7	9742	0	0	150	10	20	20	with 1.33% Zinc	47	0	0	0	
				or 3737 11	150	10	20	20		with 1.33% Zinc	0	0	0	0	
	#9A	11.3	2000	5226	17	150	10	20	20	with 1.33% Zinc	0	0	0	0	
	#11	20.0	5625	8798	52	250	10	20	20	with 0.8% Zinc	0	0	0	0	2.0
Alfalfa New Seeding	Spooner 3	4.3		3333				NONE			0	0	0	0	2.0
				or 0		300	5	10	30	with 0.6% Boron	0	0	0	0	2.0
Grass 1st Cut	#1	10.0		4135	12			NONE			0	0	26	0	1.0
				or 0		200	23	0	30		0	0	40	0	
	#3	10.8	7986	0				NONE			6	0	0	0	
Grass 2nd Cut	#1	10.0		0		200	23	0	30		0	0	0	0	
	#3	10.8		3755	12			NONE			0	0	0	0	

Table 4a-14.   Plan Summary from a Sample Plan (Pennsylvania State University Cooperative Extension, 1997).

Manure Summary Table

Manure Source	Generated on the Farm	Used on the Farm	Exported from the Farm
liquid dairy	523,000 gal	523,000 gal	0 gal
uncollected solid dairy	263 tons	263 tons	0 tons
collected solid dairy	175 tons	175 tons	0 tons
solid poultry	1,860 tons	0 tons	1,860 tons

Nutrient Application Rates by Crop Group

Crop Group	Acres	Starter Fertilizer Nutrients (lbs per acre)			Planned Manure Application Rate/ac.	When Manure Applied (incorp. time)	Additional Chemical Fertilizer Nutrients Applied		
		N	P_2O_5	K_2O			N	P_2O_5	K_2O
Corn, grain (liquid manure)	32	10	20	10	9,000 gal	spring (2-4 days)	0	0	0
Corn, grain (liquid manure)	18	10	20	10	9,000 gal	fall (2-4 days)	50	0	0
Corn, silage (liquid manure)	12	20	20	10	6,000 gal	fall (2-4 days)	0	0	0
Corn, silage (solid manure)	9	20	20	10	20 tons	fall/spring (2-4 days)	0	0	20
Alfalfa (new)	21	10	20	10	0	–	0	40	230
Alfalfa	53	0	0	0	0	–	0	120	200

– All numbers rounded off recognizing the built-in variation in figures used.

– **Manure application is restricted in the following areas:**

   a) within 100 feet of the farm well (field A-13) and the neighbor's well (field A-7), where surface flow is towards the well (unless the manure is incorporated within 24 hours of application, in which case manure application rates and supplemental fertilizer needs may need to be adjusted)

   b) within 100 feet of Little Fishing Creek when the ground is frozen, snow-covered, or saturated (fields A-2 and A-3)

   c) within the grassed waterway when the ground is frozen, snow-covered, or saturated (fields A-1 and A-2)

responsive ground water systems if significant changes in nitrogen application are accomplished across the watershed.

☐ In a pilot program in Butler County, Iowa, 48 farms managing 25,000 acres reduced fertilizer nitrogen use by 240,000 pounds by setting realistic yield goals based on soils, giving appropriate crop rotation and manure credits, and some use of the pre-sidedress soil nitrate test (Hallberg et al., 1991). Other data from Iowa showed that in some areas fields had enough potassium and phosphorus to last for at least another decade (Iowa State University, 1991b).

☐ In Garvin Brook, Minnesota, fertilizer management on corn resulted in nitrogen savings of 29 to 49 pounds per acre from 1985 to 1988 (Wall et al., 1989). In this Rural Clean Water Program (RCWP) project, fertilizer management consisted of split applications and rates based upon previous yields, manure application, previous crops, and soil test results.

☐ Baker (1993) concluded that the downward trends in total and soluble phosphorus loads from Lake Erie tributaries for the period from the late 1970s to 1993 indicate that agricultural controls have been effective in reducing soluble phosphorus export. Tributary nitrate concentrations increased, however, possibly due to adoption of conservation tillage, which enhances water percolation into the soil, and the extensive use of tile drainage systems in the watersheds.

☐ Berry and Hargett (1984) showed a 40% reduction in statewide nitrogen use over 8 years following introduction of improved fertilizer recommendations in Pennsylvania. Findings from the RCWP project in Pennsylvania indicated that, for 340 nutrient management plans, overall recommended reductions (corn, hay, and other crops) were 27% for nitrogen, 14% for phosphorus, and 12% for potash (USDA–ASCS, 1992a). Producers achieved 79% of the recommended nitrogen reductions and 45% of the recommended phosphorus reductions. In the same project area, Hall (1992) documented 8 to 32% decreases in median nitrate concentrations in ground water samples following decreases of 39–67% in N application rates under nutrient management.

☐ Base flow concentrations of dissolved nitrate-nitrite from a 909-acre subwatershed under nutrient management decreased slightly relative to a 915-acre paired subwatershed in the Little Conestoga Creek watershed in Pennsylvania, suggesting that nutrient management had a positive impact on water quality (Koerkle et al., 1996). Nutrient applications in the 909-acre treated subwatershed (study site) decreased in the period 1986-1989 by about 30% versus the period 1984-1986 (pre-implementation) as 85% of the land was placed under nutrient management. Less than 10% of the land was under nutrient management in the 915-acre untreated subwatershed (control site). The study was extended for two years to improve upon the findings, but implementation at the control site resulted in nutrient management on 40% of agricultural land, while implementation for the study site stood at 90% (Koerkle and Gustafson-Minnich, 1997). Nitrogen applications for the period 1989-1991 were about 7% less than for the period 1984-1986 at the study site, a much smaller decrease than the 30% decrease reported for the period 1986-1989. Nutrient application data were not available for the control site. The lack of statistically significant reductions in dissolved nitrate-nitrite for the period 1989-1991 versus 1984-1986 is interpreted as an indication that a reduction in nitrogen input of 30% (as achieved in 1986-1989) is needed to cause a 0.5 mg/L decrease in dissolved nitrate-nitrite.

A related study in the Conestoga River headwaters, Pennsylvania, showed that nutrient management caused statistically significant decreases in nitrate concentrations in ground water (Hall et al., 1997). Changes in nitrogen applications to the contributing areas of five wells were correlated with nitrate concentrations in the well water on a 55-acre crop and livestock farm in carbonate terrain. Lietman et al. (1997) showed that terracing decreased suspended-sediment yield as a function of runoff, but also increased nitrate-nitrite yields in runoff, and increased nitrate concentrations in ground water at 4 of the wells on a 23.1-acre site.

☐ A 6-year study in the 403-acre Brush Run Creek watershed in Pennsylvania showed that monthly and annual base flow loads of total nitrogen, dissolved nitrite-nitrate, total ammonia plus organic nitrogen, and total and dissolved phosphorus and orthophosphorus decreased during the 3-year period when nutrient management was implemented (Langland and Fishel, 1996). However, stormflow discharges of total nitrogen and total phosphorus increased by 14 and 44%, respectively, while nitrogen and phosphorus applications were reduced by 25 and 61%. Fewer storms were sampled during two of the three years under nutrient management due to a significant decrease in precipitation during the growing seasons. Maximum total nitrogen concentrations were 21 mg/L above the tile drains before nutrient management, and 2,400 mg/L in the tile drains before nutrient management (Langland and Fishel, 1996). Median concentrations of total nitrogen and dissolved nitrite-nitrate were reduced from 3.3 and 1.2 mg/L, respectively, to 2.5 and 0.90 mg/L when nutrient management was applied above the tile drains. Nutrient management in this tile-drained watershed resulted in a 14% decrease in nitrogen and 57% decrease in phosphorus applied as commercial and manure fertilizer.

❒ In Vermont, research suggested that a newly introduced, late spring soil test resulted in about a 50% reduction in the nitrogen recommendation compared to conventional technologies (Magdoff et al., 1984). Research in New York and other areas of the nation documented fertilizer use reductions of 30 to 50% for late spring versus preplant and fall applications, with yields comparable to those of the preplant and fall applications (Bouldin et al., 1971).

❒ Improved nutrient management on a case-study group of 8 United States Department of Agriculture (USDA) Demonstration Projects (DP) and 8 Hydrologic Unit Area (HUA) Projects resulted in reported nitrogen application reductions ranging from 14 to 129 lb/ac and phosphorus

Table 4a-15. Reported changes in average annual nutrient application rates on land with practice adoption in 19 USDA Demonstration and Hydrologic Unit Area Projects, 1991-1995.

Project	Purpose[1]	Nitrogen Reductions (lb/ac)	Phosphorus Reductions (lb/ac)
AL HUA	N, P	129	106
IN HUA	N, P	21	30
MI HUA	N, P	41	18
NY HUA	N, P	14	21
UT HUA	P	—	0
DE HUA	N, P	118	96
IL HUA	N, P	117	36
OR HUA	N	52	—
MD DP	N, P	43	42
NC DP	N, P	72	n/a
WI DP	N, P	78	18
FL DP	N, P	14	3
MN DP	N, P	30	21
NE DP	N	21	—
TX DP	N, P	21	18
CA DP	N, P	47	11

1 Nutrients to be controlled as project objective: N=nitrogen, P=phosphorus

—- = data not applicable

n/a = data not available

Source: Meals, D.W., J.D. Sutton, and R.H. Griggs. 1996. *Assessment of Progress of Selected Water Quality Projects of USDA and State Cooperators.* USDA–NRCS, Washington, D.C.

application reductions of 0 to 106 lb/ac (Table 4a-15). The case study group included both animal and crop agriculture and both irrigated and non-irrigated cropland.

Additional results from evaluations of practice effectiveness may exist for specific practices in particular regions. Potential sources of such documentation include the USDA MSEA/ADEQ (Management Systems Evaluation Areas/ Agricultural Systems for Environmental Quality) Programs (http:// www.nps.ars.usda.gov/) and the US EPA Section 319 National Monitoring Program (http://h2osparc.wq.ncsu.edu/319index.html).

A summary of the literature findings regarding the effectiveness of nutrient management in controlling nitrogen and phosphorus is given in Table 4a-16.

Table 4a-16. Relative effectiveness[a] of nutrient management (Pennsylvania State University, 1992b).		
Practice	Percent Change in Total Phosphorus Loads	Percent Change in Total Nitrogen Loads
Nutrient Management[b]	-35	-15

a Most observations from reported computer modeling studies
b An agronomic practice related to source management; actual change in contaminant load to surface and ground water is highly variable.

Factors in Selection of Management Practices

The movement of available nutrients to surface and/or ground waters depends on the properties of the nutrients involved, climate, soil and geologic characteristics, and land management practices such as crops grown, fertilizer applications, erosion control, and irrigation water management. These factors determine which specific strategies and practices should be selected to reduce nutrient movement in a given situation. Land management practices such as selection of fertilizer formulation or rate and method of application can be controlled, while environmental factors such as climate cannot. Other factors, such as crop selection and farming equipment, are governed to varying degrees by economic considerations and may therefore limit nutrient management options in some cases.

> Effective nutrient management will not transfer problems from surface to ground water, or vice versa.

Care should be taken that practices to control surface runoff do not increase the risk of ground water contamination, and vice versa. In general, practices that increase the efficiency of nutrient use and thereby reduce availability of nutrients for loss are the first line of defense in nutrient management. Control of detachment and transport of nutrients in the particulate phase and of runoff and leaching of soluble forms may be achieved with other practices or management measures, including erosion and sediment control and irrigation water management.

The characteristics of the agricultural operation are critical considerations in selection of appropriate practices for nutrient management. Specific nutrient management practices will differ markedly, for example, between a large grain farm, where all nutrients are supplied by purchased fertilizer and can be applied by precision farming methods, and a small dairy farm, where nutrients are supplied by animal waste, legumes, and purchased fertilizer, and exact nutrient balance is difficult to achieve. The equipment and facilities available to the producer, such as manure or fertilizer application equipment and the type of waste storage system influence both the form of the nutrients and the producer's ability to efficiently manage the nutrients.

Climatic and other environmental conditions such as soils and geology are key determinants in the selection of practices. For example, the need for irrigation to grow crops in the Columbia Basin of Washington places a premium on careful scheduling of fertigation to protect ground water below sandy soils (Annandale and Mulla, 1995), whereas the yield variability in midwestern claypan soils makes "on-the-go" changes in fertilizer application rates essential to maximizing the efficiency of N uptake (Kitchen et al., 1995). In addition, local environmental factors, such as the presence of sensitive or protected waterbodies, may require additional practices such as buffer strips or vegetative filter strips to reduce delivery of nutrients lost from agricultural land.

Local and regional agricultural economies and land use mix can also be important factors in selecting nutrient management practices. In livestock agriculture, the available land base with respect to animal populations may limit the potential for full use of manure nutrients on farm land and require efforts to export manure from an area in order to follow a nutrient management plan. Proximity to residential and urban centers can offer opportunities for exporting manure nutrients, but may also limit some forms of nutrient management due to odor problems or other perceived nuisances.

Finally, a range of issues such as the availability of soil, manure, and plant testing services; the availability of nutrient management consultants; the opportunity for producer training; the availability of rental equipment for specialized operations; and State, Tribal, and local laws and regulations may all affect the selection of best management practices for any given location.

Cost and Savings of Practices

Costs

In general, most of the costs documented for this management measure are associated with technical assistance to landowners to develop nutrient management plans. Some costs are also involved in ongoing nutrient management activities such as soil, manure, and plant tissue testing. Technical assistance in nutrient management is typically offered by universities, farm service dealers, and independent crop consultants. Rates vary widely depending on the extent of the service and type and value of the crop. Fees can range from about $5 per acre for basic service up to $30 per acre for extensive consultation on high-value crops (NAICC, 1998).

Typical nutrient management costs for Vermont dairy farms begin with a $150 fixed charge for a nutrient management plan. There is an additional $6 per acre for corn land, which includes record-keeping for manure, fertilizer, and pesticide applications, soil analysis for each field, manure test, and a PSNT; cost for grassland is $4 per acre, which includes the same services as for corn fields except the PSNT (Stanley, 1998).

In Pennsylvania, where state law requires extensive nutrient management planning, charges for development of a plan range from $400 to $900. Specific costs vary from around $3 to $4 per acre for a "generic" plan without soil sampling or weed and insect control recommendations, up to $8 to $12 per acre for a complete plan with full scouting (Craig, 1998).

In Maryland, again subject to a recent state law requiring all farms to have nutrient management plans, average costs across the state are about $3 per acre, which includes writing the plan, technical recommendations on fertilization and waste management, maps, and record-keeping (Maryland Dept. of Agriculture, 1998). Soil and manure testing are additional costs, at $2 to $5 per analysis.

Charges listed by an Illinois crop consultant range from $5 to $15 per acre for services including scaled maps, manure analysis, soil testing, and site specific recommendations for fertilizer and manure applications (Cochran, 1998).

A Wisconsin agronomic service charges $5 to $8 per acre for nutrient management services that include farm aerial maps; identification of fields with manure spreading restrictions; soil test reports; animal inventory with manure analysis; written plans for each field specifying crop to be grown, previous crop grown, fertilizer recommendations, legume and manure credits, manure application rates, and record-keeping sheets; and regular field scouting (Polenske, 1998).

In Nebraska, a crop consulting service charges $5 per acre for basic soil fertility and pest and water management, another $4 per acre for precision-farming GPS grid samples, plus a separate soil analysis charge (Michels, 1998).

Savings

In many instances landowners can actually save money by implementing nutrient management plans. For example, Maryland estimated (based on the over 750 nutrient management plans that were completed prior to September 30, 1990) that plan recommendations would save the landowners an average of $23 per acre per year (Maryland Dept. of Agriculture, 1990). This average savings may be high because most of the 750 plans were for farms using animal waste. Savings for farms using commercial fertilizer may be less.

In the South Dakota RCWP project, the total cost (1982–1991) for implementing fertilizer management on 46,571 acres was $50,109, or $1.08 per acre (USDA–ASCS, 1991a). In the Minnesota RCWP project, the average cost for fertilizer management for 1982–1988 was $20 per acre (Wall et al., 1989). Assuming a cost of $0.15 per pound of nitrogen, the savings in fertilizer cost due to improved nutrient management on Iowa corn was about $2.25 per acre as rates dropped from 145 pounds per acre in 1985 to about 130 pounds per acre in 1989 and 1990 (Iowa State University, 1991a).

USDA/NRCS *Comprehensive Nutrient Management Planning Technical Guidance*, December 1, 2000.

The goal of the NRCS *Comprehensive Nutrient Management Planning Technical Guidance* is to promote voluntary actions that will minimize water pollution from the production areas of animal feeding operations (AFOs) and the land application of manure and organic by-products. To accomplish this goal, NRCS envisions that AFOs will develop and implement technically sound, economically feasible, and site-specific Comprehensive Nutrient Management Plans (CNMP) using a conservation planning process.

The document explains that conservation planning is a natural resource problem-solving process, that integrates ecological (natural resource), economic, and production considerations meeting both the operator's objectives and the public's resource protection needs. This approach emphasizes identifying desired future conditions, improving natural resource management, minimizing conflict, and addressing problems and opportunities. The plan will help AFO owners and operators manage manure and organic by-products by combining conservation practices and management activities into a conservation system that, when implemented, will protect or improve water quality.

The guidance identifies six elements that must be considered when developing a CNMP. These elements include:

1. **Manure and Wastewater Handling and Storage**

2. **Land Treatment Practices**

3. **Nutrient Management**

4. **Record Keeping**

5. **Feed Management**

6. **Other Utilization Activities**

The specific criteria that each of these elements should address is presented in the guidance. The guidance also states that practices in CNMPs should meet requirements of NRCS Field Office Technical Guide conservation practice standards.

The technical guidance also provides information on the expertise required to prepare CNMPs. As a minimum, the three elements that address Manure and Wastewater Handling and Storage, Land Treatment Practices, and Nutrient Management must be developed by certified specialists. Because of the diversity and complexity of specific skills associated with each element of the CNMP, it is envisioned that most individuals will pursue "certification" for only one of the elements. Therefore, to develop a CNMP could require the interaction of three separate certified specialists, each addressing only one element. NRCS envisions that a certified conservation planner, assisting the AFO owner/operator, would facilitate the CNMP development process, with "certified specialists" developing the detailed specifics associated with the element they are certified to produce.

The CNMP Technical Guidance is available at *www.policy.nrcs.usda.gov/scripts/lpsis.dll/H/H_180_600_E5.htm*

4B: Pesticide Management

Management Measure for Pesticides

To reduce contamination of ground and surface water from pesticides:

1. Inventory pest problems, previous pest control measures, and cropping history.

2. Evaluate the soil and physical characteristics of the site including mixing, loading, and storage areas for potential leaching or runoff of pesticides. If leaching or runoff is found to occur, steps should be taken to prevent further contamination.

3. Use integrated pest management (IPM) strategies that

   ❑ apply pesticides only when an economic benefit to the producer will be achieved (i.e., applications based on economic thresholds) and

   ❑ apply pesticides efficiently and at times when runoff losses are least likely.

4. When pesticide applications are necessary and a choice of registered materials exists, consider the persistence, toxicity, runoff potential, and leaching potential of products in making a selection.

5. Periodically calibrate pesticide application equipment.

6. Use anti-backflow devices on the water supply hose, and other safe mixing and loading practices such as a solid pad for mixing and loading, and various new technologies for reducing mixing and loading risks.

> Six general principles guide safe pesticide management.

Management Measure for Pesticides: Description

The goal of this management measure is to reduce contamination of ground and surface water from pesticides. The basic concept of the pesticide management measure is to foster effective and safe use of pesticides without causing degradation to the environment. The most effective approach to reducing pesticide pollution of waters is, first, to release a lesser quantity of and/or less toxic pesticides into the environment and, second, to use practices that minimize the movement of pesticides to ground and surface water (Figure 4b-1). In addition, pesticides should be applied only when an economic benefit to the producer will be achieved. This usually results in some reduction in the amount of pesticides being applied to the land, plants, or animals, thereby enhancing the protection of water quality and possibly reducing production costs as well.

> Pesticide management consistent with this management measure is based on pesticide application only when an economic benefit is anticipated.

The pesticide management measure identifies a series of steps or thought processes that producers should use in managing pesticides. First, the pest problems, previous pest control measures, and cropping history should be evaluated for pesticide use and water contamination potential. Second, the physical characteristics of the soil and the site, including mixing, loading, and storage areas, should be evaluated for leaching and/or runoff potential. Integrated pest management (IPM) strategies should be used to minimize the amount of pesticides applied. In rare cases, IPM practices may not be available for some

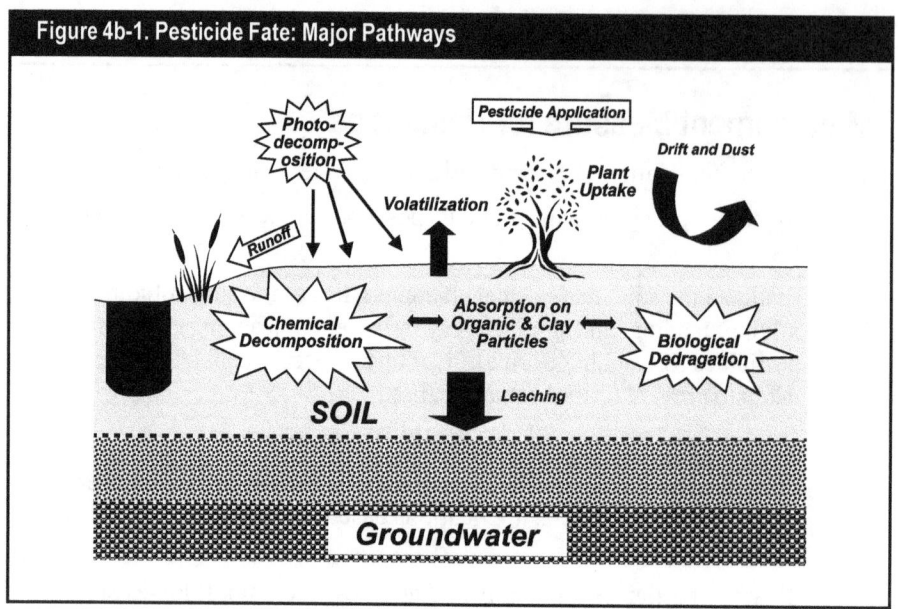

Figure 4b-1. Pesticide Fate: Major Pathways

commodities or in certain regions. An effective IPM strategy should call for pesticide applications only when an economic benefit to the producer will be achieved and not on a routine schedule. In addition, pesticides should be applied efficiently and at times when runoff and leaching losses are unlikely.

Pesticide labels must be followed.

When pesticide applications are necessary and a choice of materials exists, producers are encouraged to choose the most environmentally benign pesticide products. State Cooperative Extension Service specialists and Natural Resources Conservation Service field staff may be able to assist producers in this selection process.

Users must apply pesticides in accordance with the requirements on the label of each pesticide product. Label instructions include the following: allowable use rates; whether the pesticide is classified as "restricted use" for application only by certified and trained applicators; safe handling, storage, and disposal requirements.

At a minimum, effective pest management requires evaluating past and current pest problems and cropping history; evaluating the physical characteristics of the site; applying pesticides only when an economic benefit to the producer will be achieved; applying pesticides efficiently and at times when runoff losses are unlikely; selecting pesticides (when a choice exists) that are the most environmentally benign; using anti-backflow devices on hoses used for filling tank mixtures and on chemigation systems; and providing suitable mixing, loading, and storage areas. Other factors which may influence pesticide management decisions include long-term pest management, resistance management, nutrient management, and soil conservation.

Calibrating equipment saves money and reduces damage to the environment.

Pest management practices should be updated whenever the crop rotation is changed, pest problems change, or the type of pesticide used is changed. Application equipment should be calibrated and inspected for wear and damage frequently and repaired when necessary. Anti-backflow devices should also be inspected and repaired on a regular basis.

Pesticides: An Overview

What are pesticides?

Agricultural pesticides are chemicals which are used to protect crops against damaging organisms. They are generally divided into four categories according to the target pests:

Insecticides are targeted at insect pests. There are many kinds of insecticides in use today. They may be applied to the soil to protect roots, seeds, or seedlings. They may also be applied to the crop to protect stems, leaves, or fruit. Some of the most common insecticides include chlorpyrifos, diazinon, and carbaryl. Many insecticides kill the insects by disrupting their nervous system, resulting in paralysis and death. Unfortunately, they can have the same effect on non-target insects or fish and animals if enough of the applied product drifts or washes from the field.

Herbicides are used to control weeds in crops. Up to 80% of all pesticides sold are herbicides and they are used in most crop production systems. Weed control is one of the most effective practices to increase yields. Herbicides can be selective, killing the weeds but not the crop, such as atrazine in corn or trifluralin in soybeans. Other herbicides, such as glyphosate or paraquat, are non-selective, killing all plants they contact except those genetically engineered to be resistant to that particular herbicide or those that have developed resistance due to selection by the herbicide. Many herbicides have relatively low toxicity to insects, fish, or animals because they target specific enzyme systems found only in plants (Stevens and Sumner, 1991). This is particularly true for newer herbicides.

Fungicides are used to control fungi which cause disease in crops. They are applied to seeds, to soil, or to the crop to prevent or slow disease when conditions are favorable for the fungus. Fungicides are used primarily on high-value food crops and in turf and ornamental plant maintenance. They generally kill the fungal spores before they can germinate and infect the plant. Fungicides such as benomyl, metalaxyl, and chlorothalonil are used for a wide variety of crops, turf, and ornamental plants.

Nematicides are targeted at nematodes which infect plant roots and stunt or kill the crop. They are always applied to the soil as that is where the target occurs. Nematicides are generally non-selective, killing most everything they contact in the soil.

Why are pesticides used in agriculture?

Pests have affected crop production since man first started planting seeds. Crop damage from insects, fungi, and weeds can reduce yields and crop quality or even kill the crop in some cases. As a result, farmers have always sought ways to reduce this damage. Pest control using chemicals such as sulfur or plant extracts has been around for thousands of years. The first synthetic pesticides were discovered in the late 1930s and early 1940s and thousands have been developed since.

Pesticide use became widespread in part because the early results were so promising. Pests which farmers had battled for centuries seemed to be eliminated quickly and easily with these sprays. In many cases, less labor was required to produce a crop since hand or mechanical weeding was no longer necessary. As a result, yields increased and more acres could be managed by a farmer.

What are the risks associated with pesticides?

One problem which became evident in the early years of pesticide application was that pests developed resistance to the chemicals; this in turn devastated crops. When large areas are regularly sprayed with a pesticide, a population of pests resistant to the applied chemical can develop. It was learned later that this problem can be reduced by spraying only when necessary and using different pesticides when possible.

Another problem was the effect of pesticides on non-target organisms, which were inadvertently exposed through the food chain. Many of the first pesticides were persistent in the environment and accumulated in animals which consumed contaminated insects or fish. As a result of this problem, most modern pesticides are much less persistent and do not accumulate in the food chain.

There are several potential problems caused by pesticides reaching surface or ground water. The most severe occurrences involve acute toxicity. Acute toxicity occurs when negative effects are seen after exposure to relatively high doses of a pollutant over a short period of time, measured in hours or days. An amount of pesticide reaching a water body and killing fish or other nontarget species would be an example of acute toxicity. Most cases of pesticide acute toxicity are caused by insecticides which drift or wash from fields soon after application. As noted above, insecticides tend to be much more acutely toxic than other pesticides.

The most widespread problem is the occurrence of pesticides in surface and ground water used for drinking water. Because this may result in many people being exposed to the pesticide through their drinking water, there are concerns about chronic toxicity in these groups. Chronic exposure is when the exposure occurs over many years at concentrations which cause no outward effects, but which may increase cancer or other disease risks. Studies have shown that it is highly improbable that the types and concentrations of pesticides found in drinking water pose significant risks. However, most agree that it is prudent to minimize or eliminate pesticide occurrence in drinking water supplies.

The U.S. Geological Survey's (USGS) National Water Quality Assessment Program (NAWQA) has shown widespread herbicide occurrence in agricultural streams and shallow ground water. The presence of insecticides was also frequently detected in streams draining high insecticide use watersheds. The concentrations of these pesticides were measured at levels well below EPA drinking water standards 99% of the time. However, water quality standards are based on exposure to a single chemical or pesticide. In the NAWQA studies, where pesticide contamination of waters was found, there were generally two or more pesticides present (USGS, 1999).

In recent years, research on pesticides in water supplies, including the NAWQA studies, has included the study of pesticide degradation products. Degradation products are the compounds found in the environment as a result of the natural breakdown of the original pesticide or parent compound. They are usually less toxic than the original pesticide. While this document does not directly address pesticide breakdown products or their effects, the issue is an emerging concern and will likely receive more attention in the future.

Pesticide Movement into Surface and Ground Water

Pesticides can reach ground and surface water in a number of ways. Surveys of ground and surface water have found pesticides in many areas of the country. The extent of the contamination is often well defined, but the source or sources of contamination can be quite elusive in some cases. Figures 4b-2 to 4b-4 illustrate the major environmental fates of pesticides and are indicative of how difficult it is to quantitatively assess pesticide fate. However, the sources and problems associated with ground and surface water contamination are described in the following section.

Movement to Surface Water

Importance of pesticide contamination of surface water: About half of the population in the United States gets its water from surface sources. Therefore, pesticide contamination of surface water is of great concern to many. Several studies have shown that water supply reservoirs in the Midwest routinely exceed the health limits for pesticides, although these levels often only occur briefly in late spring after the main application season.

Losses of pesticides to runoff generally range from ≤1 to 5% of applied amounts, depending on various factors. Losses are usually greatest in the 1 to 2 weeks after application, and are highly dependent on storm events. Often, pesticide residues are only detectable in the first storm event after application.

Pesticides can enter surface water from the atmosphere in the form of drift or rainfall. Drift into surface waters can be serious locally if the pesticide is highly toxic to aquatic organisms, as in the case of many insecticides. Rain and fog have been shown to contain pesticide residues, particularly during the spring planting season. However, neither drift nor rain are major contributors to surface water contamination when compared to runoff.

Most pesticide contamination of streams, lakes, and estuaries occurs as a result of runoff from agricultural and urban areas. Runoff carries with it a mix of suspended soil particles and any pesticides which were either attached to the particles or dissolved in surface moisture just before runoff began. The amount of pesticide loss due to runoff is affected by the following factors:

Good soil and water management are also essential for effective pesticide management.

Rain Intensity — Heavy downpours result in minimal infiltration and maximum runoff. If soil is already moist prior to a rainfall event, then runoff will be greater since the soil's capacity to store additional water is reduced.

Surface Conditions — Recently tilled soil and soil with good ground cover have the most resistance to runoff, since water infiltrates relatively easily and the surface is "rough" enough to impede the flow of water. Maximum runoff potential occurs during the month after planting, since the soil is exposed and the crop has not grown large enough to intercept rain and reduce its ability to detach and transport soil particles. Reduced tillage practices that maintain residue on the surface will decrease runoff relative to conventional tillage practices that leave the soil bare and smooth at planting.

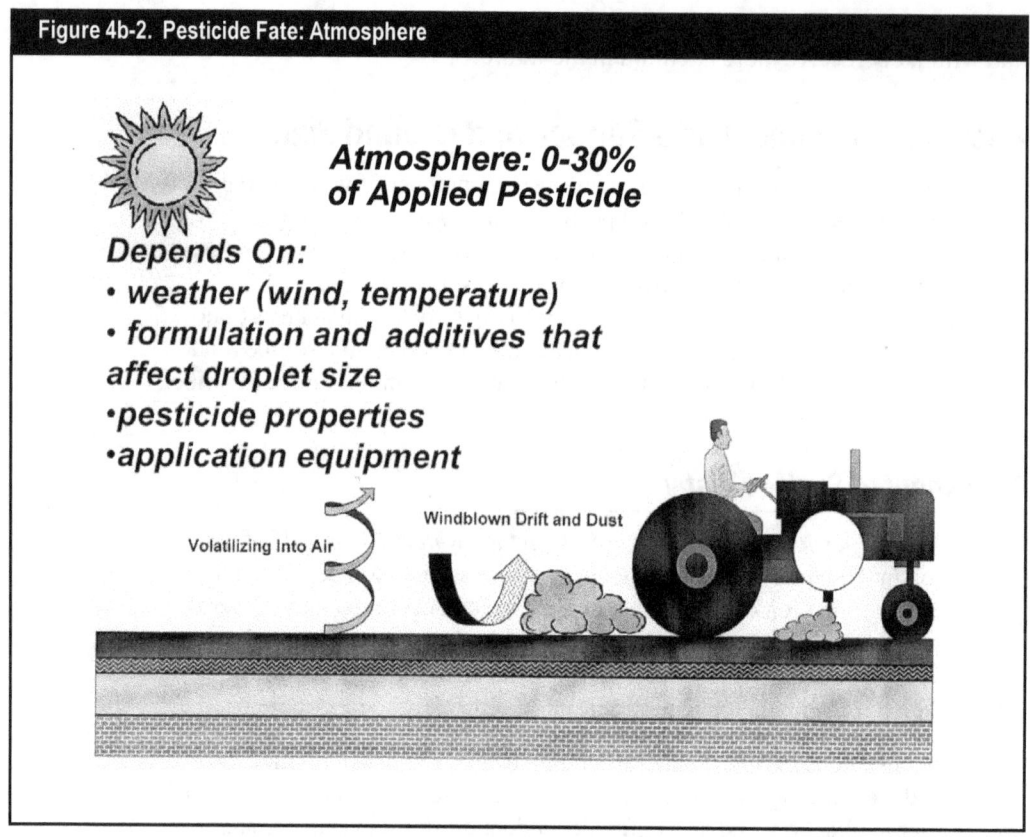

Figure 4b-2. Pesticide Fate: Atmosphere

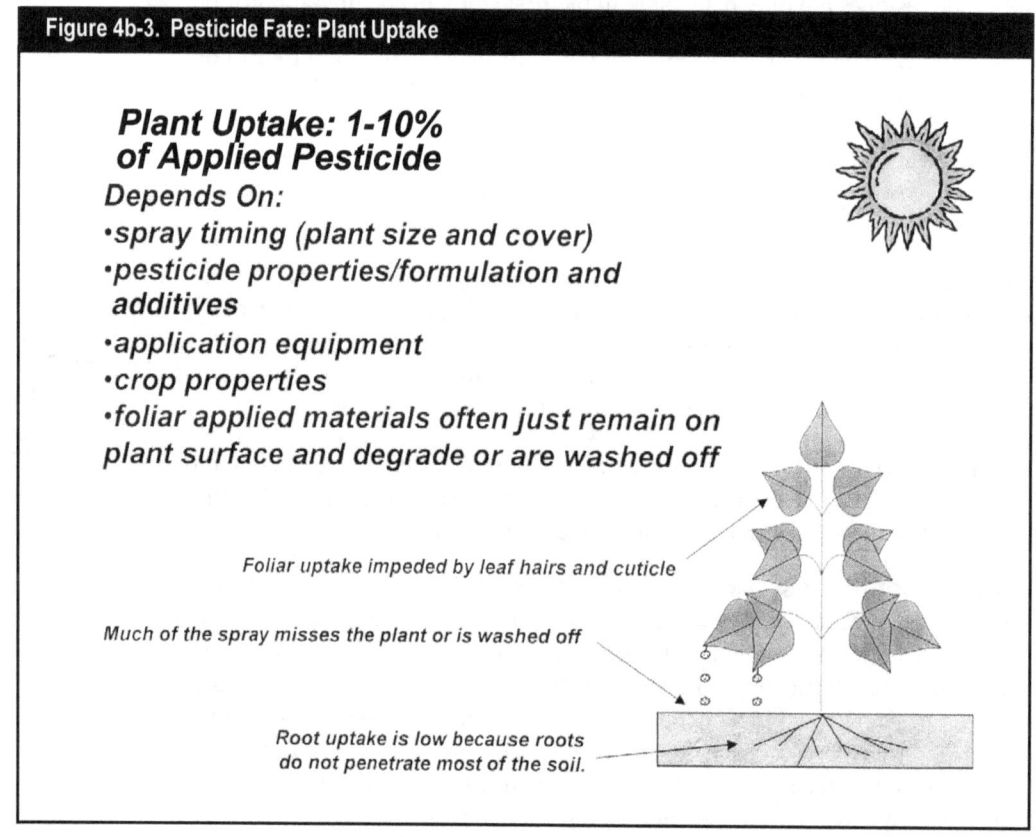

Figure 4b-3. Pesticide Fate: Plant Uptake

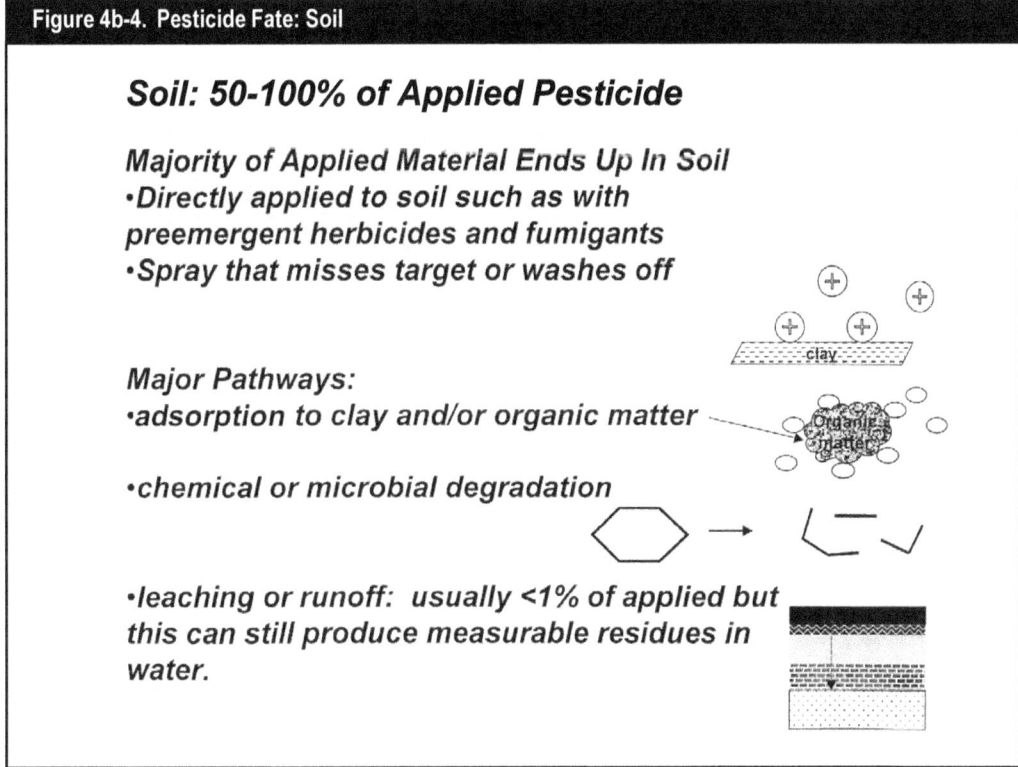

Figure 4b-4. Pesticide Fate: Soil

Soil: 50-100% of Applied Pesticide

Majority of Applied Material Ends Up In Soil
•Directly applied to soil such as with preemergent herbicides and fumigants
•Spray that misses target or washes off

Major Pathways:
•adsorption to clay and/or organic matter

•chemical or microbial degradation

•leaching or runoff: usually <1% of applied but this can still produce measurable residues in water.

Length of Slope and Percent Slope — Steeper and longer field slopes increase runoff energy, and the transport of soil and adsorbed pesticides.

Rate and Method of Application — Pesticides tilled or injected into the soil are less likely to be lost in runoff, although the disturbance of the soil by tilling or injection may increase soil (and attached pesticides) losses. Large losses of foliar pesticides in runoff can result if a heavy downpour occurs soon after application. Higher application rates will also generate higher pesticide concentrations in runoff.

Timing — If a runoff event occurs soon after the pesticide is applied, substantial losses can occur.

Vegetated Buffers — The beneficial effects of grassed buffers can be quite substantial, with reductions of pesticide movement into adjoining streams of up to 80 to 90%. The combination of infiltration, reduced overland flow rates, and adsorption in these zones can be quite effective in keeping pollutants in field runoff from being delivered to waterways.

It is important to emphasize that buffers function only under conditions of overland or sheet flow. Pesticides in runoff which moves through a buffer in a ditch or channel have little opportunity to degrade or adsorb before delivery to surface water.

Pesticide Degradation in Surface Waters — Once pesticides enter surface water, their rate of degradation slows considerably compared to degradation rates in soils. A portion of the pesticide may adsorb to the sediment and remain there until a flood event moves the sediment back into the moving water. This cycle of deposition and re-suspension is one of the mechanisms responsible for the presence of low levels of pesticides long after the application season.

Movement to Ground Water

Importance of pesticides in ground water: Approximately half of the U.S. population drinks water from wells; therefore, ground water protection is very important. Once a pesticide reaches ground water, it is very slow to degrade or flush out, so prevention is very important.

Movement of pesticides into ground water can occur through leaching after normal applications or by more direct pathways not related to normal uses (i.e. spills and direct contamination):

❏ **Leaching** — Pesticides can be moved downward toward ground water as rain or irrigation water percolates through the soil. Such a leaching process is controlled by the properties of the pesticide, the properties of the soil, the weather, and hydrologic loading.

Pesticide Properties: There are hundreds of pesticides and each one has a unique set of properties which determine if it is more or less likely to contaminate ground water. The most important are:

Persistence: measured in amount of time required for 50% to be degraded (half-life). The more persistent a chemical, the more likely it will find its way into ground water.

Adsorption: measured by how much of the chemical binds to soil, when shaken in water, as opposed to that which dissolves in water. The greater the adsorption ability of a pesticide, the less likely it will leach through the soil.

Application Rate and Method: measured in amount of active ingredient applied per acre. Pesticides requiring higher application rates may have an increased chance of leaching into ground water. Pesticides applied to growing crops are less likely to have the opportunity to leach than those applied to the soil.

Soil Properties: Pesticides often are applied to, or wash into, soils, where they may be adsorbed, degraded, or leached into shallow ground water. The properties of the soil that most influence these processes are discussed below. In addition to the soil properties listed here, any management practice (e.g., tillage) that impacts the properties or structure of soil has the potential to affect the movement of pesticides to ground water.

Organic Matter: measured as a fraction of the soil by weight. Most pesticides bind tightly to organic matter in soil so higher organic matter contents reduce the risk of leaching.

Clay: measured as a fraction of the soil by weight. Clay can bind many pesticides and it tends to reduce or slow the movement of percolating water. These two effects combined result in lower leaching risk with increasing clay content.

pH: measured on a scale of 0-14, with most soils falling in the 5-8 range. Generally, lower pH values will reduce leaching of pesticides and increase their rate of degradation.

Depth to Ground Water: not exactly a soil property but often closely related. The farther pesticide residues have to leach to reach ground water, the greater the chance of biological or chemical degradation. Although degradation rates decline rapidly below the root zone, most pesticides will degrade slowly as they move toward the ground water table.

Weather: The degradation and movement of pesticides in soil is highly influenced by the weather. Warmer or cooler temperatures will speed up or slow down degradation, respectively.

Hydrologic Loading: The addition of water to areas of pesticide application is key to the transport of pesticides toward ground water. Precipitation or irrigation in excess of evapotranspiration rates and soil water holding capacity can move pesticides deeper into the soil profile and increase the likelihood of pesticides leaching into ground water aquifers.

❐ **Spills** — Although some soils are very good at adsorbing and degrading applied pesticides, high concentrations of pesticides which result from spills overwhelm all these processes. Highly contaminated soils can be a long-term source of contamination because percolating water will continue to carry the pesticide into the ground water. Although the movement of pesticide residues is through leaching, a spill is still considered a point source.

❐ **Direct Contamination** — Ground water can be contaminated directly in many ways. Some of the most serious include backsiphoning, surface water movement into wells, or drainage into limestone channels or sink-holes. These contamination problems can almost always be prevented. Once they occur, however, the point of entry becomes a point source for contamination. A plume of contamination moves slowly away from the source and can spread to contaminate many downgradient wells.

Well contamination is often the result of a lack of proper backflow prevention devices or poor well construction. Problems such as a poor or absent casing, lack of grouting, location in a low spot where water accumulates, or capping below the soil surface are all invitations for contaminated surface water to enter the well. High nitrates and bacterial contamination are often associated with these problems.

Pesticide Management Practices and Their Effectiveness

The practices set forth below have been found by EPA to be representative of the types of practices that can be applied successfully to achieve the management measure described above. Additional information about individual practices, their purpose, and how they work is presented in Appendix A.

1. *Inventory current and historical pest problems, cropping patterns, and use of pesticides for each field.*

   The purpose of this procedure is to assist the grower in evaluating the potential for water contamination at the site and to determine IPM strategies which may be applied to the operation. Much of this information is important for many aspects of farm operation beyond pollution prevention. This inventory can be accomplished by using a farm and field map, and by compiling the following information for each field:

❑ *Crops to be grown and a history of crop production.* Certain IPM strategies, such as crop rotation, require this information.

❑ *Information on soil types.* Different soils can have very different susceptibility to either runoff or leaching losses of applied pesticides.

❑ *The exact acreage of each field.* This information can be used to check application rates as well as yields.

❑ *Records on past pest problems, pesticide use, and other information for each field.* By keeping these records, the grower can evaluate options for pest management such as crop rotations and alternative pesticides.

2. ***Evaluate the soil and physical characteristics of the site including mixing, loading, and storage areas for potential for the leaching and/or runoff of pesticides.*** The most important types of features for evaluation include:

❑ *Sinkholes, drainage wells, abandoned wells, and karst topography which allow direct access to ground water.* These allow surface water carrying sediment, bacteria, and pesticides to quickly enter and contaminate the ground water.

❑ *Proximity to surface water.* Pesticides should not be used directly adjacent to surface water because of the high potential for pesticide contamination from runoff and drift. An untreated buffer around the surface water will provide a measure of protection.

❑ *Runoff potential.* Steeper slopes, heavier soils, and conventional tillage all increase the runoff potential for a field. Greater amounts of organic matter and clay increase the ability of the soil to bind the pesticide. Conservation tillage tends to increase infiltration and decrease the amount of runoff, further reducing potential pesticide losses.

❑ *Aerial drift.* Fields with their longer dimension at 90 degrees to the prevailing wind direction will have lower drift potential than those parallel to the wind.

❑ *Soils with a high risk of erosion.* Cropping practices such as no-till can greatly reduce the runoff potential for pesticides on steep slopes with heavy soils.

❑ *Soils with poor adsorptive capacity.* Low organic matter (<1%) and clay content reduces the ability of the soil to bind applied pesticides and prevent them from leaching through to ground water.

❑ *Highly permeable soils.* Often soils with poor adsorptive capacity also have high sand contents which allow water to percolate rapidly through them. This allows any pesticides present to move quickly downward before they are degraded by the more abundant microbes in the surface horizons.

❑ *Shallow aquifers.* A shorter distance between the application zone in the surface soil to the aquifer means less opportunity for binding and degradation of the pesticide.

❏ *Wellhead protection areas.* Private wells should have a 100-foot buffer in which no pesticides or fertilizers are applied. Public water supply wells may require a larger buffer. The buffer minimizes the risk of agricultural chemicals leaching into the ground water immediately adjacent.

3. ***Use IPM strategies to minimize the amount of pesticides applied,*** including:

❏ *Scouting fields for pest problems.* Most universities have scouting guides for farmers which will provide guidance for procedures appropriate to their area. Often county extension staff provide training for scouting, or a farmer may be able to hire a consultant to provide this service. Many agricultural retailers also provide scouting services as a part of their pesticide application contracts. The key is to know how and where to look for pests and their correct identification. For weeds, a farmer may rely on problems from the previous year or he may walk a specified length of row to count weed seedlings. For insects, a sweep net may be brushed through the crop and the insects identified and counted to estimate the potential for crop damage.

❏ *Determine the economic threshold for pests.* This is also information that is usually available from local extension offices. The expected value of the crop and the anticipated losses caused by the pest are estimated against the cost of an application before any sprays occur.

❏ *Use varieties of crops resistant to pests.* Resistant varieties usually require fewer pesticide applications.

❏ *Use crop rotation.* Crop rotations interrupt pest buildup by eliminating the host plants or by allowing the application of pesticides which reduce pest populations. An example is a corn-soybean rotation, in which broadleaf weeds are more easily controlled in the corn crop and grass weeds are more easily controlled in the soybean crop.

❏ *Foster biological controls.* Identifying the pest properly and recognizing beneficial insects is key. If a spray is necessary, select a pesticide which is the most specific to the pest and least toxic to non-target species. Natural enemies can be introduced and their habitats preserved. Pheromones can be used to monitor populations, disrupt mating, or attract predators or parasites.

❏ *Use of improved tillage practices* such as ridge tillage.

❏ *Use of cover crops* in the system to promote water use and reduce deep percolation of water that contributes to leaching of pesticides into ground water.

❏ *Destruction of pest breeding, refuge, and overwintering sites* (this may result in loss of crop residue cover and an increased potential for erosion).

❏ *Use of mechanical destruction of weed seed through the use of tillage techniques.* Erosion control goals must also be considered when tillage alternatives are being examined.

❏ *Diversification of habitat.* The abundance of pests is greatly influenced by the environment created by the farmer. Monocultures create a simple environment in which pests may have little or no competition or predators. Having a broad array of plant species as crops and in borders diversifies the habitat and dampens pest populations.

❑ *Use of trap crops.* A species or variety of plant which is more attracted to pests than the main crop can be planted earlier or in an adjacent area. This will concentrate the pests in a smaller area where they can be controlled with a pesticide, thus avoiding a wider pesticide application.

❑ *Use of allelopathic characteristics of crops.* There is evidence that some crops can naturally inhibit the growth of pest populations. For example, a rye cover crop may reduce weed populations in subsequent crops.

❑ *Use of timing of field operations* (planting, cultivating, irrigation, and harvesting) to minimize application and/or runoff of pesticides.

❑ *Use of efficient application methods, e.g., spot spraying and banding of pesticides.* Often pest problems occur primarily in one portion of the field, allowing for targeted pesticide application. Banding may provide protection of the crop without the entire area being sprayed.

4. ***When pesticide applications are necessary and a choice of material exists, consider the persistence, toxicity, and runoff and leaching potential of products along with other factors, including current label requirements, in making a selection.*** This is a complex area and most pesticide users will not have much of the information necessary to make such judgements. The leaching potential for many pesticides has been estimated in several ways and are in general agreement with each other. One example is the PLP, or Pesticide Leaching Potential, which is an index of persistence and leaching characteristics of each chemical (Table 4b-1).

    Table 4b-1 may be useful as a starting point, but other information may be available from State agencies, NRCS, or universities.

    Users must apply pesticides in accordance with the instructions on the label of each pesticide product and, when required, must be trained and certified in the proper use of the pesticide. Labels include a number of requirements including allowable use rates; classification of pesticides as "restricted use" for application only by certified applicators; safe handling, storage, and disposal requirements; and other requirements. Users should contact their state and/or federal pesticide program with questions concerning specific requirements.

    Grower practices can have significant impact on the movement of pesticides into surface water. Tillage practices, incorporation, and filter strips all provide significant reductions in pesticide movement from fields to surface water in most cases (Tables 4b-2, 4b-3). Generally, practices which slow runoff, increase infiltration, and trap sediment tend to reduce pesticide losses.

5. ***Maintain records of application of restricted use pesticides (product name, amount, approximate date of application, and location of application of each such pesticide used) for a 2-year period after use, pursuant to the requirements in section 1491 of the 1990 Farm Bill.***

Table 4b-1. Typical pesticide leaching potential (PLP) index values calculated using commonly reported pesticide properties, and estimated fraction hitting the soil for six example herbicides (NCCES, 1994).

Common Name	Trade Name	Application Method[a]	PLP Index[b]
Herbicides:			
Acifluoren	Blazer	f	40
Alachlor	Lasso	s	52
Ametryn	Evik	s	50
		f	46
Amitrole	Amitrole-T	f	53
Asulam	Asulox	f	51
Atrazine	AAtrex	f, ph7	56
		s, ph7	60
		s, ph5	52
		s, ph7, noncrop	66
		s, ph5, noncrop	57

[a]s = soil application and f = foliar application of pesticide. pH is given where differences have a known effect and data are available. Noncrop indicates difference in rates, usually higher than crop uses.

[b]PLP values range from 0 (no leaching potential) to 100 (maximum leaching potential).

Source: North Carolina Cooperative Extension Service. 1994. Soil Facts: Protecting Groundwater in North Carolina, a Pesticide and Soil Ranking System. North Carolina State University. AG-439-31.

Table 4b-2. Effect of BMPs on pesticide losses compared to conventional tillage or no filter strips.

Practice	Range of Reductions	Average	Reference
Ridge Till	-33 – 65	30	Baker and Johnson, 1979
No-Till	-98 – 9	51	Baker and Johnson, 1979
	29 – 100	77	Glenn and Angle, 1987
	64 – 100	86	Hall et al., 1991
	85 – 99	92	Hall et al., 1984
	6 – 41	21	Franti et al., 1995
	41	—	Seta et al., 1993
	100	—	Isensee and Sadeghi, 1993
Contour Ridges	53 – 100	79	Ritter et al., 1974
Incorporation	26 – 75	—	Hall et al., 1983
	24 – 36	30	Baker and Laflen, 1979
	7–79	52	Franti et al., 1995
Filter Strips	28 – 31	—	Asmussen et al., 1977
	4 – 14	—	Rhode et al., 1980
	9 – 35	22	Hall et al., 1983
	40 – 72	56	Mickelson and Baker, 1993
	50 – 74	63	Misra et al., 1994
	15 – 72	45	Misra, 1994

Table 4b-3. Summary of buffer studies measuring trapping efficiencies for specific pesticides. K_{oc} values listed for each pesticide are from the NRCS Field Office Technical Guide, Section II Pesticide Property data base (USDA-NRCS, 2000).

Pesticide	K_{oc}	Study reference	Percent pesticide trapped
Highly adsorbed pesticides			
Chlorpyrifos	6,070	Boyd, et al., 1999	57–79
		Cole, et al., 1997	62–99
Diflufenican	1,990	Patty, et al., 1997	97
Lindane	1,100	Patty, et al., 1997	72–100
Trifluralin	8,000	Rhode, et al., 1980	86–96
Moderately adsorbed pesticides			
Acetochlor	150	Boyd, et al., 1999	56–67
Alachlor	170	Lowrance, et al., 1997	91
Atrazine	100	Arora, et al., 1996	11–100
		Boyd, et al., 1999	52–69
		Hall, et al., 1983	91
		Hoffman 1995	30–57
		Lowrance, et al., 1997	97
		Mickelson and Baker 1993	35–60
		Misra, et al., 1996	26–50
		Patty, et al., 1997	44–100
Cyanazine	190	Arora, et al., 1996	80–100
		Misra, et al., 1996	30–47
2,4-D	20	Asmussen, et al., 1977	70
		Cole, et al., 1997	89–98
Dicamba	2	Cole, et al., 1997	90–100
Fluormeturon	100	Rankins, et al., 1998	60
Isoproturon	120	Patty, et al., 1997	99
Mecoprop	20	Cole, et al., 1997	89–95
Metolachlor	200	Arora, et al., 1996	16–100
		Misra, et al., 1996	32–47
		Webster and Shaw 1996	55–74
		Tingle, et al., 1998	67–97
Metribuzin	60	Webster and Shaw 1996	50–76
		Tingle, et al., 1998	73–97
Norflurazon	600	Rankins, et al., 1998	65

Section 1491 requires that such pesticide records shall be made available to any Federal or State agency that deals with pesticide use or any health or environmental issue related to the use of pesticides, on the request of such agency. Section 1491 also provides that Federal or State agencies may conduct surveys and record the data from individual applicators to facilitate statistical analysis for environmental and agronomic purposes; however, in no case may a government agency release data, including the location from which the data was derived, that would directly or indirectly reveal the identity of individual producers. Section 1491 provides that in the case of Federal agencies, access to records maintained under section 1491 shall be through the Secretary of Agriculture or the Secretary's designee. This section also provides that State agency requests for access to records maintained under section 1491 shall be through the lead State agency so designated by the State.

Section 1491 includes special access provisions for health care personnel. Specifically, when a health professional determines that pesticide information maintained under this section is necessary to provide medical treatment or first aid to an individual who may have been exposed to pesticides for which the information is maintained, upon request persons required to maintain records under section 1491 shall promptly provide records and available label information to that health professional. In the case of an emergency, such record information shall be provided immediately.

Operators should consider maintaining records beyond those required by section 1491 of the 1990 Farm Bill. For example, operators may want to maintain records of all pesticides used for each field, i.e., not just restricted use pesticides. These records will be useful in setting up IPM programs and in crop rotation and management decisions. In addition, operators may want to maintain records of other pesticide management activities such as scouting records or other IPM techniques used and procedures used for disposal of remaining pesticides after application. Operators should also check with state and local agencies regarding record keeping requirements.

6. *Use only the recommended amount of pesticide for the problem you or a professional have identified and determined to merit pesticide application.*

7. *Recalibrate and repair application equipment, including chemigation equipment, at least each spray season. Use anti-backflow devices on hoses used for filling tank mixtures and on chemigation systems.* Calibration of pesticide spray equipment at least once each spray season is critical to ensuring that proper application rates are maintained.

   As replacement equipment is needed, purchase new, more precise application equipment and other related farm equipment (including improved nozzles, computer sensing to control flow rates, radar speed determination, electrostatic applicators, and precision equipment for banding and cultivating).

8. *Solid pad for mixing and loading pesticides.*

EPA's Office of Pesticide Programs Promotes Registration of Lower Risk Pesticides

Reduced risk conventional pesticides

Since 1993 EPA's Office of Pesticide Programs has encouraged pesticide companies to register lower risk pesticides. The Agency expanded this program in 1998 to further encourage replacements for organophosphate (OP) pesticides, a class of neurotoxins. EPA's Reduced-risk Initiative expedites the registration of *conventional* pesticides that the Agency believes pose less risk to human health and the environment than existing alternatives. The goal of the program is to quickly register commercially-viable alternatives to riskier pesticides such as neurotoxins, carcinogens, reproductive and developmental toxicants, and ground and surface water contaminants. Reduced risk pesticides generally have low human toxicity; low risk to non-target terrestrial and aquatic plants and animals; reduced application rates; rapid field degradation; low potential to contaminate ground or surface water; and work well with integrated pest management programs. *Biological* pesticides which also have many of these desirable characteristics are described below.

The major incentive for pesticide companies to register reduced risk conventional pesticides is a one to two year reduction in the time to get their product on the market. This allows the chemical to be introduced into the market at the earliest possible time and displace riskier alternatives as soon as possible. It also allows the registrant to recoup their investment costs sooner and gain several additional growing seasons under patent. In addition, although companies are not allowed to put a reduced-risk claim on their labels, EPA believes that companies use the reduced-risk status to marketing advantage. Some reduced risk pesticides have already gained large market shares (up to 70%) over riskier compounds.

Biological Pesticides

Office of Pesticide Programs also encourages the registration of biological pesticides. Biological pesticides are expedited in a fast-track registration process by their own working group, the Biopesticides and Pollution Prevention Division. Examples include microbial pesticides (bacteria, viruses or other microorganisms used to control pests), and biochemical pesticides, such as pheromones (insect mating attractants), insect and plant growth regulators, and hormones used as pesticides. Most biological pesticides are applied at very low rates, are highly volatile, or are applied in bait, trap, or "encapsulated" formulations and thus result in less exposure (and less likelihood of adverse effects to humans and the environment than from the use of most conventional pesticides). Among these new pesticides approved are the first plant pesticide products, which are agricultural plants that are altered to produce proteins toxic to insects that destroy crops. As with reduced risk conventional pesticides, a major incentive to pesticide companies to register biological pesticides is a reduction in the time to get their product on the market and the benefits that accrue from an earlier release date.

For more information on reduced risk pesticides, contact the EPA Reduced Risk Pesticide Coordinator, in the Registration Support Branch, Registration Division, Office of Pesticide Programs.

Factors in the Selection of Management Practices

The best way to control pests in crops is to know the crop and pest well enough to determine a control plan which maximizes crop production while minimizing environmental impacts. This is often a combination of cultural, biological, and chemical practices. Cultural controls include tillage, crop rotations, resistant varieties, and varying planting or harvest dates. Biological controls involve encouraging or introducing natural enemies of the pest and managing the crop environment to the disadvantage of the pest. Chemical controls should involve a selection process which selects a pesticide which results in the greatest economic benefit for the least environmental cost. Such a determination requires knowledge and information which are beyond the average grower. However, many states have guides to assist in pesticide selection.

Relationship of Pesticide Management Measures to Other Programs

Under the Federal Insecticide, Fungicide and Rodenticide Act (FIFRA), EPA registers pesticides on the basis of evaluation of test data showing whether a pesticide has the potential to cause unreasonable adverse effects on humans, animals, or the environment. Data requirements include environmental fate data showing how the pesticide behaves in the environment, which are used to determine whether the pesticide poses a threat to ground water or surface water. If the pesticide is registered, EPA imposes enforceable label requirements, which can include, among other things, maximum rates of application, classification of the pesticide as a "restricted use" pesticide (which indicates that a pesticide may have adverse effects on the environment and/or the applicator and restricts use to certified applicators trained to handle such pesticides), or restrictions on use practices. FIFRA allows States to develop more stringent pesticide requirements than those required under FIFRA, and some States have chosen to do this. The EPA and the U.S. Department of Agriculture Cooperative Extension Service provide assistance for pesticide applicator and certification training in each State.

Cost and Savings of Practices

Costs

In general, most of the costs of implementing the pesticide management measure are program costs associated with providing additional educational programs and technical assistance to producers to evaluate pest management needs and for field scouting during the growing season.

One of the most important IPM practices is scouting, which carries with it a cost to the producer. High and low scouting costs are given for major crops in each of the coastal regions (Table 4b-4). These costs reflect variations in the level of service provided by various crop consultants. For example, in the Great Lakes region, the relatively low cost of $4.95 per acre is based on five visits per season at the request of the producer. Higher cost services include scouting and weekly written reports during the growing seasons. Cost differences may also reflect differences in the size of farms (i.e., number of acres) and distance between farms.

The variations in scouting costs between regions and within regions also occur because of differences in the provider of the service. For example, in some states the Cooperative Extension Service provides scouting services and training at no cost or for a nominal fee. In other areas, farmer cooperatives have formed crop management associations to provide scouting and crop fertility/pest management recommendations. There are also consulting firms and agricultural retailers with scouting expertise.

Scouting costs also vary by crop type. Scouting services for high-value cash crops, such as fruits and vegetables, must be very intensive given that pest damage is permanent and may make the crop unmarketable.

Another issue regarding the cost of pesticide management practices is selection of the tillage system and direct and indirect costs associated with that system. Conservation tillage or no-till practices often rely on the use of herbicides to control weeds rather than multiple passes with a cultivator employed in conventional tillage, which mechanically destroy the weeds. When deciding between conservation versus conventional tillage, the direct costs of buying more pesticides (and specific pesticides) for no-till must be weighed against the cost of running more equipment in the field for conventional tillage. Corn production under conventional tillage requires an average of more than three passes through the field to cultivate, while no-till may only require one pass to plant and spray herbicides. Since each cultivation pass costs nearly seven dollars per acre, production costs may increase by more than $14/acre for conventional tillage compared to no-till, minus any additional costs of herbicides.

Savings

Most of the savings of implementing the pesticide management measure are associated with a reduction in the amount of pesticides used. IPM usually requires less pesticide use, thereby reducing the cost of production and increasing the profitability of the crop. In a review of 61 studies of IPM impacts on crop yield, pesticide use, and economics, pesticide use declined in seven of the eight commodities evaluated (Norton and Mullen, 1994; Table 4b-5). Some studies found increased use of pesticides with IPM due to increased awareness of pest problems, but the majority found reductions.

An additional benefit is associated with the use of no-till practices. Soil losses are reduced by up to 90% in no-till compared to conventional tillage, reducing both the indirect costs of erosion and consequent crop yield losses and also adverse environmental impacts of sedimentation of surface water bodies. Yields with conservation tillage are often reduced when a farmer first experiments with it, as it is a new practice which requires new skills and equipment. However, this situation usually changes with time. An added benefit of no-till is that considerable time is saved by only needing to work the field once instead of three or more times.

Table 4b-4. Estimated scouting costs (dollars/acre) by coastal region and crop in the coastal zone in 1992 (EPA, 1992a).

COASTAL REGION	Corn	Soybean	Wheat	Rice	Cotton	Fresh Market Vegetables[a]	Hay[b]
Northeast							
Low	5.50	NA	3.75	—	—	25.00	2.50
High	6.25	NA	4.50	—	—	28.00	2.75
Southeast							
Low	5.00	3.25	3.00	8.00	6.00	30.00	2.00
High	6.00	4.00	3.50	12.00	8.00	35.00	3.00
Gulf Coast							
Low	6.00	4.50	—	5.00	6.00	35.00	—
High	8.00	6.50	—	9.00	9.00	40.00	—
Great Lakes							
Low	4.95	4.25	3.75	—	—	—	4.75
High	5.50	5.00	4.00	—	—	—	5.25
West Coast							
Low	NA	NA	3.50	NA	6.75	32.00	NA
High	NA	NA	5.50	NA	9.30	38.00	NA

NA = not available
— = not applicable
[a] Most fresh market vegetables are produced under a regular spraying schedule.
[b] Scouting costs for hay are based on alfalfa insect inspection. The higher cost in the Great Lakes region includes pesticide and soil sampling.

Table 4b-5. Summary of results of farm-level economic evaluations of IPM programs.

Average Commodity	Percent States	Number of Studies	Percent Percent Change in Pesticide Use[a]	Change in Production Cost with IPM[a]	Percent Yield Change with IPM[a]	Change in Net Returns Per Acre[a]	Level of Risk with IPM
Cotton	TX, GA, MS, NC, SC, LA, MO, TN, AZ, NM, CA, AR	18	-15	-7	+29	+79	decreased
Soybeans	NC, VA, MD GA, IN	7	-35	-5	+6	+45	decreased
Corn	IN, IL, and 10 other states	3	+20	+3	+7	+54	—
Vegetables and Flowers	CT, CA, MA, TX, FL, OH, NY, HI	15	-43	Quality increased in 4 studies and remained the same in others			
Fruits	NY, MA, WA, NJ, CA, CT	8	-20	0	+12	+19	—
Peanuts	GA, TX, OK, NC	5	-5	-5	+13	+100	—
Tobacco	NC	2	-19	—	0	+1	—
Alfalfa	OK, WI, Northwest	3	-2	—	+13	+37	decreased
Unweighted Average[b]			-14.9	-2.8	+11.4	+47.8	decreased

[a] For those producers that adopted the specified IPM practices compared to those that did not.
[b] Weighting is not possible without an accurate accounting of the acreage affected for each commodity in each state.
Source: Norton, G.W. and J. Mullen. 1994. *Economic evaluation of integrated pest management programs: a literature review.* Va. Coop. Ext. Pub. 448-120, Virginia Tech, Blacksburg, VA 24061.

4C: Erosion and Sediment Control

Management Measure for Erosion and Sediment

Apply the erosion component of a Resource Management System (RMS) as defined in the Field Office Technical Guide of the U.S. Department of Agriculture–Natural Resources Conservation Service (see Appendix B) to minimize the delivery of sediment from agricultural lands to surface waters, *or*

Design and install a combination of management and physical practices to settle the settleable solids and associated pollutants in runoff delivered from the contributing area for storms of up to and including a 10-year, 24-hour frequency.

Management Measure for Erosion and Sediment: Description

Application of this management measure will preserve soil and reduce the mass of sediment reaching a water body, protecting both agricultural land and water quality.

This management measure can be implemented by using one of two general strategies, or a combination of both. The first, and most desirable, strategy is to implement practices on the field to minimize soil detachment, erosion, and transport of sediment from the field. Effective practices include those that maintain crop residue or vegetative cover on the soil; improve soil properties; reduce slope length, steepness, or unsheltered distance; and reduce effective water and/or wind velocities. The second strategy is to route field runoff through practices that filter, trap, or settle soil particles. Examples of effective management strategies include vegetated filter strips, field borders, sediment retention ponds, and terraces. Site conditions will dictate the appropriate combination of practices for any given situation. The United States Department of Agriculture (USDA)–Natural Resources Conservation Service (NRCS) or the local Soil and Water Conservation District (SWCD) can assist with planning and application of erosion control practices. Two useful references are the USDA–NRCS Field Office Technical Guide (FOTG) and the textbook "Soil and Water Conservation Engineering" by Schwab et al. (1993).

Resource management systems (RMS) include any combination of conservation practices and management that achieves a level of treatment of the five natural resources (i.e., soil, water, air, plants, and animals) that satisfies criteria contained in the Natural Resources Conservation Service Field Office Technical Guide (FOTG). These criteria are developed at the State level. The criteria are then applied in the provision of field office technical assistance.

The erosion component of an RMS addresses sheet and rill erosion, wind erosion, concentrated flow, streambank erosion, soil mass movements, road bank erosion, construction site erosion, and irrigation-induced erosion. National (minimum) criteria pertaining to erosion and sediment control under an RMS will be applied to prevent long-term soil degradation and to resolve existing or potential off-site deposition problems. National criteria pertaining to the water

Sedimentation causes widespread damage to our waterways. Water supplies and wildlife resources can be lost, lakes and reservoirs can be filled in, and streambeds can be blanketed with soil lost from cropland.

resource will be applied to control sediment movement to minimize contamination of receiving waters. The combined effects of these criteria will be to both reduce upland soil erosion and minimize sediment delivery to receiving waters.

The practical limits of resource protection under an RMS within any given area are determined through the application of national social, cultural, and economic criteria. With respect to economics, landowners should implement an RMS that is economically feasible to employ. In addition, landowner constraints may be such that an RMS cannot be implemented quickly. In these situations, a "progressive planning approach" may be used to ultimately achieve planning and application of an RMS. Progressive planning is the incremental process of building a plan on part or all of the planning unit over a period of time. For additional details regarding RMS, see Appendix B.

Sediment Movement into Surface and Ground Water

Sedimentation is the process of soil and rock detachment (erosion), transport, and deposition of soil and rock by the action of moving water or wind. Movement of soil and rock by water or wind occurs in three stages. First, particles or aggregates are eroded or detached from the soil or rock surface. Second, detached particles or aggregates are transported by moving water or wind. Third, when the water velocity slows or the wind velocity decreases, the soil and rock being transported are deposited as sediment at a new site.

It is not possible to completely prevent all erosion, but erosion can be reduced to tolerable rates. In general terms, tolerable soil loss is the maximum rate of soil erosion that will permit indefinite maintenance of soil productivity, i.e., erosion less than or equal to the rate of soil development. The USDA–NRCS uses five levels of erosion tolerance ("T") based on factors such as soil depth and texture, parent material, productivity, and previous erosion rates. These T levels are expressed as annual losses and range from about 1–5 tons/acre/year (2–11 t/ha/year), with minimum rates for shallow soils with unfavorable subsoils and maximum rates for deep, well-drained productive soils.

Water Erosion

Water erosion is generally recognized in several different forms. *Sheet erosion* is a process in which detached soil is moved across the soil surface by sheet flow, often in the early stages of runoff. *Rill erosion* occurs as runoff water begins to concentrate in small channels or streamlets. Sheet and rill erosion carry mostly fine-textured, small particles and aggregates. These sediments will contain higher proportions of nutrients, pesticides, or other adsorbed pollutants than are contained in the surface soil as a whole. This process of preferential movement of fine particulates carrying high concentrations of adsorbed pollutants is called *sediment enrichment*.

Gully erosion results from water moving in rills which concentrate to form larger and more persistent erosion channels. Gullies are classified as either ephemeral or classic. Ephemeral gullies occur on crop land and are temporarily filled in by field operations, only to recur after concentrated flow runoff. This filling and recurrence of the ephemeral gully can happen numerous times throughout the year if untreated. Classic gullies may occur in agricultural fields but are so large they cannot be crossed by farming equipment, are not in production nor planted

---

Sheet, rill, and gully erosion can occur on cropland fields. Streambank and streambed erosion can occur in intermittent and perennial streams.

to crops, and are farmed around. Classic gullies are characterized by headward migration and enlargement through a combination of headcut erosion and gravitational slumping, as well as the tractive stress of concentrated flows.

Streambank and streambed erosion typically increase in streams during runoff events. Within a stream, the force of moving water on bare or undercut banks causes streambank erosion. Streambank erosion is usually most intense along outside bends of streams, although inside meanders can be scoured during severe floods. Stream power can detach, move, and carry large soil particles, gravel, and small rocks. After large precipitation events, high gradient streams can detach and move large boulders and chunks of sedimentary stone. Streambank and shoreline erosion are addressed in greater detail in EPA's guidance for the coastal nonpoint source pollution control program (EPA, 1993a).

Gully and streambank erosion can move and carry large soil particles that often contain a much lower proportion of adsorbed pollutants than the finer sediments from sheet and rill erosion. Sheet and rill erosion are generally active only during or immediately after rainstorms or snowmelt. Gullies that intercept groundwork may continue to erode without storm events.

> Excessive irrigation water application can detach and transport soil particles.

Irrigation may also contribute to erosion if water application rates are excessive. Erosion may also occur from water transport through unlined earthen ditches. See the Practices for Irrigation Erosion Control discussion in Chapter 4F: Irrigation Water Management for additional information regarding erosion from irrigation.

Water erosion rates are affected by rainfall energy, soil properties, slope, slope length, vegetative and residue cover, and land management practices. Rainfall impacts provide the energy that causes initial detachment of soil particles. Soil properties like particle size distribution, texture, and composition influence the susceptibility of soil particles to be moved by flowing water. Vegetative cover and residue may protect the soil surface from rainfall impact or the force of moving water. These factors are used in the Revised Universal Soil Loss Equation (RUSLE), an empirical formula widely used to predict soil loss in sheet and rill erosion from agricultural fields, primarily crop land and pasture, and construction sites:

Revised Universal Soil Loss Equation (RUSLE)

$$A = R * K * LS * C * P$$

where

$A =$ estimated average annual soil loss (tons/acre/year)

$R =$ rainfall/runoff factor, quantifying the effect of raindrop impact and the amount and rate of runoff associated with the rain, based on long term rainfall record

$K =$ soil erodibility factor based on the combined effects of soil properties influencing erosion rates

$LS =$ slope length factor, a combination of slope gradient and continuous extent

$C =$ cover and management factor, incorporating influences of crop sequence, residue management, and tillage

$P =$ practice factor, incorporating influences of conservation practices such as contouring or terraces

---

Prediction equations such as the RUSLE and WEQ help planners make quantitative assessments of soil loss and BMP effectiveness.

RUSLE may be used as a framework for considering the principal factors affecting sheet and rill erosion: climate (R), soil characteristics (K), topography (LS), and land use and management (C and P). Except for climate, these factors suggest areas where changes in management can influence soil loss from water erosion. Although soil characteristics (K) may be changed slightly over a long period of good management practices by an increase in organic matter, it should generally not be considered changed by management.

It is important to note that the RUSLE predicts **soil loss**, not sediment delivery to receiving waters. Even without erosion control practices, delivery of soil lost from a field to surface water is usually substantially less than 100%. Sediment delivery ratios (percent of gross soil erosion delivered to a watershed outlet) are often on the order of 15–40% (Novotny and Olem, 1994). Numerous factors influence the sediment delivery ratio, including watershed size, hydrology, and topography.

Ephemeral gully erosion can be predicted by the Ephemeral Gully Erosion Model (EGEM), (http://www.wcc.nrcs.usda.gov/hydro/hydro-tools-models.html). EGEM has two major components: hydrology and erosion. The hydrology component is a physical process model that uses the soil, vegetative cover and condition, farming practices, drainage area, watershed flow length, average watershed slope, 24-hour rainfall, and rainfall distribution to estimate peak discharge and runoff volume. Estimates of peak discharge and runoff volume drive the erosion process in the model. The erosion component uses a combination of empirical relationships and physical process equations to compute the width and depth of the ephemeral gully based on hydrology outputs. The model may be used to estimate ephemeral gully erosion for a single 24-hour storm or for average annual conditions.

Erosion control in humid tropical areas like Hawaii and Puerto Rico may present special problems. Soil loss by water erosion may be drastically higher than in temperate regions, especially in areas of steep slopes (El-Swaify and Cooley, 1980). High annual rainfall and the energy of intense storms often result in high erosion rates. Sediment yields of up to 3000 t/sq km/yr from montane basins in Puerto Rico have been reported, where mass wasting contributed most of the sediment to the receiving streams (Simon and Guzman-Rio, 1990). Land clearing and changes in soil characteristics (e.g. exhaustion of soil organic matter) can result in catastrophic soil erosion in tropical regions.

Erosion control practices that succeed in temperate regions are often less effective in the tropics. Engineered practices like terracing, contour ridging, diversions, terraces, and grassed waterways are frequently overwhelmed by torrential rains (Troeh et al., 1980; Lal, 1983). Agronomic practices that conserve the soil, such as mulch farming, reduced tillage, mixed cropping with multistorey canopy structure, and strip cropping with perennial sod crops are more likely to be successful (Troeh et al., 1980; Lal, 1983). El-Swaify and Cooley (1980) reported that pineapple and sugarcane provided adequate protection from soil erosion only a few months after planting.

Wind Erosion

Wind detaches soil particles when, at one foot above the ground surface, wind velocity exceeds 12 mph. Detached soil is moved by wind in one of three ways (Figure 4c-1):

1. Soil particles and aggregates smaller than 0.05 mm in diameter may be picked up by wind and carried in *suspension*. Suspended dust may be moved great distances, but does not drop out of the air unless rain washes it out or the velocity of the wind is dramatically reduced.

2. Intermediate sized grains — 0.05 to 0.5 mm (very fine to medium sand) — move in the wind in a series of steps, rising into the air and falling after a short flight in a motion called *saltation*.

3. Soil grains larger than 0.5 mm cannot be lifted into the wind stream, but particles up to about 1 mm may be pushed along the soil surface by saltating grains or by direct wind action. This type of movement is called *surface creep*.

> Wind can erode and transport soil particles of various sizes causing damage to land and waterways.

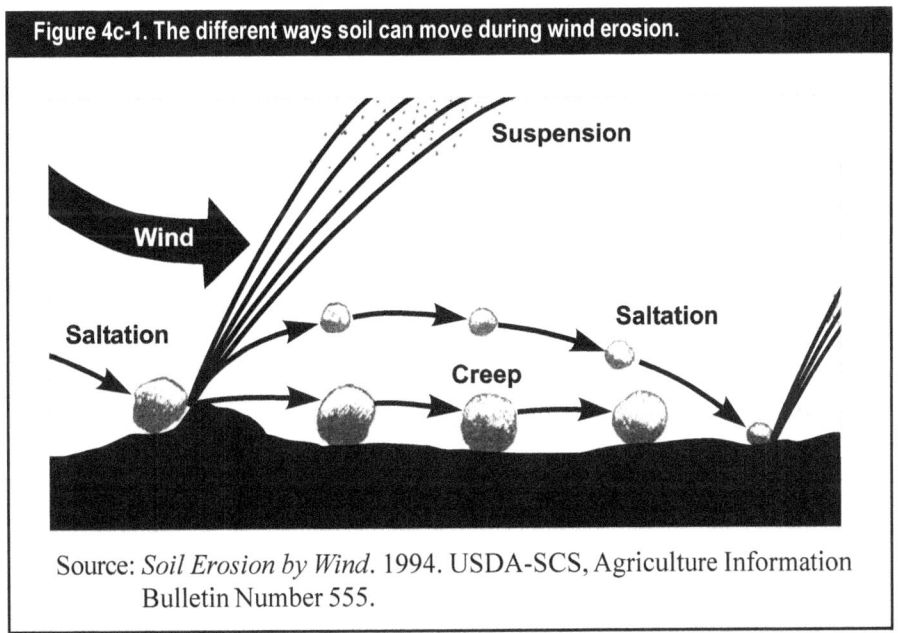

Figure 4c-1. The different ways soil can move during wind erosion.

Source: *Soil Erosion by Wind*. 1994. USDA-SCS, Agriculture Information Bulletin Number 555.

Wind erosion rates are determined by factors similar to those affecting water erosion rates, including the detachment and transport capacity of the wind, soil cloddiness, soil stability, surface roughness, residue or vegetative cover, and length of exposed area. These factors are expressed in the Wind Erosion Equation (WEQ). The WEQ is an empirical wind erosion prediction equation that is currently the most widely used method for estimating average annual soil loss by wind for agricultural fields. The equation is expressed in the general form of:

Wind Erosion Equation (WEQ)

$$E = f(I,K,C,L,V)$$

where **E** is the potential average annual soil loss (tons/acre/year), a function of:

    I, the soil erodibility index;

    K, the soil ridge roughness factor;

    C, the climate factor;

    L, the unsheltered distance across the field; and

    V, the vegetative cover.

Ground Water Protection

Although sediment movement into ground water is generally not an issue in most locations, there are places, such as areas of karst topography, where sediment and sediment-borne pollutants can enter ground water through direct links to the surface. More important from a national perspective, however, is the potential for increased movement of water and soluble pollutants through the soil profile to ground water as a result of implementing erosion and sediment control practices.

It is not the intent of this measure to correct a surface water problem at the expense of ground water. Erosion and sediment control systems can and should be designed to protect against the contamination of ground water. Ground water protection will also be provided through implementation of the nutrient and pesticide management measures.

Erosion and Sediment Control Practices and Their Effectiveness

The strategies for controlling erosion and sedimentation involve reducing soil detachment, reducing sediment transport, and trapping sediment before it reaches water. Combinations of the following practices can be used to satisfy the requirements of this management measure. The NRCS practice number and definition are provided for each management practice, where available. Additional information about the purpose and function of individual practices is provided in Appendix A.

Practices to Reduce Detachment

For both water and wind erosion, the first objective is to keep soil on the field. The easiest and often most effective strategy to accomplish this is to reduce soil detachment. Detachment occurs when water splashes onto the soil surface and dislodges soil particles, or when wind reaches sufficient velocity to dislodge soil particles on the surface.

Crop residues (e.g. straw) or living vegetative cover (e.g. grasses) on the soil surface protect against detachment by intercepting and/or dissipating the energy of falling raindrops. A layer of plant material also creates a thick layer of still air next to the soil to buffer against wind erosion. **Keeping sufficient cover on the soil is therefore a key erosion control practice.**

The implementation of practices such as conservation tillage also preserves or increases organic matter and soil structure, resulting in improved water infiltration and surface stability. In addition, creation of a rough soil surface through practices such as surface roughening will break the force of raindrops and trap water, reducing runoff velocity and erosive forces. This benefit is short-lived, however, as rainfall rapidly decreases effectiveness of surface roughness. Reducing effective wind velocities through increased surface roughness or the use of barriers or changes in field topography will reduce the potential of wind to detach soil particles. Practices which increase the size of soil aggregates increase a soil's resistance to wind erosion.

The following practices can be used to reduce soil detachment:

❑ **Chiseling and subsoiling (324):** Loosening the soil without inverting and with a minimum of mixing of the surface soil to improve water and root penetration and aeration.

❑ **Conservation cover (327):** Establishing and maintaining perennial vegetative cover to protect soil and water resources on land retired from agricultural production.

❑ **Conservation crop rotation (328):** An adapted sequence of crops designed to provide adequate organic residue for maintenance or improvement of soil tilth.

❑ **Residue Management (329):** Any tillage or planting system that maintains at least 30% of the soil surface covered by residue after planting to reduce soil erosion by water; or, where soil erosion by wind is the primary concern, maintains at least 1,000 pounds of flat, small-grain residue equivalent on the surface during the critical erosion period.

❑ **Contour orchard and other fruit area (331):** Planting orchards, vineyards, or small fruits so that all cultural operations are done on the contour.

❑ **Cover crop (340):** A crop of close-growing grasses, legumes, or small grain grown primarily for seasonal protection and soil improvement. It usually is grown for 1 year or less, except where there is permanent cover as in orchards.

❑ **Critical area planting (342):** Planting vegetation, such as trees, shrubs, vines, grasses, or legumes, on highly erodible or critically eroding areas (does not include tree planting mainly for wood products).

❑ **Seasonal Residue Management (344):** Using plant residues to protect cultivated fields during critical erosion periods.

❑ **Diversion (362):** A channel constructed across the slope with a supporting ridge on the lower side (Figure 4c-2).

❑ **Windbreak/shelterbelt establishment (380):** Linear plantings of single or multiple rows of trees or shrubs established next to farmstead, feedlots, and rural residences as a barrier to wind.

> Source area stabilization is fundamental to erosion and sediment control.

Figure 4c-2. Diversion (USDA-SCS, 1984).

☐ **Windbreak/shelterbelt renovation (650):** Restoration or preservation of an existing windbreak, including widening, replanting, or replacing trees.

☐ **Mulching (484):** Applying plant residue or other suitable material to the soil surface.

☐ **Irrigation water management (449):** Effective use of available irrigation water to manage soil moisture, reduce erosion, and protect water quality.

☐ **Prescribed Grazing (528A):** The controlled harvest of vegetation with grazing or browsing animals, managed with the intent to achieve a specified objective.

☐ **Cross wind ridges/stripcropping/trap strips (589):** Ridges formed by tillage or planting, crops grown in strips, or herbaceous cover aligned perpendicular to the prevailing wind direction.

☐ **Surface roughening (609):** Roughening the soil surface by ridge or clod-forming tillage.

☐ **Tree planting (612):** Establishing woody plants by planting or seeding.

☐ **Waste utilization (633):** Using agricultural or other wastes on land in an environmentally acceptable manner while maintaining or improving soil and plant resources.

☐ **Wildlife upland habitat management (645):** Creating, maintaining, or enhancing upland habitat for desired wildlife species.

The following additional practices, although typically applied for a different primary purpose, may have significant secondary benefits in erosion control:

☐ **Brush management (314):** The management of undesirable brush species through use of living organisms, herbicides, prescribed burning, or mechanical methods.

❑ **Irrigation System, Microirrigation (441)**: A planned irrigation system in which all necessary facilities are installed for efficiently applying water directly to the root zone of plants by means of applicators (orifices, emitters, porous tubing, or perforated pipe) operated under low pressure (Figure 4f-19).

❑ **Irrigation system - sprinkler (442)**: Distribution of water by means of sprinklers or spray nozzles to efficiently and uniformly apply irrigation water to maintain adequate soil moisture.

❑ **Pasture and hayland planting (512)**: Establishing and re-establishing long-term stands of adapted species of perennial, biannual, or reseeding forage plants.

Practices to Reduce Transport within the Field

Sediment transport can be reduced in several ways, including the use of crop residues and vegetative cover. Vegetation slows runoff, increases infiltration, reduces wind velocity, and traps sediment. Reductions in slope length and steepness reduce runoff velocity, thereby reducing sediment carrying capacity as well. Terraces and diversions are common techniques for reducing slope length. Runoff can be slowed or even stopped by placing furrows perpendicular to the slope, through practices such as contour farming that act as collection basins to slow runoff and settle sediment particles. By decreasing the distance across a field that is unsheltered from wind and by creating soil ridges or other barriers, sediment transport by wind will be reduced.

> Where conditions and opportunities permit, install practices that prevent edge-of-field sediment loss.

❑ **Contour farming (330)**: Farming sloping land in such a way that preparing land, planting, and cultivating are done on the contour. This includes following established grades of terraces or diversions.

❑ **Field windbreak (392)**: Establishment of trees in or adjacent to a field as a barrier to wind.

❑ **Grassed waterway (412):** A natural or constructed channel that is shaped or graded to required dimensions and established in suitable vegetation for the stable conveyance of runoff.

❑ **Contour stripcropping (585)**: Growing crops in a systematic arrangement of strips or bands on the contour to reduce water erosion. The crops are arranged so that a strip of grass or close-growing crop is alternated with a strip of clean-tilled crop or fallow or a strip of grass is alternated with a close-growing crop (Figure 4c-3).

❑ **Herbaceous Wind Barriers (442A)**: Herbaceous vegetation established in rows or narrow strips across the prevailing wind direction.

❑ **Field stripcropping (586)**: Growing crops in a systematic arrangement of strips or bands across the general slope (not on the contour) to reduce water erosion. The crops are arranged so that a strip of grass or a close-growing crop is alternated with a clean-tilled crop or fallow.

❑ **Terrace (600)**: An earthen embankment, a channel, or combination ridge and channel constructed across the slope (Figures 4c-4 and 4c-5).

❑ **Contour Buffer Strips (332)**: Narrow strips of permanent, herbaceous vegetative cover established across the slope and alternated down the slope with parallel, wider cropped strips.

Practices to Trap Sediment Below the Field or Critical Area

Practices are also typically needed to trap sediment leaving the field before it reaches a wetland or riparian area. Deposition of sediment is achieved by practices that slow water velocity or increase infiltration.

Trap sediment before it reaches riparian areas.

❏ **Sediment basins (350):** Basins constructed to collect and store debris or sediment.

❏ **Field border (386):** A strip of perennial vegetation established at the edge of a field by planting or by converting it from trees to herbaceous vegetation or shrubs.

❏ **Filter strip (393):** A strip or area of vegetation for removing sediment, organic matter, and other pollutants from runoff and wastewater.

❏ **Water and sediment control basin (638):** An earthen embankment or a combination ridge and channel generally constructed across the slope and minor watercourses to form a sediment trap and water detention basin.

Figure 4c-3. Stripcropping and rotations (USDA-ARS, 1987).

Contour strip cropping systems can involve up to 10 strips in a field. A strip cropping system could involve the following:

Corn (either for grain and/or silage)

Soybeans

1st year Meadow

Established Meadow (2-4 years)

Oats

Grassed waterway or diversion

Tillage systems may include two kinds in the same year such as chisel plowing for the soybean crop and moldboard plowing for the oats.

See the following figure showing typical patterns of stripcropping.

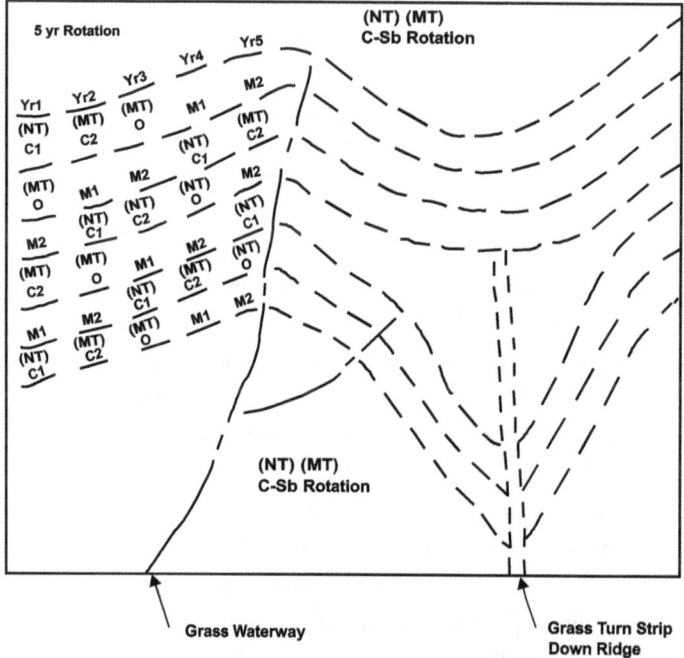

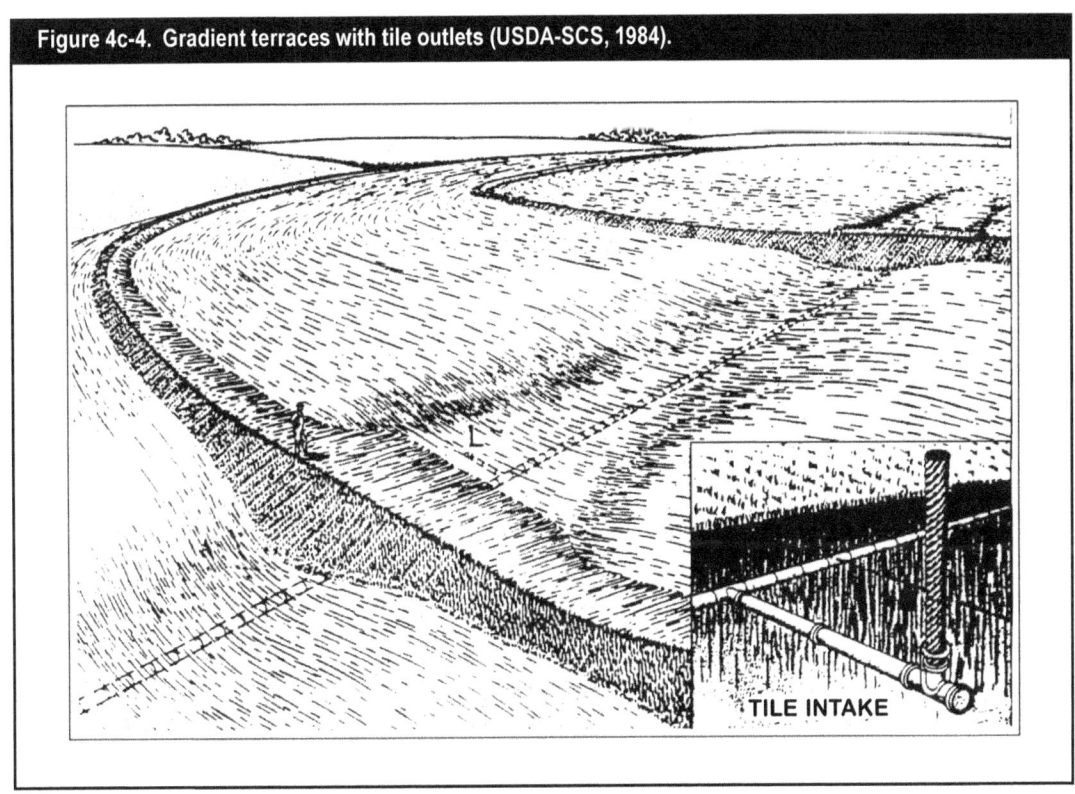

Figure 4c-4. Gradient terraces with tile outlets (USDA-SCS, 1984).

Figure 4c-5. Gradient terraces with waterway outlet (USDA-SCS, 1984).

Healthy Wetland and Riparian Areas Help Reduce Sediment Transport and Delivery

Riparian area practices can serve to repair damaged stream corridors. Assessment and remediation of runoff and sedimentation problems enhances riparian area restoration.

Properly functioning natural wetlands and riparian areas can significantly reduce nonpoint source pollution by intercepting surface runoff and subsurface flow and by settling, filtering, or storing sediment and associated pollutants. Wetlands and riparian areas typically occur as natural buffers between uplands and adjacent water bodies. Loss of these systems allows a more direct contribution of nonpoint source pollutants to receiving waters; degraded wetlands and riparian areas may even become pollutant sources. Thus, natural wetlands and riparian areas should be protected and should not be used as designated erosion control practices. Their nonpoint source control functions are most effective as part of an integrated land management system focusing on nutrient, sediment, and erosion control practices applied to upland areas.

Management measures for protection of the full range of functions for wetlands and riparian areas are discussed in *Nonpoint Source Pollution Guidance for Wetlands, Riparian Areas, and Vegetated Treatment Systems* (EPA, 2001 draft). Protection of wetlands and riparian areas should allow for both nonpoint source pollution control and maintenance of other benefits of these natural aquatic systems, e.g. wildlife habitat. **The Management Measure for Protection of Wetlands and Riparian Areas** states:

> Protect from adverse effects wetlands and riparian areas that are serving a significant NPS abatement function and maintain this function while protecting other existing functions of these wetlands and riparian areas as measured by characteristics such as vegetative composition and cover, hydrology of surface water and ground water, geochemistry of the substrate, and species composition.

Examples of implementation practices for protecting wetlands and riparian areas include:

> Identify existing functions of those wetlands and riparian areas with significant NPS control potential when implementing NPS management practices. Do not alter wetlands or riparian areas to improve their water quality functions at the expense of their other functions.

> Use appropriate preliminary treatment practices such as vegetated treatment systems or detention or retention basins to prevent adverse impacts to wetland functions that affect NPS pollutant abatement from hydrologic changes, sedimentation, or contaminants.

Practices specifically designed to repair or protect wetlands and streambanks from erosion include:

- ❑ **Wildlife wetland habitat management (644)**: Creating, maintaining, or enhancing wetland habitat for desired wildlife species.

- ❑ **Grade stabilization structure (410):** A structure used to control the grade and head cutting in natural or artificial channels.

- ❑ **Streambank and Shoreline Protection (580):** Using vegetation or structures to stabilize and protect banks of streams, lakes, estuaries, or excavated channels against scour and erosion.

❑ **Stream Channel Stabilization (584):** Stabilizing the channel of a stream with suitable structures.

❑ **Use exclusion (472):** Excluding animals, people, or vehicles from an area, primarily by means of fencing.

❑ **Riparian forest buffer/herbaceous cover (391A/390):** Establishing an area of trees, shrubs, grasses, or forbs adjacent to and up-gradient from water bodies.

❑ **Control of streambank erosion** on agricultural land requires techniques different from those used to treat upland sheet and rill erosion. The force of flowing water in a river or stream is a very important process causing streambank erosion. Protection of the slope faces on channel banks, especially those already undergoing active erosion, from the force of flowing water is the key control principle. Techniques may be divided into two general categories: bioengineering (vegetative) and structural. Vegetative methods are generally preferred, unless structural methods are more cost-effective.

Soil bioengineering uses live or dead plant materials, in combination with natural and synthetic support materials, for slope stability, erosion reduction, and vegetative establishment. It should be noted that soil bioengineering measures depending on growth of living vegetation also require livestock exclusion to protect the growing plants from grazing and trampling.

Specific bioengineering practices include:

• *Live staking*: insertion and tamping of live, rootable vegetative cuttings into the ground to create a living root mat that stabilizes the soil.

• *Live fascines and brushlayering*: placement of bundles of branch cuttings (usually of willow) in shallow trenches or benches on bare streambanks to rapidly establish protective vegetation.

• *Tree/shrub planting*: planting of rooted cuttings and tree or shrub seedlings on shaped streambanks and in the riparian zone.

• *Trench packing*: filling of a gully with woody brush to provide a barrier to retard water flow and accumulate sediment.

• *Brushrolls, brushmattresses, brush boxes*: bundles of brush of varying configurations staked against the base of an eroding streambank as a barrier to slow water flow and to settle and accumulate sediment.

Structural practices can protect streambank soils from the erosive force of streamflow, help retain eroding soil, or influence the direction or velocity of streamflow with durable nonliving materials. When using hardened structures like those below, care must be taken to avoid causing additional problems within the stream channel (e.g., channelization, incision):

• *Riprap*: rock dumped or placed along a sloped streambank to armor the bank against the force of flowing water.

• *Revetments*: structures such as timber cribbing backfilled with gravel, anchored trees, gabions, or bulkheads applied to the streambank to hold back eroding material as well as to protect from flowing water.

- Streamflow deflectors: sills, bars, or groins of logs, rock, or concrete projecting out from the bank into the stream to redirect the streamflow away from an eroding bank.

For further information on controlling streambank erosion, refer to Chapter 6: "Management Measures for Hydromodification: Channelization and Channel Modification, Dams, and Streambank and Shoreline Erosion," in *Guidance Specifying Management Measures for Sources of Nonpoint Pollution in Coastal Waters*, EPA 840-B-92-002, 1993. *Stream Corridor Restoration: Principles, Processes, and Practices,* from the Federal Interagency Stream Restoration Working Group (ISBN-0-934213-59-3), (FISRWG, 1998), also contains valuable information on streambank erosion, as well as restoration.

Practice Effectiveness

The available information shows that erosion and sediment control practices in-field can be used to greatly reduce the quantity of eroding soil on agricultural land, and that edge-of-field practices can effectively reduce sediment transport. The benefits of this management measure include preservation of productive agricultural soils and significant reductions in the mass of sediment and associated pollutants (e.g., phosphorus, some pesticides) entering water bodies.

The effectiveness of sediment control practices depends on several factors, including:

❑ The contaminant (e.g. sediment, phosphorus) to be controlled;

❑ The nature of the soil particles to be controlled;

❑ The types of practices or controls being considered;

❑ Site-specific conditions (e.g. crop rotation, topography, tillage, harvesting method); and

❑ Operation and maintenance.

Management practices or systems of practices must be designed for site-specific conditions to achieve desired effectiveness levels. Management practice systems include combinations of practices that provide source control of the contaminant(s) as well as control or reductions in edge-of-field losses and delivery to receiving waters. Table 4c-1 provides a gross estimate of practice effectiveness (i.e., "average" changes in runoff and pollutant loads due to the addition of the practice(s) at sites where erosion control practices are generally lacking) as reported in research literature. Even within relatively small watersheds, extreme spatial and temporal variations are common. Because of this variation, the actual effectiveness of practices at a specific site may differ considerably from the gross estimates given in Table 4c-1.

> Although some sites are challenging, detailed local information combined with sound erosion control knowledge and experience should result in an effective system plan for erosion and sediment control.

Table 4c-1. Relative Gross Effectiveness[a] of Sediment[b] Control Measures Pennsylvania State University, 1992b).				
Practice Category[c]	**Runoff Volume**	**Total[d] Phosphorus**	**Total[d] Nitrogen**	**Sediment**
		(% reduction)		
Reduced Tillage Systems[e]	reduced	45	55	75
Diversion Systems[f]	reduced	30	10	35
Terrace Systems[g]	reduced	70	20	85
Filter Strips[h]	reduced	75	70	65

a Actual effectiveness depends on site-specific conditions. Values are not cumulative between practice categories.

b Includes data where land application of manure has occurred.

c Each category includes several specific types of practices.

d Total phosphorus includes total and dissolved phosphorus; total nitrogen includes surface-delivered organic-N, ammonia-N, and nitrate-N.

e Includes practices such as conservation tillage, no-till, and crop residue use.

f Includes practices such as grassed waterways and grade stabilization structures.

g Includes several types of terraces with safe outlet structures where appropriate.

h Includes all practices that reduce contaminant losses using vegetative control methods.

Conservation tillage is now promoted widely by a large number of groups and organizations because it is both profitable and effective in controlling erosion. For example, researchers at Louisiana State University have shown that the use of no-till with or without a cover crop (2-6 tons of soil loss per acre per year) is much more effective at controlling erosion on cotton fields than is use of conventional tillage with or without a cover crop (13-16 tons per acre per year) (Zeneca, 1994). It is reported that the top three reasons soybean farmers adopt no-till are reduced soil erosion, increased profit potential, and time and labor savings (Alesii, 1998). The percentage of soybeans planted in no-till has increased from 1992 to 1997 at an average annual rate of 11.6 percent, ranging from 4 percent (Minnesota) to 25 percent (North Dakota) in the Upper Midwest (CTIC, 1997). According to some of the leading authorities on conservation tillage, the economic and environmental benefits of farming with conversation tillage are simply too numerous to ignore (CTIC, ca. 1997). CTIC reported that, on average, no-till resulted in 93 percent less erosion and 69 percent less water runoff than moldboard plowing.

Factors in the Selection of Management Practices

Two fundamental options exist to minimize water and wind erosion from agricultural land and the delivery of sediment to receiving waters: (1) Controlling soil loss from fields or streambanks by reducing detachment and transport of sediment, and (2) Encouraging deposition of eroded sediment to prevent delivery to surface waters. Different management strategies are employed with the different options. Preventing initial soil loss (option "(1)") is generally the most desirable option because it not only minimizes the delivery of sediment to receiving waters but also provides an agronomic benefit by preserving soil resources. Option "(2)" minimizes the delivery of sediment to receiving waters, but does not necessarily provide the agronomic benefits of upland erosion control. In addition, practices encouraging sediment deposition require mainte-

Site conditions, cost, and maintenance requirements are considered for practice selection. Local demonstrations are also needed to refine practices and encourage adoption.

nance to retain their effectiveness over time. In some cases, for example, management or economic constraints may prevent full installation of all practices needed to adequately reduce field soil loss, and additional practices to prevent delivery of eroded sediment may be needed. In other cases, even if field soil loss can be reduced to "T" level, additional practices may be needed to prevent delivery of sediment to critical or sensitive water bodies. Using one or both of these options, planners have the flexibility to address erosion and sediment problems in a manner that best reflects State, local, and land owner/operator needs and preferences.

Management practices for a given site should not result in undue economic impact on the operator. Many of the practices that could be used to implement this measure may already be encouraged or required by Federal, State, or local programs (e.g., filter strips or field borders along streams) or may otherwise be in use on agricultural fields. By building upon existing erosion and sediment control efforts, the time, effort, and cost of implementing this measure will be reduced.

It should be noted that basic erosion control measures will not always provide adequate control of nutrients, pesticides, or other sediment-attached pollutants. Erosion control practices tend to be most effective on larger particles, which tend to carry a lower proportion of adsorbed pollutants than do finer particles like clays. Many erosion control practices or structures may not effectively control the majority of pollutants that are attached to fine soil particles. If pollutants attached to soil particles are the primary concern, practices specifically designed to control fine sediments should be applied.

Conversely, some nutrient or irrigation management practices may contribute to erosion control, even though their primary purpose is not erosion control. Waste utilization, for example, may help reduce soil erodibility by both water and wind through improvements in soil organic matter content. Improved irrigation water management may help reduce wind erosion potential by maintaining adequate soil moisture during critical periods.

Continued performance of this measure will be ensured through supporting maintenance operations where appropriate. Although some practices are designed to be effective and withstand a design storm, they may suffer damage when larger storms occur. It is expected that damage will be repaired after such storms and that practices will be inspected periodically. To ensure that practices selected to implement this measure will continue to function as designed and installed, some operational functions and maintenance will be necessary over the life of the practices.

Most structural practices for erosion and sediment control are designed to operate without human intervention. Management practices such as conservation tillage, however, do require some attention each time they are used. Field operations should be conducted with practices like contouring or terraces in mind to ensure that the practices or structures are not damaged or destroyed by the operations. For example, non-selective herbicides should not be applied to areas of permanent vegetative cover that are used as part of erosion control practices, such as waterways and filter strips.

Structural practices such as diversions, grassed waterways, and filter strips may require grading, shaping, and reseeding. Trees and brush should not be allowed to grow on berms, dams, or other structural embankments. Cleaning of sediment retention basins will be needed to maintain their original design capacity and trapping efficiency.

Filter strips and field borders must be maintained to prevent channelization of flow and the resulting short-circuiting of filtering mechanisms. Reseeding of filter strips may be required on a frequent basis. Periodic removal of vegetative growth will help keep filter strips actively growing and remove nutrients and other potential pollutants that have been taken up by the plants or attached to the vegetative growth. Grazing and other livestock activities should be managed to avoid damage to vegetation cover, especially near streams.

Finally, conditions sometimes occur when serious wind erosion is imminent or has just begun, and immediate action is needed to protect soil and crops. Several emergency techniques can lessen or slow wind erosion. Emergency measures are not as effective as long-term planned erosion control; they are last resort options and should not be relied on for primary erosion control or continued use. The following emergency control methods can reduce damage from anticipated wind erosion (Smith et al., 1991).

- ❏ Emergency tillage to produce surface roughness, ridges, and clods
- ❏ Addition of crop residue
- ❏ Application of manure
- ❏ Irrigation to increase soil moisture
- ❏ Temporary, artificial wind barriers
- ❏ Soil additives or spray-on adhesives

Choice of specific methods depends on severity of erosion, soil type, crop type and growth stage, and equipment available.

Cost and Savings of Practices

Costs

Both national and selected State costs for a number of common erosion control practices are presented in Table 4c-2. The variability in costs for practices can be accounted for primarily through differences in site-specific applications and costs, differences in the reporting units used, and differences in the interpretation of reporting units.

The cost estimates for control of erosion and sediment transport from agricultural lands in Table 4c-3 are based on experiences in the Chesapeake Bay Program.

Savings

It is important to note that for some practices, such as conservation tillage, the net costs often approach zero and in some cases can be negative because of the savings in labor and energy. In fact, it is reported that cotton growers can lower their cost per acre by $24.32 due to lower fixed costs associated with conservation tillage (Zeneca, 1994).

Reliable and current information on cost of initial investment, along with annualized cost throughout practice life, helps planners and farmers make sound decisions.

Table 4c-2. Representative costs of selected erosion control practices.

Practice	Unit	Range of Capital Costs[1]	References
Diversions	ft	1.97 - 5.51	Sanders et al., 1991 Smolen and Humenik, 1989
Terraces	ft a.s.[2]	3.32 - 14.79 24.15 - 66.77	Smolen and Humenik, 1989 Russell and Christiansen, 1984
Waterways	ft ac a.e.[3]	5.88 - 8.87 113 - 4257 1250 - 2174	Sanders et al., 1991 Barbarika, 1987; NCAES, 1982; Smolen and Humenik, 1989 Russell and Christiansen, 1984
Permanent Vegetative Cover	ac	69 - 270	Barbarika, 1987; Russell and Christiansen, 1984; Sanders et al., 1991; Smolen and Humenik, 1989
Conservation Tillage	ac	9.50 - 63.35	NCAES, 1982; Russell and Christiansen, 1984; Smolen and Humenik, 1989

1 Reported costs inflated to 1998 dollars by the ratio of indices of prices paid by farmers for all production items, 1991=100.
2 acre served
3 acre established

[Note: 1991 dollars from CZARA were adjusted by +15%, based on ratio of 1998 Prices Paid by Farmers/1991 Prices Paid by Farmers, according to USDA National Agricultural Statistics Service, http://www.usda.gov/nass/sources.htm, 28 September, 1998]

Table 4c-3. Annualized cost estimates and life spans for selected management practices from Chesapeake Bay Installations[a] (Camacho, 1991).

Practice	Practice Life Span (Years)	Median Annual Costs[b] (EAC[c])($/acre/yr)
Nutrient Management	3	2.40
Strip-cropping	5	11.60
Terraces	10	84.53
Diversions	10	52.09
Sediment Retention Water Control Structures	10	89.22
Grassed Filter Strips	5	7.31
Cover Crops	1	10.00
Permanent Vegetative Cover on Critical Areas	5	70.70
Conservation Tillage[d]	1	17.34
Reforestation of Crop and Pasture[d]	10	46.66
Grassed Waterways[e]	10	1.00/LF/yr
Animal Waste System[f]	10	3.76/ton/yr

a Median costs (1990 dollars) obtained from the Chesapeake Bay Program Office (CBPO) BMP tracking data base and Chesapeake Bay Agreement Juristictions' unit data cost. Costs per acre are for acres benefited by the practice.
b Annualized BMP total cost including O&M, planning, and technical assistance costs.
c EAC = Equivalent annual cost: annualized total; costs for the life span. Interest rate = 10%.
d Government incentive costs.
e Annualized unit cost per linear foot of constructed waterway.
f Units for animal waste are given as $/ton of manure treated.

4D: Animal Feeding Operations (AFOs)

Management Measure for Animal Feeding Operations

Animal feeding operations (AFOs) should be managed to minimize impacts on water quality and public health. To meet this goal, management of AFOs should address the following eight components:

1. *Divert clean water.* Siting or management practices should divert clean water (run-on from uplands, water from roofs) from contact with feedlots and holding pens, animal manure, or manure storage systems.

2. *Prevent seepage.* Buildings, collection systems, conveyance systems, and storage facilities should be designed and maintained to prevent seepage to ground and surface water.

3. *Provide adequate storage.* Liquid manure storage systems should be (a) designed to safely store the quantity and contents of animal manure and wastewater produced, contaminated runoff from the facility, and rainfall from the 25-year, 24-hour storm and (b) consistent with planned utilization or utilization practices and schedule. Dry manure, such as that produced in certain poultry and beef operations, should be stored in production buildings, storage facilities, or otherwise covered to prevent precipitation from coming into direct contact with the manure.

4. *Apply manure in accordance with a nutrient management plan that meets the performance expectations of the nutrient management measure.*

5. *Address lands receiving wastes.* Areas receiving manure should be managed in accordance with the erosion and sediment control, irrigation, and grazing management measures as applicable, including practices such as crop and grazing management practices to minimize movement of nutrient and organic materials applied, and buffers or other practices to trap, store, and "process" materials that might move during precipitation events.

6. *Recordkeeping.* AFO operators should keep records that indicate the quantity of manure produced and its utilization or disposal method, including land application.

7. *Mortality management.* Dead animals should be managed in a way that does not adversely affect ground or surface waters.

8. *Consider the full range of environmental constraints and requirements.* When siting a new or expanding facility, consideration should be given to the proximity of the facility to (a) surface waters; (b) areas of high leaching potential; (c) areas of shallow groundwater; and (d) sink holes or other sensitive areas. Additional factors to consider include siting to minimize off-site odor drift and the land base available for utilization of animal manure in accordance with the nutrient management measure. Manure should be used or disposed of in ways that reduce the risk of environmental degradation, including air quality and wildlife impacts, and comply with Federal, State and local law.

> Animal Feeding Operations should be designed and operated to avoid waste discharge by having engineered runoff controls, waste storage, waste utilization, and nutrient management.

USDA–EPA Unified National Strategy for Animal Feeding Operations

USDA-EPA Unified National Strategy for Animal Feeding Operations

Animal feeding operations (AFOs) can pose a number of risks to water quality and public health, mainly because of the amount of animal manure and wastewater they generate. To minimize water quality and public health impacts from AFOs and land application of animal waste, the U.S. Department of Agriculture (USDA) and the U.S. Environmental Protection Agency (EPA) released the Unified National Strategy for Animal Feeding Operations on March 9, 1999. The Strategy sets a national performance expectation that all AFO owners and operators develop and implement technically sound and economically feasible site-specific **Comprehensive Nutrient Management Plans (CNMPs)** by 2009.

A CNMP identifies actions that will be implemented to meet clearly-defined nutrient management goals at an agricultural operation. AFO owners and operators may seek technical assistance for the development, implementation and review of CNMPs from qualified specialists.

The following components may be contained in a CNMP:

❑ **Feed Management:** reducing nutrients in manure by modifying animal diets

❑ **Manure Handling and Storage:** proper handling and storage of manure

❑ **Land Application of Manure:** utilizing the nutrients and organic matter in manure while minimizing the risk to water quality and public health

❑ **Land Management:** installing best management practices to minimize movement of potential pollutants to surface or ground water

❑ **Record Keeping:** recording the quantity of manure produced and how the manure was utilized

❑ **Other Utilization Options:** finding alternative uses or markets (e.g., composting, sale to other farmers, power generation) for manure when land application is not feasible

Voluntary and regulatory programs serve complementary roles in providing AFO owners and operators and the animal agricultural industry with the assistance and certainty they need to achieve individual business and personal goals, and in ensuring protection of water quality and public health. For the vast majority of AFOs, voluntary efforts will be the principal approach to assist owners and operators in developing and implementing site-specific CNMPs and in reducing water pollution and public health risks associated with AFOs. While CNMPs are not required for AFOs participating only in voluntary programs, they are strongly encouraged as the best possible means of managing potential water quality and public health impacts from these operations.

Impacts from certain higher risk AFOs are addressed through National Pollutant Discharge Elimination System (NPDES) permits under the authority of the Clean Water Act. AFOs that meet certain specified criteria in the NPDES regulations are referred to as concentrated animal feeding operations or CAFOs. NPDES permits will require CAFOs to develop CNMPs and to meet other conditions that minimize the threat to water quality and public health and otherwise ensure compliance with the requirements of the Clean Water Act.

The Strategy identifies three categories of CAFOs that are priorities for the regulatory program:

❏ **Significant Manure Production:** large facilities (i.e., greater than 1000 animal units)

❏ **Unacceptable Conditions:** facilities that discharge through a man-made conveyance to waters or allow animals direct contact with waters

❏ **Significant Contributors to Water Quality Impairment:** facilities that are significantly contributing to the impairment of a waterbody

In addition, the Unified AFO Strategy addresses strategic issues to be addressed by the agencies. The discussion of each strategic issue identifies several action items that the agencies intend to pursue in implementing the Strategy. Some of these actions are listed below.

❏ Assure the availability of qualified specialists from the public or private sectors to assist in the development and implementation of CNMPs

❏ Review USDA's practice standards and revise as necessary

❏ Develop a CNMP guidance

❏ Strengthen and improve existing EPA regulations for CAFOs

❏ Coordinated research, technical innovation, and technology transfer activities

❏ Provide compliance assistance and establish a single point information center

❏ Promote the involvement of the animal agriculture industry in CNMP adoption

❏ Coordinate data sharing while protecting the relationship of trust between USDA and farmers and providing regulatory authorities with information that is useful in protecting water quality and public health

❏ Develop an approach for measuring the effectiveness of efforts to minimize the water quality and public health impacts of AFOs

> **For additional information on the Strategy, see** http://cfpub.epa.gov/npdes/
> home.cfm?program_id=7

AFOs, CAFOs, and CZARA

Existing regulatory definitions of AFOs and *concentrated animal feeding operations (CAFOs)* are given at 40 *CFR* 122.23 and Part 122, Appendix B (as revised February 12, 2003). These regulations define an AFO as a facility that meets the following criteria:

1. Animals (other than aquatic animals) have been, are, or will be stabled or confined and fed or maintained for a total of 45 days or more in any 12-month period, and

2. Crops, vegetation forage growth, or post-harvest residues are not sustained in the normal growing season over any portion of the lot or facility.

As described in Chapter 1, EPA published guidance specifying management measures for sources of nonpoint pollution in coastal waters as required under section 6217(g) of CZARA. With regard to the management measures for livestock operations (EPA, 1993a), EPA defined a *confined animal facility* as a lot or facility that meet the same two criteria (1 and 2) specified above for AFOs. AFOs include the areas used to grow or house the animals, areas used for processing and storage of product, manure and runoff storage areas, and silage storage areas.

The subset of AFOs within the section 6217 coastal management areas that are subject to the CZARA management measures for confined animal facilities is determined by the number of head at the operation and whether or not the operation is designated as a CAFO. Those facilities that are required by Federal regulation 40 CFR 122.23(c) to apply for and receive discharge permits, are *NOT* covered by section 6217 since they are CAFOs. CAFOs are defined generally as an AFO that:

❑   Confines the number of animals presented in the second column of Table 4d-1: or

❑   Confines the number of animals presented in the third column of Table 4d-1 and discharges pollutants:

- Into waters of the U.S. through a man-made ditch, flushing system, or similar man-made device; or

- Directly into waters of the U.S. that originate outside of and pass over, across, or through the facility or otherwise come into direct contact with the animals confined in the operation.

In addition, 40 CFR 122.23(c) provides that the Director of a National Pollutant Discharge Elimination System (NPDES) permit program may designate any AFO as a CAFO upon determining that it is a significant contributor of water pollution. AFOs containing fewer than the number of head listed in Table 4d-1 for small confined animal facilities are not subject to the CZARA management measures for confined animal facilities. Figure 4d-1 shows the relationship between AFOs, CAFOs, and large and small confined animal facilities under the NPDES and CZARA programs. Operators of confined animal facilities should contact their state or federal NPDES permitting authority for information on permit application procedures.

It is important to note that in December 2002 EPA finalized revised regulations for concentrated animal feeding operations under 40 CFR 122. The final regulations changed some of the definitions. Readers are encouraged to contact EPA's

Table 4d-1. Comparison of CAFO and AFO Size Difinitions under the NPDES and CZARA Programs.				
Animal Type	Defined as a CAFO by Size and must have a NPDES Permit	Defined as CAFO by Size and Site Conditions* and must have a NPDES Permit	Large Animal Feeding Operations under CZARA (that do not have a NPDES Permit)	Small Animal Feeding Operations under CZARA
	Number of Head			
Beef cattle or heifers	≥1,000	< 1,000 & ≥300	≥300	51 - 299
Veal calves	≥1,000	< 1,000 & ≥300	ND**	ND
Mature dairy cattle	≥700	< 700 & ≥200	≥70	20 - 69
Swine	≥2,500 (each 55 lbs or more)	< 2,500 & ≥750 (each 55 lbs or more)	≥200	100 - 199
Swine	≥10,000 (each under 55 lbs)	< 1,000 & ≥300 (each under 55 lbs)	ND	ND
Turkeys	≥55,000	<55,000 & ≥16,500	≥13,750	5,000 - 13,749
Chickens with liquid manure handling	≥30,000	< 30,000 & ≥9,000	ND	ND
Chickens (except laying hens) with dry manure handling	≥125,000	< 125,000 & ≥37,500	≥15,000 (all broilers)	5,000 - 14,999 (all broilers)
Laying hens with dry manure handling	≥82,000	< 82,000 & ≥25,000	≥15,000 (all laying hens)	5,000 - 14,999 (all laying hens)
Horses	≥500	< 500 & ≥150	≥200	100 - 199
Sheep or lambs	≥10,000	< 10,000 & ≥3,000	ND	ND
Ducks with liquid manure handling	≥5,000	< 5,000 & ≥1,500	ND	ND
Ducks with dry manure handling	≥30,000	< 30,000 & ≥10,000	ND	ND

*AFOs are defined as CAFOs if they have the number of animals shown above AND have a man-made ditch or pipe that carries manure or wastewater from the operation to surface waters OR the animals come into contact with surface water running through the area where they are confined.
**Not defined.

Office of Wastewater Management (www.epa.gov/owm) or their state NPDES permitting authority for the latest information on the final CAFO regulations.

Management Measure for Animal Feeding Operations: Description

The water quality problems associated with confined animal facilities result from accumulated animal wastes, facility wastewater, and storm runoff, all of which may be controlled under this management measure (Figure 4d-2). The goal of this management measure is to minimize the discharge of contaminants in facility wastewater, runoff, and seepage to ground water, while at the same time preventing any other negative environmental impacts such as increased air pollution. Accumulated animal wastes include manure, litter, or other waste products that are deposited within the confinement area and are periodically removed by scraping, flushing, or other means and can be conveyed to a storage or treatment facility. Facility wastewater is water generated in the operation of an animal facility as a result of animal or poultry watering; washing, cleaning, or

Management of Soil Phosphorus Levels to Protect Water Quality

Phosphorus in Agriculture

Phosphorus (P) is important to and used extensively in both the crop production and confined livestock segments of agriculture, making it one of the most common elements used in agriculture today.

One of the most important functions of P in plants is the storage and transfer of energy. Phosphorus is essential for seed production, promotes increased root growth, stalk strength and early plant maturity, and aids in resistance to root rot diseases and winter kill.

In the confined livestock segment, producers use P as a diet supplement, in addition to the P already contained in feeds, to improve animal performance. To avoid excessive buildup of soil-P on the lands surrounding confined animal operations, consideration must be given to the amount of land available to absorb P from livestock.

Environmental Impacts

In areas of intense crop and livestock production, continued inputs of fertilizer and manure P in excess of crop requirements have led to a build-up of soil P levels. This increases the potential for nonpoint source (NPS) runoff to carry excess phosphorus to surrounding streams and lakes.

Phosphorus is usually the limiting nutrient in freshwater aquatic systems. When excess phosphorus enters streams and lakes, creating P concentrations between the critical values of 0.01 and 0.02 ppm (Sawyer, 1947; Vollenweider, 1968), accelerated eutrophication occurs. Eutrophication, a natural process that usually occurs over a long period of time, is characterized by increased aquatic plant growth, oxygen depletion, and pH variability. It eventually leads to a decline in plant species quality and adverse food chain effects (Sharpley et al., 1994), all of which may reduce water quality.

Transport Mechanism

Phosphorus enters the soil through mineral dissolution, desorption from clay and mineral surfaces, and biological conversion from organic materials to inorganic forms. As rainfall or irrigation water interacts with a thin layer of surface soil, P is either moved into agriculture runoff through dissolution from the soil and plant material, or is transported by erosion, remaining either attached to soil or in vegetation. The dissolved P is immediately available for uptake by aquatic biota (bioavailable), while the particulate P is available only after all of the dissolved P is consumed. Once bioavailable P moves from the field into receiving waters, it can contribute to eutrophication (Wood et al., 1998).

Another mechanism for P transport occurs when large accumulations of P occupy all available sites on the soil surface, causing additional P to leach downward through the soil column. When this leaching is followed by lateral movement of water under the soil surface, especially under high water table conditions, dissolved P may be added to the surface waters.

Soil Testing

The prime goal of soil testing methods have also been developed and tested to determine if they might more accurately predict the runoff and drainage P levels. Some of the most promising new methods are:

(1) Breeuswma et al., 1995 – developed to determine the degree of P saturation in soils

(2) Chardon et al., 1996 – using an iron oxide coated filter paper strip as an "infinite sink" to measure the amount of P in soils that is subject to runoff or leaching

(3) Pote et al., 1996 – using distilled water to extract readily desorbable soil P, simulating the rapid release of P to runoff water.

Soil test extractants now used for phosphorus in the U.S. (Kamprath and Watson, 1980)		
Soil Test Category	**Common Soil Test**	**Regions in the U.S. Where Commonly Used**
Dilute concentrations of strong acids: Solvent nature of acids primarily extracts Al and Fe bound P, plus some Ca-P. Best for soils with pH < 7.0	Mehlich 1	Southeast and Mid-Atlantic
Dilute concentrations of strong acids plus a complexing ion: Extractants remove P by solvent action of acids and complexing ability of fluoride ion for Al-P. Best on acidic soils.	Bray P1 Mehlich 3	Bray: North Central and Midwest Mehlich 3: Widespread use in U.S.
Dilute concentrations of weak acids: Anion replacement	Morgan and Modified Morgan	Northeast
Buffered Alkaline Solutions: Extract P by hydrolysis of cations binding P. Precipitate $CaCO_3$ from calcareous, alkaline, and neutral soils, reducing Ca and increasing P concentrations in solution, making P more accurately and easily measured.	Olsen AB-DTPA	West and Northwest

Other Control Options

One reason for high phosphorus levels in runoff from fields fertilized with poultry or swine litter is that these animals lack phytase enzymes, making most of the phytate P (65% of total P) in corn and soybeans unavailable to these animals. In order for normal growth and development, other forms of P must be added to the diet. This addition of inorganic P results in much higher levels of P in manure.

Phytase products - One way to reduce the level of inorganic P fed to these animals, thus lowering the level of P in manure, is to add phytase enzyme to the feed aiding the breakdown of phytate P.

Low phytic acid or high available P (HAP) corn - Another way to reduce the amount of additional P needed in the animal diets, thus reducing amounts of P in manure, is to feed the animals a corn hybrid containing lower amounts of phytate P or higher amounts of available P.

While some studies have shown that P levels in runoff decrease with the use of these products in livestock diets, more comprehensive research must be done before any conclusions can be drawn.

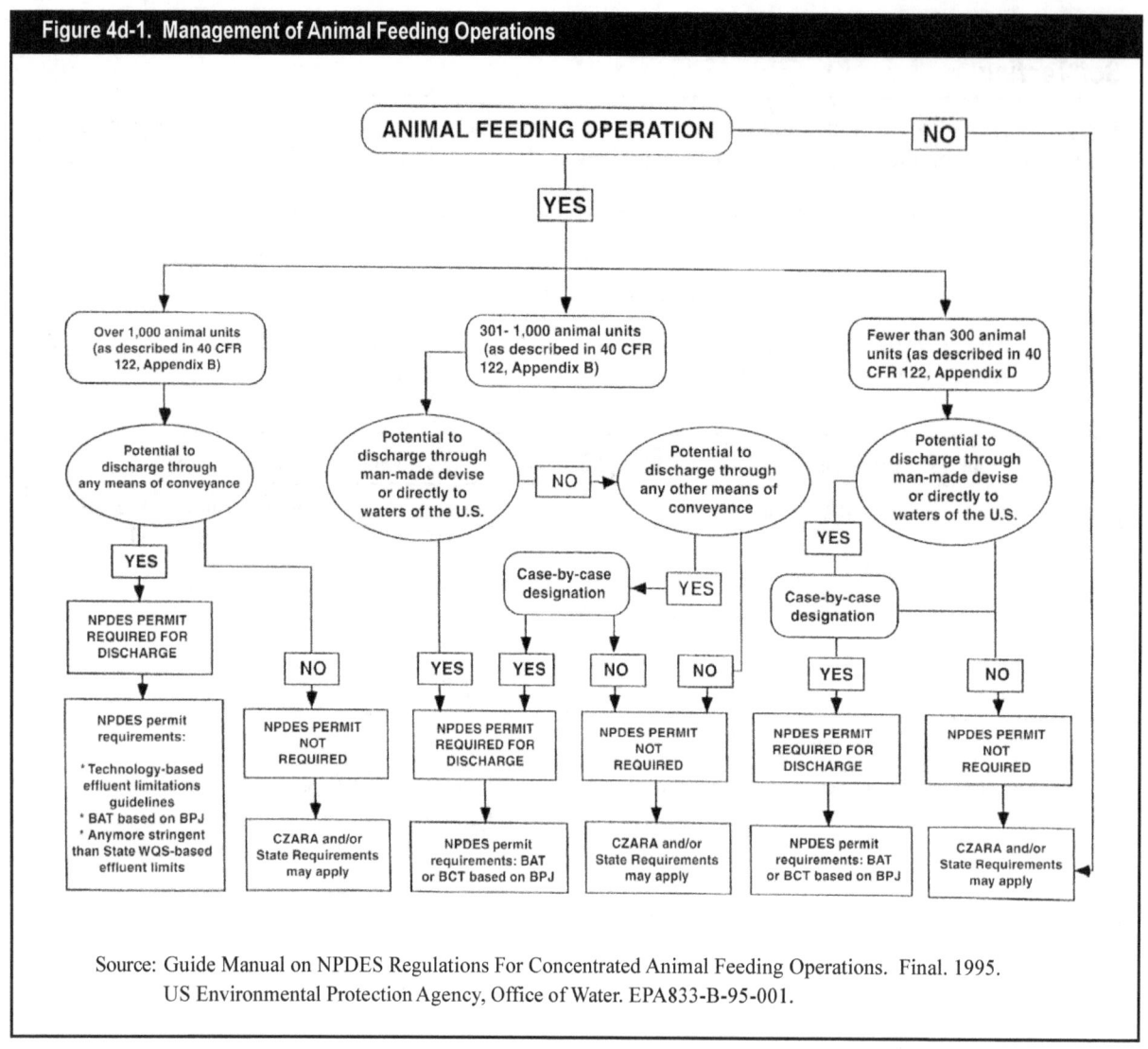

Figure 4d-1. Management of Animal Feeding Operations

Source: Guide Manual on NPDES Regulations For Concentrated Animal Feeding Operations. Final. 1995.
US Environmental Protection Agency, Office of Water. EPA833-B-95-001.

flushing pens, barns, manure pits, or other facilities; washing or spray cooling of animals; and dust control. Animal lot runoff includes any precipitation (rain or snow) that comes into contact with manure, feed, litter, or bedding and may potentially leave the facility either by overland flow or by infiltration.

Implementation of this management measure greatly reduces the volume of runoff, manure, and facility wastewater reaching a water body due to structural practices such as solids separation basins in combination with vegetative practices and other techniques that reduce runoff while also protecting ground water. The measure can be implemented by using practices that divert clean runoff water from upslope sites and roofs away from the facility, thereby minimizing the amount of contaminated water to be stored and managed. Accumulated animal wastes should be protected as much as possible from runoff and stored in such a way that any runoff water, seepage, or leachate can be captured and managed with runoff and wastewater. Runoff water and facility wastewater should be routed through a settling structure or debris basin to remove solids, and then stored in a pit, pond, or lagoon for application on agricultural land in accordance with the Nutrient Management Measure. If manure is managed as a

Diverting clean water from upslope areas and roof runoff away from the animal lot and waste storage structure can reduce waste volume and storage requirements.

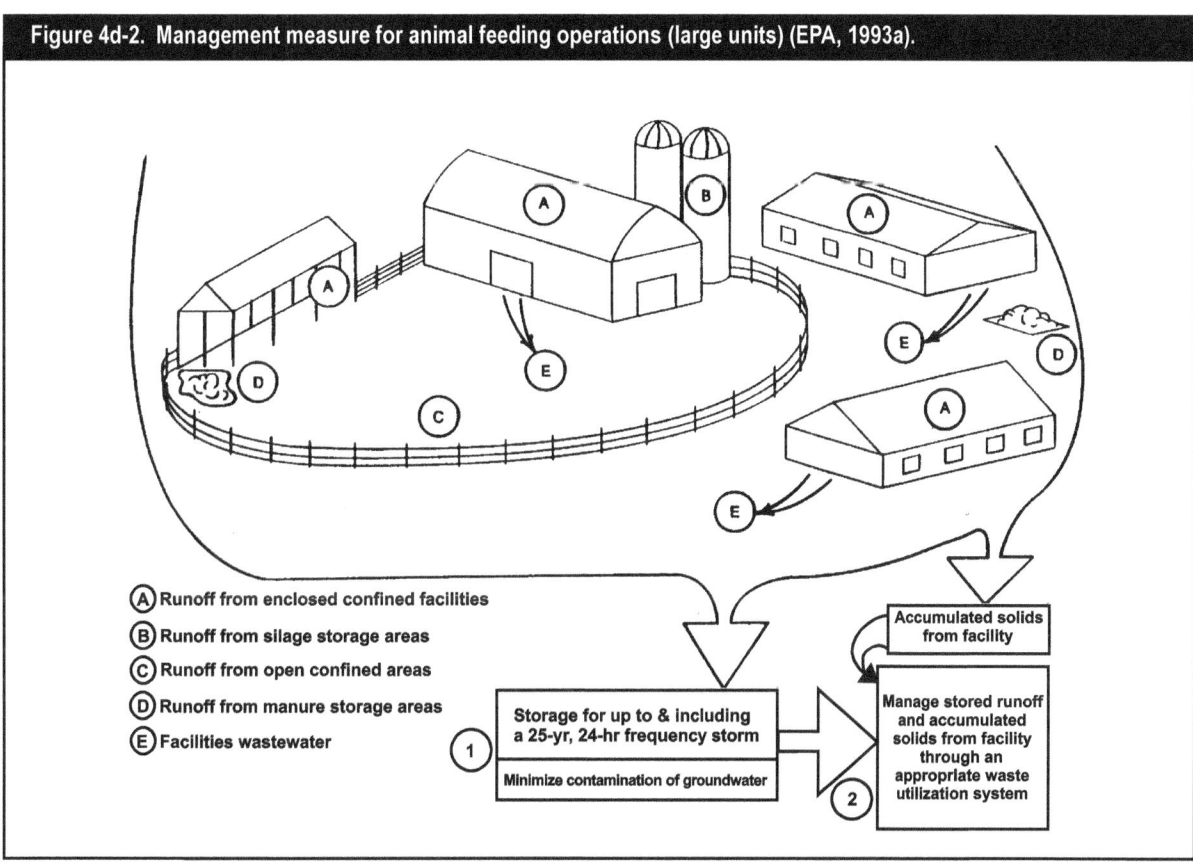

Figure 4d-2. Management measure for animal feeding operations (large units) (EPA, 1993a).

(A) Runoff from enclosed confined facilities

(B) Runoff from silage storage areas

(C) Runoff from open confined areas

(D) Runoff from manure storage areas

(E) Facilities wastewater

(1) Storage for up to & including a 25-yr, 24-hr frequency storm

Minimize contamination of groundwater

Accumulated solids from facility

(2) Manage stored runoff and accumulated solids from facility through an appropriate waste utilization system

liquid, all manure, runoff, and facility wastewater can be stored in the same structure and there is no need for an additional debris basin. In some areas, certain systems may be preferred over others due to competing environmental concerns (e.g., liquid systems may raise concerns regarding air quality), and innovative alternatives that achieve the management measure goals should be considered.

This management measure is consistent with, yet more specific than the CZARA management measure for large confined animal facilities, and it goes beyond the expectation for small confined animal facilities under CZARA by calling for storage. This does *NOT* change, however, the performance expectations for either large or small facilities that are subject to the CZARA management measures.

Contaminant Movement from Animal Feeding Operations into Surface and Ground Water

The concentration of livestock production and housing in large systems has resulted in large accumulations of animal wastes with the potential to contribute nutrients, suspended solids, pathogens, oxygen-demanding materials, and heavy metals to surface and ground waters. Animal operations can also be a source of atmospherically transported pollutants, particularly ammonia, via volatilization (Harper and Sharpe, 1997). The pollution potential of such accumulations is influenced by the number and type of animals in the operation, the facilities and practices used to collect and store the wastes, and the methods chosen to manage the wastes (e.g., application to the land).

Animal feeding operations have the potential to contribute large pollutant loads to waterways. Because they may be located near streams and water supplies, AFOs require well-planned and maintained systems of practices to minimize human health and aquatic ecosystem impacts.

Movement to Surface Waters

The volume of runoff from animal facilities is influenced by several major factors including: (1) *water inputs*, dependent on rain storm intensity and duration, time since last runoff, snowpack accumulation and melting, and runoff entering from outside the facility and (2) *runoff generation* from impervious surfaces such as roofs and paved areas. While precipitation inputs cannot usually be managed, the diversion of clean water from upgradient areas, and the reduction and diversion of runoff from impervious areas (e.g. installation of roof gutters on facility buildings) to avoid contact with pollutants can affect the volume of runoff that needs to be controlled. In regions of the country with very high rainfall, some animal facilities are entirely roofed to prevent precipitation from coming into contact with animal wastes and to minimize the total volume of stored wastes that must be managed.

The pollutant load carried in runoff from animal facilities is affected by several additional factors, including: (1) pollutants *available* for transport in the facility; (2) the rate and path of runoff-*movement* through the facility; and (3) passage of runoff through *settling or filtering* practices before exiting the facility. Management activities like scraping manure from pavement areas or proper storage of feeds and bedding can significantly reduce the availability of pollutants for transport. Structures such as detention basins can affect pollutant transport by regulating runoff movement and increasing settling within the facility. Vegetated filter strips, riparian buffers, or other vegetated areas located around animal facilities can reduce delivery of pollutants to surface waters by infiltrating, settling, trapping, or transforming nutrients, sediment, and pathogens in runoff leaving the facility.

The ranges in concentrations of pollutants from some typical sources on a dairy farm are shown in Table 4d-2. The total pounds of pollutants that could come from a typical 100-cow dairy is shown in Table 4d-3. These values were obtained by multiplying the concentrations by the typical volume in Table 4d-2. The pounds per year from these concentrated sources may be small but represent significant pollutant sources if not controlled. Each farm is different, as shown by the range in concentration and amount of pollutants from the various sources. Some of the variation is under the control and management of the farmer and their day-to-day operations, while some of it is due to the type and layout of the facility.

Facility wastewater volumes and pollutant loads are controlled primarily through the design and operation of the facilities involved in watering, washing, and cleaning. Frequency of wash-downs and the volume of water used, for example, will influence both total volume of wastewater to be managed and the concentrations of pollutants in the wastewater. In dairy milking center wastewater, both volume and concentrations of pollutants in the wastewater are controlled by the type of milking and plumbing systems and the formulation of cleaning compounds used.

An important part of the management of milking center waste is to reduce the volume of water and the amount of material that must be handled. The amount of waste can be affected by management as shown by the variability of both the flows and the concentrations in Tables 4d-2 and 4d-3. Reducing the volume of wastewater to be treated will reduce the cost of wastewater treatment. Energy

Table 4d-2. Waste characteristics from dairy farms (Wright, 1996).

Potential Pollutant Source	Biochemical Oxygen Demand[a] ppm	Nitrogen ppm	Phosphorus ppm	Volume gallons per 100 cows[b]
Milking Center Waste	400-10,000	80-900	25-170	73,000
Silage Leachate	12,000-90,000	4,400[c]	500[c]	105,000
Barnyard Runoff	1,000-10,000	50-2,100	5-500	80,000
Dairy Manure	20,000[c]	5,600[c]	900[c]	660,000
Domestic Waste	150-250	20-30	5-10	365,000

[a] 5 day BOD
[b] yearly volumes assuming: 2 gallons/cow/day milking center waste
bunk silo, 25% DM, no drainage water, 36" precipitation
70 ft²/cow, 36" precip., scraped daily, good solid retention
22,000 LB/cow/yr. milk production, 18 gal./cow/day
10 people producing 100 gal./day/person

[c] Typical values

Table 4d-3. Annual waste production on a typical[b] 100 cow dairy (Wright, 1996).

Potential Pollutant Source	Biochemical Oxygen Demand[a] lb.	Nitrogen lb.	Phosphorus lb.
Milking Center Waste	250-6,100	50-550	15-100
Silage Leachate	10,500-79,000	3900[c]	440[c]
Barnyard Runoff	670-6,700	30-1,400	3-330
Dairy Manure	110,000[c]	31,000[c]	5,000[c]
Domestic Waste	450-760	60-90	15-30

[a] 5 day BOD
[b] yearly volumes assuming: 2 gallons/cow/day milking center waste
bunk silo, 25% DM, no drainage water, 36" precipitation
70 ft²/cow, 36" precip., scraped daily, good solid retention
22,000 LB/cow/yr. milk production, 18 gal./cow/day
10 people producing 100 gal./day/person

[c] Typical values

savings for reduced pumping costs and water heating can also be realized. Using only the amount of cleaners that are necessary and using low phosphorus detergents can significantly decrease the amount of phosphorus in the wastewater. Using automated systems appropriately and water treatment where needed can result in a cost savings. Manure reduction methods for milking centers are shown in Table 4d-4 and methods for phosphorus reduction are described in Table 4d-5.

Table 4d-4. Manure reduction methods and costs for milking centers (Wright, 1996).

Manure Reduction Methods	Reduction Potential	Estimated Cost
Schedule the cleaning of alleys and holding areas to minimize the amount of manure tracked into parlors	High	<$300 to >$1,200
Scrape the cow platforms before hosing down parlors	High	<$300
Don't install drains in the cow platform	High	>$1,200
Slope the floors of the parlor to facilitate scraping to the holding area	High	>$300
Install deep traps in drains	Low	$300-$1,200
Keep traffic from manure areas out of the milk house	Low	<$300

Table 4d-5. Phosphorus reduction methods and costs (Springman, 1992).

Phosphorus Reduction Methods	Reduction Potential	Estimated Cost
Install water softener and/or increase softening time	High	<$1,200
Install an iron filter if needed	Low	<$300
Install automatic, programmable CIP dispensing system	Medium	>$1,200
Use low or no phosphorus containing detergents and acid rinses	High	<$300
Reuse CIP detergent and/or acid rinse water	Medium	>$300
Install water conservation methods in CIP	Medium	>$300

Movement to Ground Water

Implementation of some surface runoff controls may increase the potential for movement of water and soluble pollutants through the soil profile to the ground water. The intent of this measure is not to address a surface water problem at the expense of ground water. Facility wastewater and runoff control systems can and should be designed to protect against the contamination of ground water. Ground water protection will also be provided by minimizing seepage of stored, contaminated water to ground water, and by implementing the nutrient and pesticide management measures.

Most parts of AFOs are either paved or highly compacted, and therefore relatively impervious. Thus, in most cases, threats to ground water by infiltration at the feedlot are low, and most actions for ground water protection will occur on land application sites and should be approached through the Nutrient Management Measure. There are, however, a few important concerns within the feedlot and storage areas. Unpaved feedlots and earthen impoundments are generally believed to be "self-sealing" through compaction or with fine organic matter and bacterial cells after a few months of operation. The rate and effectiveness of sealing varies with waste and soil type. Cattle manure generally seals better than swine waste; fine-textured soils generally seal more quickly and effectively than do more porous soils. This sealing, however, is neither immediate nor 100% effective. Significant leaching of pathogens or soluble pollutants such as nitrate may occur early in the life of a facility, and the very slow seepage after "sealing" may still pose a long-term threat to ground water. Additional sealing by compaction, soil additives, or impermeable membranes is often required over porous soils or fractured bedrock. Whenever possible, liners made of clay or synthetic materials should be used in the original design and construction of the facility. Construction with concrete or use of closed storage tanks are effective means of preventing seepage.

Vegetated filter strips located within or adjacent to the facility may sometimes represent an additional ground water concern. When such areas receive a high pollutant load and infiltration occurs, ground water levels of nitrate may be increased. While it may not be necessary to implement a nutrient management plan on the vegetative control practices themselves, ground water should be protected by taking care to not exceed the capacity of the practices to assimilate nutrients.

Finally, wells within the facility represent a direct path to ground water and may be vulnerable to direct contamination by runoff water or by accidental spills of wastes. Wells are a particular concern where drinking water may be threatened by nitrates, bacteria, viruses, or other pathogens. Care should be taken to protect wells from routine or accidental contamination. Wells should be properly cased, grouted, and sealed, and abandoned wells should be properly filled and sealed. Participation in Farm*A*Syst, a voluntary farmstead pollution risk assessment program, is an excellent way to identify ways to prevent contamination of wells (Jackson et al., undated).

Animal Feeding Operation Management Practices and Their Effectiveness

Management Practices

One of the most important considerations in preventing water pollution from AFOs is the location of the facility. For new facilities and expansions to existing facilities, consideration should be given to siting the facility:

- ❏ Away from surface waters;
- ❏ Away from areas of shallow ground water;
- ❏ Away from areas with high leaching potential;
- ❏ Away from sinkholes and other critical or sensitive areas;
- ❏ To avoid odor drift to homes, churches, and communities; and
- ❏ In areas where adequate land is available; to apply animal wastes in accordance with the nutrient management measure.

Combinations of the following practices can be used to satisfy the requirements of this management measure. The Natural Resources Conservation Service (NRCS) practice number and definition are provided for each management practice, where available. Additional information about the purpose and function of individual practices is provided in Appendix A. In some emergency situations, such as extreme animal mortality or structure failure, certain management methods such as commercial rendering, incineration, or approved burial sites may be necessary.

> A large set of management practices are available to custom fit most facilities for an effective pollution prevention system.

Practices to Divert Clean Water

❑ **Diversions (362):** A channel constructed across the slope with a supporting ridge on the lower side.

❑ **Field Border (386):** A strip of perennial vegetation established at the edge of a field by planting or by converting it from trees to herbaceous vegetation or shrubs.

❑ **Filter strip (393):** A strip or area of vegetation for removing sediment, organic matter, and other contaminants from runoff and wastewater.

❑ **Grassed waterway (412):** A natural or constructed channel that is shaped or graded to required dimensions and established in suitable vegetation for the stable conveyance of runoff.

❑ **Lined waterway or outlet (468):** A waterway or outlet having an erosion-resistant lining of concrete, stone, or other permanent material. The lined section extends up the side slopes to a designed depth. The earth above the permanent lining may be vegetated or otherwise protected.

❑ **Roof runoff management (558):** A facility for controlling and disposing of runoff water from roofs.

❑ **Terrace (600):** An earthen embankment, a channel, or combination ridge and channel constructed across the slope.

Practices for Waste Storage

❑ **Dikes (356):** An embankment constructed of earth or other suitable materials to protect land against overflow or to regulate water.

❑ **Sediment basin (350):** A basin constructed to collect and store debris or sediment.

❑ **Water and sediment control basin (638):** An earth embankment or a combination ridge and channel generally constructed across the slope and minor water courses to form a sediment trap and a water detention basin.

❑ **Waste storage facility (313):** A waste impoundment made by constructing an embankment and/or excavating a pit or dugout, or by fabricating a structure.

❑ **Waste treatment lagoon (359):** An impoundment made by excavation or earth fill for biological treatment of animal or other agricultural wastes.

Practices for Waste Management

❑ **Constructed wetlands (656):** A wetland that has been constructed for the primary purpose of water quality improvement.

❑ **Heavy use area protection (561)**: Protecting heavily used areas by establishing vegetative cover, by surfacing with suitable materials, or by installing needed structures.

❑ **Waste utilization (633)**: Using agricultural wastes or other wastes on land in an environmentally acceptable manner while maintaining or improving soil and plant resources.

❑ **Composting facility (317)**: A facility for the biological stabilization of waste organic material.

❑ **Application of manure and/or runoff water to agricultural land**: Manure and runoff water are applied to agricultural lands and incorporated into the soil in accordance with the Nutrient Management Measure.

Practices for Mortality Management

❑ **Composting facility (317)**: A facility for the biological stabilization of waste organic material.

Practice Effectiveness

The effectiveness of practices to control contaminant losses from confined livestock facilities depends on several factors including:

❑ The contaminants to be controlled and their likely pathways in surface, subsurface, and ground water flows;

❑ The types of practices and how these practices control surface, subsurface, and ground water contaminant pathways; and

❑ Site-specific variables such as soil type, topography, precipitation characteristics, type of animal housing and waste storage facilities, method of waste collection, handling and disposal, and seasonal variations. The site-specific conditions must be considered in system design, thus having a large effect on practice effectiveness levels.

The gross effectiveness estimates reported in Table 4d-6 simply indicate summary literature values. For specific cases, a wide range of effectiveness can be expected depending on the value and interaction of the site-specific variables cited above. When runoff from storms up to and including the 24-hour, 25-year frequency storm is stored, there should be no release of pollutants from an AFO via surface runoff. Rare storms of a greater magnitude or sequential storms of combined greater magnitude may produce runoff, however.

Table 4d-7 shows reductions in pollutant concentrations that are achievable with solids separation basins that receive runoff from small barnyards and feedlots. Concentration reductions may differ from the load reductions presented in Table 4d-6 since loads are determined by both concentration and discharge volume. Solids separation basins combined with drained infiltration beds and vegetated filter strips (VFS) provide additional reductions in contaminant concentrations. The effectiveness of solids separation basins is highly dependent on site variables. Solids separation; basin sizing and management (clean-out); characteristics of VFS areas such as soil type, land slope, length, vegetation type, vegetation quality; and storm amounts and intensities all play important roles in the performance of the system.

Table 4d-6. Relative gross effectiveness[a] (load reduction) of animal feeding operation control measures (Pennsylvania State University, 1992b).

Practice[b] Category	Runoff Volume	Total[d] Phosphorus (%)	Total[d] Nitrogen (%)	Sediment (%)	Fecal Coliform (%)
Animal Waste Systems[e]	reduced	90	80	60	85
Diversion Systems[f]	reduced	70	45	NA	NA
Filter Strips[g]	reduced	85	NA	60	55
Terrace System	reduced	85	55	80	NA
Containment Structures[h]	reduced	60	65	70	90

NA = not available.
[a] Actual effectiveness depends on site-specific conditions. Values are not cumulative between practice categories.
[b] Each category includes several specific types of practices.
[d] Total phosphorus includes total and dissolved phosphorus; total nitrogen includes organic-N, ammonia-N, and nitrate-N.
[e] Includes methods for collecting, storing, and disposing of runoff and process-generated wastewater.
[f] Specific practices include diversion of uncontaminated water from confinement facilities.
[g] Includes all practices that reduce contaminant losses using vegetative control measures.
[h] Includes such practices as waste storage ponds, waste storage structures, waste treatment lagoons.

Table 4d-7. Concentration reductions in barnyard and feedlot runoff treated with solids separation.

Site Location	Constituent Reduction (%)			
	TS	COD	Nitrogen	TP
Ohio - basin only[a,b]	49-54	51-56	35	21-41
Ohio - basin combined w/infiltration bed[a]	82	85	—	80
VFS[b]	87	89	83	84
Canada - basin only[c]	56	38	14(TKN)	—
Canada - basin w/VFS[c]	(High 90's in fall and spring)			
Ilinois - basin w/VFS[d]	73		80(TKN)	78

[a] Edwards et al., 1986.
[b] Edwards et al., 1983.
[c] Adam et al., 1986.
[d] Dickey, 1981.

Constructed wetlands have been developed and evaluated for animal waste treatment. These constructed wetlands use the same plants, soils and microorganisms as natural wetlands to remove contaminants, nutrients and solids from the wastewater. Constructed wetlands have been used for years to treat municipal wastewater, industrial wastewater, and stormwater. More recently, they have been used for animal wastewater treatment. A literature review cited in Constructed Wetlands and Wastewater Management for Confined Feeding Operations published by the Gulf of Mexico program (Alabama Soil and Water Conservation Committee et al., 1997) identified 68 different sites using constructed wetlands to treat wastewater from confined animal feeding operations.

Overall, the wetlands reduced the concentration of wastewater constituents such as 5-day biochemical oxygen demand, total suspended solids, ammonia nitrogen, total nitrogen, and total phosphorus. Table 4d-8 shows the average treatment performance.

Of the 68 sites identified, 46 were at dairy and cattle feeding operations. The herd sizes ranged from 25 to 330, with an average of 85 head. Dairy wastewater often included water from milking barns and from feeding/loafing yards with varying characteristics. Cattle feeding wastewater typically came from areas where animals were confined. Usually, dairy and cattle wastewaters were pretreated or diluted before being discharged to constructed wetlands.

Swine operations accounted for 19 of the wetland systems in the study. Swine wastes were collected using flush water from solid floor barns and paved lots, or they were collected directly from slatted floors in farrowing or nursery barns. In many cases, the wastewater was pretreated in lagoons and then discharged to a wetland system to further reduce concentrations to a level that could be applied to the land.

Constructed wetland systems which provided high levels of nitrogen removal for swine wastewater was recently reported by Rice et al. 1998. Three sets of two 3.6 x 33.5 m wetlands received lagoon liquid from a 2600-pig nursery operation. In these wetlands, mass reduction of total nitrogen was 94% when the low nitrogen loading rate of 3 kg/ha specified for advanced treatment for stream discharge was used. However, discharge requirements for nitrogen and phosphorus could not consistently be achieved at this low loading rate, so the goal was changed to determine the maximum loading and nitrogen removal that could be achieved. At the current loading rate of approximately 25 kg/ha/day, the mean nitrogen removal efficiency was 87%. The nitrogen loading rates and mass removal efficiencies for these investigated loading rates are shown in Table 4d-9.

It was determined that there was not enough nitrate in the wetlands for denitrification; hence, treatment experiments were also conducted with nitrified wastewater, for which the nitrogen removal rate was 4 to 5 times higher than when non-nitrified wastewater was added. Also, wetlands with plants were more effective than those with bare soil. These results suggest that vegetative wet-

Table 4d-8. Summary of average performance of wetlands treating wastewater from confined animal feeding operations[a].

Wastewater Constituent	Average Concentration (mg/L)[b]		Average Reduction (%)
	Inflow	Outflow	
5-Day biochemical oxygen demand (BOD$_5$)	263	93	65
Total suspended solids (TSS)	585	273	53
Ammonium nitrogen (NH$_4$-N)	122	64	48
Total nitrogen (TN)	254	148	42
Total phosphorus (TP)	24	14	42

[a] Data from the Livestock Wastewater Treatment Wetland Database (LWDB), which includes wetland systems at dairy, cattle, swine, poultry, and aquaculture sites (Knight et al., 1996).
[b] Average concentration is based on a hydraulic loading rate of 1.9 inches per day (50,000 gallons per day per acre [gpd/ac]). Averages were calculated from data for 30 to 86 systems.
mg/L = milligrams per liter

Table 4d-9. Nitrogen loading rates and mass removal efficiencies for the constructed wetlands, Duplin Co., NC (June 1993–November 1997) (Rice et al., 199).		
Nitrogen	**System**	**% Mass Removal**
3 kg/ha/day	Rush/bulrush	94
	Cattails/bur-reed	94
8 kg/ha/day	Rush/bulrush	88
	Cattails/bur-reed	86
15 kg/ha/day	Rush/bulrush	85
	Cattail/bur-reed	81
25 kg/ha/day	Rush/bulrush	90
	Cattail/bur-reed	84

% Mass Removal = % mass reduction of N (NH_3-N + NO_3-N) in the effluent with respect to the nutrient mass inflow.

lands with nitrification pretreatment is a viable treatment alternative for the removal of large quantities of nitrogen from swine wastewater.

Major conclusions of these studies were that wetlands by themselves cannot remove sufficient amounts of nitrogen and phosphorus to meet stream discharge requirements but do show promise for high rates of nitrogen mass removal. Since wetlands are nitrate limited, the mass removal rate can be increased by nitrifying the wastewater prior to wetland application. With nitrification pre-treatment, wetlands have the potential to annually remove more than 14,000 kg N/ha. By sequencing nitrification and denitrification unit processes, advanced wastewater treatment levels can be achieved. Such systems could provide a safer alternative to anaerobic lagoons, with reduced ammonia volatilization and odor.

Operation and Maintenance

Appropriate operation and maintenance are critical to achieving the full environ-mental benefits of this management measure. Holding ponds and treatment lagoons should be operated such that the design storm volume is available for storage of runoff. Facilities filled to or near capacity should be pumped. Solid separation basins should be pumped or cleaned out according to design specifi-cations. Pollutant loads can be reduced by managing manure to prevent or minimize accumulation on open lots.

It is appropriate to evaluate the waste management capabilities and interests of the grower, herdsman, or stock manager. Factor this information into the daily and periodic site operation requirements for facility design.

Diversions will need periodic reshaping and should be free of trees and brush growth. Gutters and downspouts should be inspected annually and repaired when needed. Established grades for lot surfaces and conveyance channels should be maintained at all times.

Channels should be free of trees and brush growth. Periodic cleaning of debris basins, holding ponds, and lagoons will be needed to ensure that design volumes are maintained. Clean water should be excluded from the storage structure unless it is needed for further dilution in a liquid system.

It is appropriate to evaluate the waste management capabilities and interests of the grower, herdsman, or stock manager. Factor this information into the daily and periodic site operation requirements for facility design.

Infiltration areas or vegetative filter areas need to be maintained in permanent vegetative cover, with vegetation harvested when conditions permit. Where possible, runoff should be alternated between two infiltration areas to provide alternating use and rest periods.

To protect ground water, it is important to avoid disturbing the manure-soil seal when cleaning or emptying a feedlot, barnyard, or waste storage structure.

Factors in the Selection of Management Practices

The first priority in the selection of management practices should be clean water diversion. Diverting as much precipitation, snowmelt, and overland flow as possible away from the facility before the water can come into contact with wastes will greatly reduce the volumes of contaminated runoff and wastewater requiring later management. Once all clean water sources are diverted, facility runoff and wastewater should be collected and conveyed to the management systems. Simple facilities may have a single outlet that makes collection relatively easy; large facilities with complex topography and layout may require regrading, curbs, diversions, dikes, channels, or pipes to effectively collect and convey runoff and wastewater.

Proper design and construction are essential to the performance of settling basins, storage structures, and filter strips. Management practices and components must be physically compatible with the functional layout of the facility itself. Impoundments should always be located so that gravity flow can be employed; however, clean water or runoff should be diverted from the site as a precaution. It is also desirable to position buildings and waste treatment systems so that prevailing winds do not immediately transport dust and odors to sensitive areas. Distance and topography play a major role in determining what portions of the site will receive direct land application of waste or irrigation of lagoon liquid. State and local NRCS offices, Cooperative Extension Service offices, State agriculture departments, State Land Grant Universities, and the American Society of Agricultural Engineers are good sources of information for size and layout requirements for management practices.

Wastewater management systems must protect water, soil and air quality. Therefore, consideration also needs to be directed to storage, treatment and land application techniques that minimize odor and ammonia volatilization. Nitrogen loss during land application of manure by ammonia volatilization for various waste management techniques is shown in Table 4d-10. Concerns also exist regarding uncontrolled methane released from animal waste because it is considered to be an important factor in gases that cause global warming. Odor has become one of the major concerns of the general public and livestock producers. Therefore, techniques to reduce in-house odors, such as alternative manure collection and emptying techniques and dietary studies which reduce waste volume and odor have received increased attention. Major soil quality concerns include the buildup of phosphorus. Concern also exists about other constituents that accumulate in the soil, such as copper and zinc. Therefore, management practices should be selected that are both compatible with a given facility and protective of water, air and soil quality.

Soil and manure testing data must be considered along with fertilizer recommendations to be sure that the proper amount of manure is applied to land. Land

Table 4d-10. Nitrogen volatilization losses during land application of manure (percent of nitrogen applied that is lost within 4 days of application).

Application method	Type of waste	Percent of nitrogen lost
Broadcast	Solid	15 to 30
	Liquid	10 to 25
Broadcast with immediate cultivation	Solid	1 to 5
	Liquid	1 to 5
Injection	Liquid	0 to 2
Drag-hose injection	Liquid	0 to 2
Sprinkler irrigation	Liquid	15 to 35

This table shows typical nitrogen losses due to volatilization—evaporation into the air. Remember, practices that reduce volatilization losses will also reduce surface runoff losses.

Source: Hirschi et al., 1997, adapted from Livestock Waste Facilities handbook, MWPS-18, 3rd edition, 1993. ©MidWest Plan Service, Ames, IA 50011-3080.

application techniques which minimize ammonia volatilization and thus loss of fertilizer value need to be employed. These techniques will also protect air quality so that ammonia volatilization and odor are minimized. Calibration methods to assist in the proper land application of manure are given in Table 4d-11.

The management of stored runoff and accumulated solids through an appropriate waste utilization system can be achieved under a range of options, including land application, composting, biogas generation, recycling as feedstuffs, aquaculture, and biomass production (Hauck, 1995). Early efforts to conserve animal waste nutrients and other valuable components for fertilizer are directing renewed interest to conserve and process waste into value-added products. These strategies involve using manure and dead animals in conjunction with other materials such as sawdust, soybean and corn products, culled sweet potatoes, soybean hulls, and other organic waste products processed by rendering, extrusion, fluid bed cook-dehydration procedures and other techniques to produce value-added products. Crab bait is one successful value-added byproduct produced from animal waste at the North Carolina State University Animal and Poultry Waste Processing Center which has successfully utilized these waste nutrients and reduced the use of bait fish. Any stored water, accumulated solids, processed dead animals, or manure should be applied in accordance with the Nutrient Management Measure.

Cost of Practices

Construction costs for control of runoff and manure from confined animal facilities are provided in Table 4d-12. The annual operation and maintenance costs average 4% of construction costs for diversions, 3% of construction costs for settlement basins, and 5% of construction costs for retention ponds (DPRA, 1992). Annual costs for repairs, maintenance, taxes, and insurance are estimated to be 5% of investment costs for irrigation systems (DPRA, 1992).

Table 4d-11. Calibration methods (some common ways to calculate the application rate of manure spreaders) (Hirschi et al., 1997).

Manure source	What you need to know	Calculations
Liquid manure in a tank	1. Tank load size (gallons of manure) 2. Acreage over which manure is spread at even rate	$\dfrac{\text{gallons}}{\text{acreage}} = \begin{array}{l}\text{application rate}\\ \text{(gallons per acre)}\end{array}$
Liquid manure in spreader: volume method	1. Spreader load size (gallons of manure) 2. Distance driven and width spread (feet)	$\dfrac{\text{gallons} \times 43,560}{\text{distance} \times \text{width}} = \begin{array}{l}\text{application rate}\\ \text{(gallons per acre)}\end{array}$
Liquid manure in spreader: weight method*	1. Spreader load size (pounds of manure) 2. Distance driven and width spread (feet)	$\dfrac{\text{pounds} \times 5,248}{\text{distance} \times \text{width}} = \begin{array}{l}\text{application rate}\\ \text{(gallons per acre)}\end{array}$
Solid manure in spreader: spreader volume method**	1. Spreader struck-level load size (bushels of manure) 2. Distance driven and width spread (feet)	$\dfrac{\text{bushels} \times 1,688}{\text{distance} \times \text{width}} = \begin{array}{l}\text{application rate}\\ \text{(tons per acre)}\end{array}$
Solid manure in spreader: plastic sheet weight method	1. Pounds of manure on the sheet after drive-over 2. Square footage of plastic sheet	$\dfrac{\text{pounds} \times 21.78}{\begin{array}{l}\text{square footage}\\ \text{of plastic sheet}\end{array}} = \begin{array}{l}\text{application rate}\\ \text{(tons per acre)}\end{array}$
Shortcut method #1 with plastic sheet: for lighter application rates (use a 9' x 12' sheet)	1. Pounds of manure on the sheet after drive-over	$\text{pounds} \div 5 = \begin{array}{l}\text{application rate}\\ \text{(tons per acre)}\end{array}$
Shortcut method #2 with plastic sheet: for heavier application rates (use a 4'8" x 4'8" sheet or 87" x 3' sheet)	1. Pounds of manure on the sheet after drive-over	$\begin{array}{l}\text{pounds of manure}\\ \text{collected on the sheet}\end{array} = \begin{array}{l}\text{application rate}\\ \text{(tons per acre)}\end{array}$

*The calculation for this method assumes that a gallon of manure will weigh a certain number of pounds. An average figure is used.

**The calculation for this method assumes that a bushel of manure will weigh a certain number of pounds. An average figure is used.

Table 4d-12. Costs for runoff control systems (DPRA, 1992; USDA, 1998).

Practice [a]	Unit	Cost/Unit Construction in 1997 Dollars[b, c, d]
Diversion	foot	2.38
Irrigation		
- Piping (4-inch)	foot	2.35
- Piping (6-inch)	foot	3.02
- Pumps (10 hp)	unit	2,350
- Pumps (15 hp)	unit	2,690
- Pumps (30 hp)	unit	4,030
- Pumps (45 hp)	unit	4,700
- Sprinkler/gun (150 gpm)	unit	1,180
- Sprinkler/gun (250 gpm)	unit	2,350
- Sprinkler/gun (400 gpm)	unit	4,300
- Contracted service to empty retention pond	1,000 gallon	3.68
Infiltration[e]	acre	2980
Manure Hauling	mile per 4.5-ton load	2.64
Dead Animal Composting Facility	cubic foot	5.96
Retention Pond		
- 241 cubic feet in size	cubic foot	3.08
- 2,678 cubic feet in size	cubic foot	1.48
- 28,638 cubic feet in size	cubic foot	0.72
- 267,123 cubic feet in size	cubic foot	0.37
Settling Basin		
- 53 cubic feet in size	cubic foot	5.08
- 488 cubic feet in size	cubic foot	3.27
- 5,088 cubic feet in size	cubic foot	2.04
- 49,950 cubic feet in size	cubic foot	1.29

a Expected lifetimes of practices are 20 years for diversions, settling basins, retention ponds, and filtration areas and 15 years for irrigation equipment.
b Table is derived from DPRA estimates presented in an earlier edition adjusted by USDA price indices.
c Table does not present annualized costs.
d Costs for pumps, sprinklers, and infiltration are rounded to the nearest 10 dollars.
e Does not include land costs.

Sources:
* DPRA. Draft Economic Impact Analysis of Coastal Zone Management Measures Affecting Confined Animal Facilities, DPRA, Inc., Manhattan, KS, 1992.
* United States Department of Agriculture (USDA), Agricultural Prices - 1997 Summary, National Agricultural Statistics Service, July 1998.

4E: Grazing Management

Grazing Management Measure

Manage rangeland, pasture, and other grazing lands to protect water quality and aquatic and riparian habitat by:

1. improving or maintaining the health and vigor of selected plant(s) and maintaining a stable and desired plant community while, at the same time, maintaining or improving water quality and quantity, reducing accelerated soil erosion, and maintaining or improving soil condition for sustainability of the resource. These objectives should be met through the use of one or more of the following practices:

   a. maintain enough vegetative cover to prevent accelerated soil erosion due to wind and water;

   b. manipulate the intensity, frequency, duration and season of grazing in such a manner that the impacts to vegetative and water quality will be positive;

   c. ensure optimum water infiltration by managing to minimize soil compaction or other detrimental effects;

   d. maintain or improve riparian and upland area vegetation;

   e. protect streambanks from erosion;

   f. manage for deposition of fecal material away from water bodies and to enhance nutrient cycling by better manure distribution and increased rate of decomposition; and,

   g. promote ecological and stable plant communities on both upland and bottom land sites.

2. excluding livestock, where appropriate, and/or controlling livestock access to and use of sensitive areas, such as streambanks, wetlands, estuaries, ponds, lake shores, soils prone to erosion, and riparian zones, through the use of one or more of the following practices:

   a. use of improved grazing management systems (e.g., herding) to reduce physical disturbance of soil and vegetation and minimize direct loading of animal waste and sediment to sensitive areas;

   b. installation of alternative drinking water sources;

   c. installation of hardened access points for drinking water consumption where alternatives are not feasible;

   d. placement of salt and additional shade, including artificial shelters, at locations and distances adequate to protect sensitive areas;

   e. provide stream crossings, where necessary, in areas selected to minimize the impacts of the crossings on water quality and habitat; and,

   f. use of exclusionary practices, such as fencing (conventional and electric), hedgerows, moats and other practices as appropriate

   and

> The restoration or protection of designated water uses (e.g. fisheries) is the goal of BMP systems designed to minimize the water quality impact of grazing and browsing activities on pasture and range lands.

3. achieving either of the following on all rangeland, pasture, and other grazing lands not addressed above:

   a. apply the planning approach of the U.S. Department of Agriculture (USDA), Natural Resources Conservation Service (NRCS) to implement the grazing land components in accordance with one or more of the following from NRCS: a Grazing Land Resource Management System (RMS); National Range and Pasture Handbook (USDA-NRCS, 1997b); and NRCS Field Office Technical Guide, including NRCS Prescribed Grazing 528A;

   b. maintain or improve grazing lands in accordance with activity plans or grazing permit requirements established by the Bureau of Land Management, the National Park Service, or the Bureau of Indian Affairs of the U.S. Department of Interior, or the USDA Forest Service; or other federal land manager.

Management Measure for Grazing: Description

The management measure is intended to be applied to activities on rangeland, irrigated and non-irrigated pasture, and other grazing lands used by domestic livestock. This management measure applies to both public and private range and pasture lands. A grazing management plan/system should be used to plan and achieve implementation of this management measure.

The goals of this management measure are to protect water quality and quantity and sensitive areas. The grazing management plan/system is the primary mechanism through which these goals are achieved. A grazing management plan/system may include management strategies and practices such as herding, alternative water sources, livestock exclusion, and conservation of range, pasture, and other grazing lands. Grazing management systems are intended to achieve specified objectives and ensure "proper use." Proper use can be defined as grazing managed so that the total vegetation available is grazed at a time and intensity that does not degrade the existing-riverine/aquatic-riparian-upland systems or in the case of degraded rangelands, inhibit system response to a more desirable state (adapted from Platts, 1990). As such, a clear understanding of plants and their ecology are key to good grazing management.

It is recognized that livestock exclusion is more practicable on pasture than rangeland in many cases, but livestock exclusion can be used for the protection of water quality in key sensitive areas on rangelands. In grazing systems, major environmental improvements can be achieved by minimizing livestock access to streambanks and riparian areas during periods of streambank instability and regrowth of key riparian vegetation.

To meet the objectives of the management measure, a comprehensive management system should be employed to manage the entire grazing area. This grazing area may include uplands, riparian areas, and wetlands. Special attention should be given to grazing management in riparian and wetland areas due to their sensitivity to disturbance and the tendency of many grazing animals to favor these areas for foraging and loafing. *Riparian areas* are defined by Mitsch and Gosselink (1986) and Lowrance et al. (1988) as:

> vegetated ecosystems along a water body through which energy, materials, and water pass. Riparian areas characteristically have a high water table and are subject to periodic flooding and influence from the adjacent water body.

Riparian area and wetland protection strategies should be integrated with upland management strategies. The health of the riparian and wetland ecosystems, receiving waterbody quality, and stream base flow levels are often dependent on the use, management and condition of adjacent uplands. Proper management of uplands can reduce grazing pressure on riparian areas and also increase forage productivity due to increased water table height and stream base flow. Increased forage productivity and overall upland health can result in increased economic benefits to the landowner or grazing management entity.

This management measure also contains recommendations under 3a and 3b that USDA/NRCS methodologies and guidance and/or other federal agency require-ments should be employed in addition to the management elements listed in 1a-g and 2a-f to provide the requisite level of natural resource protection. Resource management systems (RMS) include any combination of conservation practices and management that achieves a level of treatment of the five natural resources (i.e., soil, water, air, plants, and animals) that satisfies criteria contained in the Natural Resources Conservation Service (NRCS) Field Office Technical Guide (FOTG). The rangeland and pasture components of a RMS address erosion control, proper grazing, adequate pasture stand density, and rangeland condition. National (minimum) criteria pertaining to rangeland and pasture under an RMS are applied to achieve environmental objectives, conserve natural resources, and prevent soil degradation.

Recommendations for Grazing Management in Riparian Areas

- ❑ Tailor the grazing approach to the specific riparian area under consideration.
- ❑ Incorporate management of riparian areas into the overall management plan for the whole operation.
- ❑ Select a season or seasons of use so grazing occurs, as often as possible, during periods compatible with animal behavior and conditions in the riparian area.
- ❑ Control the distribution of livestock within the targeted pasture.
- ❑ Ensure adequate residual vegetative cover.
- ❑ Provide adequate regrowth time and rest for plants
- ❑ Be prepared to play an active role in managing riparian areas.

Source: *Best Management Practices for Grazing Montana,* Montana Watershed Coordination Council's Grazing Practices Work Group, 1999.

Grazing and Pasturing: An Overview

In addressing nonpoint source pollution concerns, producers must balance production and water quality objectives. This section explores some of the production-oriented resources management decisions confronting livestock producers.

Livestock can obtain their needed nutrients through feed supplied to them in a confined livestock facility, through forage, or through a combination of forage and feed supplements. Forage systems can be pasture-based or rangeland-based.

It is important for the reader to be aware of the difference between rangeland and pasture. *Rangeland* refers to those lands on which the native or introduced vegetation (climax or natural potential plant community) is predominantly grasses, grasslike plants, forbs, or shrubs suitable for grazing or browsing. Rangeland includes natural grassland, savannas, many wetlands, some deserts, tundra, and certain forb and shrub communities. *Pastures* are those improved lands that have been seeded, irrigated, and fertilized and are primarily used for the production of adapted, domesticated forage plants for livestock. Other grazing lands include grazable forests, native pastures, and crop lands producing forage.

The major differences between rangeland and pasture are the kind of vegetation and level of management that each land area receives. In most cases, range supports native vegetation that is extensively managed through the control of livestock rather than by agronomy practices, such as fertilization, mowing, or irrigation. Rangeland also includes areas that have been seeded to introduced species (e.g., clover or crested wheatgrass) but are managed with the same methods as native range. For both rangeland and pasture, the key to good grazing practice is vegetative management, i.e., timing of grazing should be managed to ensure adequate vegetative regrowth and soil stability.

Pastures are represented by those lands that have been seeded, usually to introduced species (e.g., legumes or tall fescue) or in some cases to native plants (e.g., switchgrass or needle grass), and which are intensively managed using agronomy practices and control of livestock. Permanent pastures are typically based on perennial warm-season (e.g., bermudagrass) or cool-season (e.g., tall fescue) grasses and legumes (e.g., warm-season alfalfa, cool-season red clover), while temporary pastures are generally plowed and seeded each year with annual legumes (e.g., warm-season lespedezas, cool-season crimson clover) and grasses such as warm-season pearl millet and cool-season rye (Johnson et al., 1997). Plant selection for pastures should be based upon consideration of climate, soil type, soil condition, drainage, livestock type and expected forage intake rates, and the type of pasture management to be used. Management of pH and soil fertility is essential to both the establishment and maintenance of pastures (Johnson et al., 1997). In some climates (e.g., Georgia), overseeding of summer perennials with winter annuals is done to provide adequate forage for the period from mid-winter to the following summer.

Factors Affecting Animal Performance on Grazed Lands

The manager of a forage system must be concerned with care and management of the livestock, control of noxious plants, and the quality of forage (McGinty,

1996). Both forage quality and forage intake must be managed to ensure the performance, or quality, of livestock on pasture and grazing lands.

Forage quality

Forage quality is generally measured in terms of its nutritional value and digestibility. Nutritional value can be assessed based on the amount of protein, phosphorus, and energy the plants contain (Ruyle, 1993). The nutritional value of rangeland forage varies with season (e.g., higher in spring and summer), and differs among forage types. For example, protein availability from grasses decreases rapidly as the grasses mature, while shrubs are good sources of protein even at full maturity. The protein content of forbs (e.g., weeds, wildflowers) falls between that of grasses and shrubs. Grasses are generally considered to be good sources of energy, shrubs are good energy sources before fruit development, and the value of forbs is intermediate between that of grasses and shrubs for livestock.

Rangeland condition also affects the nutritive value of forage plants, with better rangeland condition yielding more digestible plants (Ruyle, 1993). Other factors affecting the quality of forage include the plant parts eaten (e.g., leaves versus stem), the presence of secondary compounds (e.g., lignin, tannins, terpenes) in the plants (Lyons et al., 1996b), and pests (Johnson et al., 1997). The stocking rate and the type of grazing system can affect grazing animal nutrition as well. Over-stocking will cause a shift toward less productive and less palatable forage plants, resulting in decreased forage intake due to less total forage and less desirable forage (Lyons et al., 1996b). The preservation of some of the forage on grazed lands is necessary to protect the resource, but forage quality may suffer if too much old growth is maintained. Closely-grazed forage is generally good for animal performance since it results in younger forage that is higher in nutrient value and more digestible (Johnson et al., 1997). The quality of regrowth in pastures is improved with intensive grazing, but the rate of regrowth, and therefore the yield, is reduced (Cannon et al., 1993). Grazing management decisions should allow for plant vigor and regrowth and maintenance of soil stability. Growing season factors should be considered when evaluating the potential for plant regrowth.

Many practitioners currently use forage utilization or stubble height as a management tool to gauge the acceptable level of grazing. Stubble height measurements can be used successfully as one component of a comprehensive grazing management strategy. Stubble height measurements are a good tool to help practitioners begin to focus on stream ecology and forage availability for animal production. However, the exclusive and continuing use of stubble height as the only or primary indicator of riparian health can be problematic. As a result stubble height measurements are sometimes improperly used. Stubble height measurements often are conducted at the wrong time or intervals, in the wrong places, and based on measurements of the wrong plant species. To properly use stubble height as an effective grazing management tool, stubble height must be measured frequently during the grazing period to ensure that adequate vegetative cover and soil stability are maintained at the end of both the growing season and grazing period. The proper use of stubble height measurements can benefit animal production and help ensure the stability of the riparian area, however, the practicality and expense of frequent stubble height measurements may be burdensome, and, as a result, this technique may be improperly applied.

In Oregon, it is recommended that pastures be grazed from about 2,400 to 2,800 pounds of dry matter growth per acre down to about 1,500-1,600 pounds of dry matter growth per acre, maintaining a height of 2-6 inches for clover and grasses (Cannon et al., 1993). Guidelines for Texas ranchers recommend minimum stubble height and plant residue as follows: 1.5 inches and 300-550 pounds per acre for short grass; 4-6 inches and 750-1,000 pounds per acre for mid-grass; and 8-10 inches and 1,200-1,500 pounds per acre for tall grass (McGinty, 1996). However, these stubble height strategies may oversimplify the complexity and site specificity of herbage dynamics under grazing, and it has been argued that these assessments are qualitative, subjective, and not truly quantitative (Scarnecchia, 1999).

The Montana Watershed Coordination Council's Grazing Practices Work Group publication, *Best Management Practices for Grazing Montana* (1999) recommends that rangeland managers set target levels for grazing use based on animals' nutritional needs balanced against the need to maintain a healthy plant community. This approach is based on setting target levels for key species and evaluating on a site level basis rangeland condition and trends. As a general rule of thumb, the Council advises that the planned grazing target should be to use no more than 50-60% of the key species.

Forage intake

Forage intake generally increases as forage quality increases (Lyons et al., 1995). As illustrated in Figure 4e-1, forage intake increases with digestibility since digestion creates room for additional forage. Livestock do not generally stop eating once their nutrient requirements are met. Because of this, ranchers cannot assume that higher quality forage alone will result in adequate resource protection. Grazing management systems will still be needed to protect the

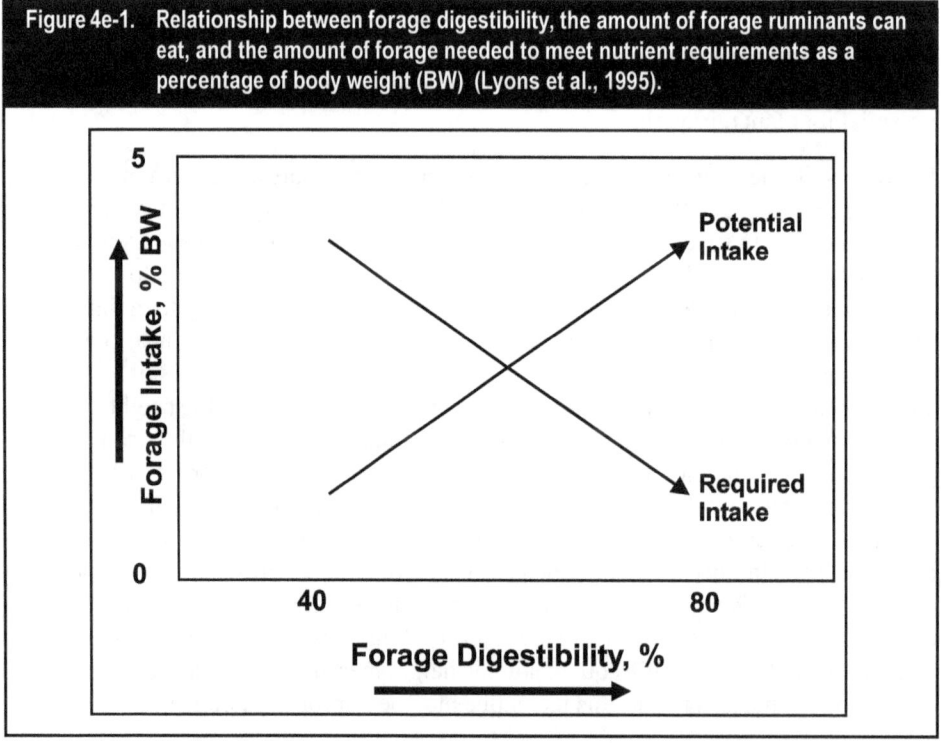

Figure 4e-1. Relationship between forage digestibility, the amount of forage ruminants can eat, and the amount of forage needed to meet nutrient requirements as a percentage of body weight (BW) (Lyons et al., 1995).

resource from improper grazing. With low-quality forage, more forage is needed to meet nutrient needs, but the lower digestibility makes it much more difficult for the livestock to meet their nutrient needs since the forage does not pass through the rumen as quickly.

Forage intake is also affected by herbivore species and size, foraging behavior (e.g., preference for certain forage types, preference for specific areas), physiological status, animal production potential, supplemental feed, forage availability, and environmental factors (Lyons et al., 1995). Smaller herbivore species (e.g., sheep) have greater intake rates when measured as a percentage of live weight than do larger species (e.g., beef cattle). Sheep and goats tend to be more selective of the plants they graze than are cattle, and tend to have higher forage intake rates due to their consumption of a readily digestible mixture of grass, forbs, and browse (young twigs, leaves, and tender shoots of plants or shrubs suitable for animal consumption). Horses may consume up to 70 percent more forage than a cow of similar size due mostly to the rapid passage rate of horses.

The forage selected by herbivore species varies, and is determined largely by their mouth parts and the anatomy of their digestive systems (Lyons et al., 1996a). For example, horses eat more grass than cattle, sheep, and goats as a percentage of their annual diet, while goats eat the most browse, and sheep eat the greatest share of forbs. Diet also varies across season within a given species. Browse constitutes 34 percent of the diet of Texas-raised goats in spring and 53 percent in fall and winter, while forbs account for 6 percent of the diet of cattle in fall and 25 percent in spring. Management strategies should control animal distribution and plant harvest timing to counter the effects of preference (Platts, 1990).

The importance of physiological status is evidenced by the fact that lactating animals generally have a higher nutrient demand and greater forage intake rate than animals that are dry, open, or pregnant (Lyons et al., 1995). In fact, an animal can eat 35 to 50 percent more when lactating than when dry, open, or pregnant. Highly productive cows early in lactation require the highest quality feed to maintain production (Cannon et al., 1993). Thus, the good farm manager gives high priority to the provision of adequate forage to lactating dairy herds in order to avoid a drop in milk production.

Producers may need to provide feed nutrient supplements to ensure suitable livestock production on rangeland (Ruyle, 1993) and other grazing lands. Protein supplements are often given to livestock grazing on low-protein forage, and the quantity and timing of the supplemental feeding can affect forage intake (Lyons et al., 1995). For example, supplemental protein can increase forage intake to a point, beyond which forage intake is reduced with increasing supplemental protein.

Forage availability is often measured in terms of stocking rates, or the number of animals that use a unit of land for a specified period of time (White, 1995; Sedivec, 1992). Forage growth and production can vary greatly over any given land area, as seasons change, and as a function of weather conditions, so matching stocking rates with forage availability is dependent upon assumptions regarding forage production. Further, since forage intake is dependent upon forage quality, it becomes necessary to carefully monitor forage quality and quantity to determine if stocking rates need to be adjusted. A general rule-of-thumb for grazing is to allow livestock to use 50 percent of the forage (Sedivec, 1992). USDA encourages development of a feed, forage, livestock balance sheet to assist in management of grazing lands, and provides procedures and worksheets to assist managers (USDA-NRCS, 1997b).

An alternative approach to addressing forage availability in management decisions is based on the concept of a forage allowance, which is the weight of forage allocated per unit of animal demand at any instance (Cropper, 1998). Forage allowance is expressed as a percentage of live body weight or as pounds of forage per animal per day, and generally averages 2.5-3% for beef and sheep, 2% for horses, and 3-4% for lactating cows (Cropper, 1998). Research has shown that forage intake increases with forage allowance, reaching a maximum level at a forage allowance of about 6.5% of herd live weight (Figure 4e-2). Forage utilization rate, however, decreases as forage allowance is increased, meaning that more forage is potentially wasted since it is not consumed by livestock. With knowledge of the number of animals on the pasture, the percentage of forage intake derived from the pasture, forage intake per animal, and the desired forage utilization rate, one can manage forage and livestock to achieve desired animal performance without wasting or degrading pasture (Cropper, 1998).

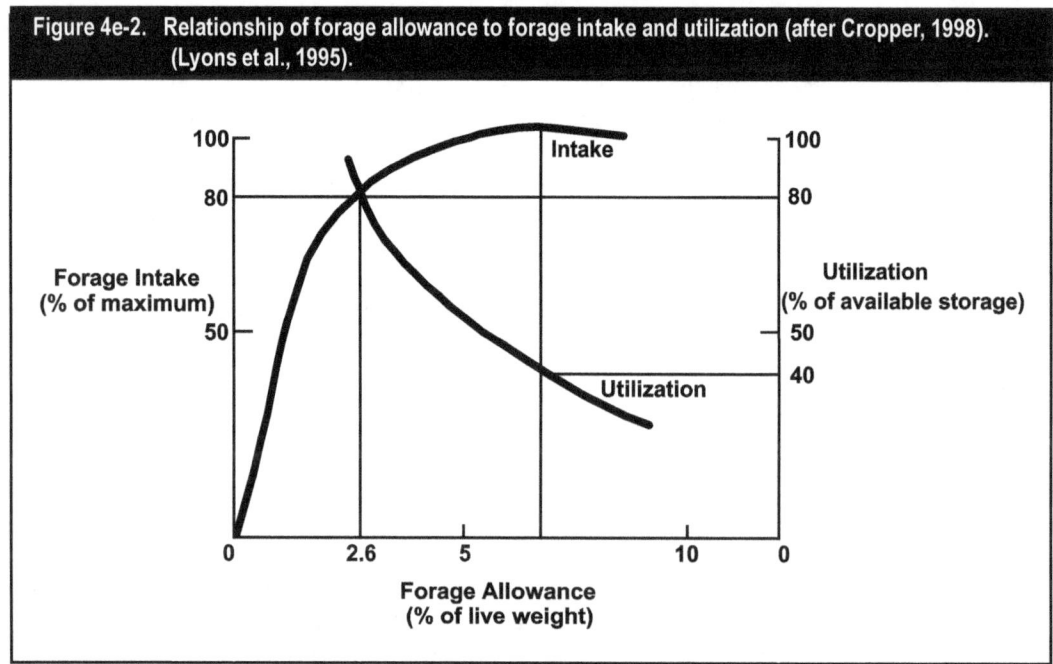

Figure 4e-2. Relationship of forage allowance to forage intake and utilization (after Cropper, 1998). (Lyons et al., 1995).

Environmental factors, including air temperature, soil moisture, and snowcover, also affect forage intake. Each species of herbivore has a temperature-based comfort zone, the thermoneutral zone, within which forage intake is not affected (Lyons et al., 1995). Above and below the thermoneutral zone, however, intake may increase or decrease depending upon outside conditions.

There is also a need to assess and compensate for wildlife forage utilization when managing livestock to protect water quality. In many areas, wildlife consumes a significant portion of available forage and wildlife ungulates (i.e., mammals with hooves) may have a major impact on riparian areas and woody vegetation. Land managers should take these impacts into account when planning and managing grazing management programs and setting grazing use levels for each grazing unit.

Because of the many sources of variability in forage quality, forage availability, and forage intake, the rancher faces a significant challenge in providing an appropriate mix of forage to ensure that livestock receive adequate nutrition throughout the year.

Water

Water is essential to the survival, growth, and productivity of livestock. Insufficient water supply will result in reductions in feed intake, production, and profits (Faries et al., 1998). High salinity, high nitrate and nitrite levels, bacterial contamination, excessive growth of blue-green algae, and spills of petroleum, pesticides, and fertilizers are the water quality problems that most affect livestock production.

Research in Missouri has shown that water consumption of pastured beef cow-calf pairs increased almost linearly as the temperature increased from 50 degrees to 95 degrees Fahrenheit (Gerrish, 1998). At 50 degrees F, water consumption was approximately 6 gallons per day, increasing to about 24 gallons per day at 95 degrees F. Cattle in Texas drink from 7 to 16 gallons per day, while horses (8-12 gallons per day) and sheep and goats (1-4 gallons per day) drink less (McGinty, 1996). Dry cows drink 8-10 gallons of water per day, while cows in their last three months of pregnancy need up to 15 gallons of water per day (Faries et al., 1998). The frequency with which livestock seek water varies, ranging from 3-5 times per day for beef cows in the Midwest, to less frequent visits in drier climates (Gerrish, 1998). A recent study showed that distance from water supply had a large effect on water consumption, as cows within 800 feet of water drank 15 percent more water than cows further than 800 feet from water (Gerrish, 1998). The maximum distance that livestock will travel to water in Texas ranges from 0.5 miles in rough terrain to 2.0 miles in smooth, flat terrain (McGinty, 1996).

Minerals

Sodium, chloride, and other minerals are essential to the bodily functions of animals, and livestock on the rangeland should consume about 20 pounds of salt per year (Schwennesen, 1994). Well managed vegetation can provide the needed minerals for healthy animals, but mineral supplements can benefit animals if they are developed to meet local deficiencies. Livestock are attracted to salt and other mineral supplements, and will remain with it as long as it remains, making mineral supplements a very useful grazing land management tool. By placing measured quantities of minerals at various locations throughout the year, livestock operators can manage the location of livestock to control grazing, help manage the grazing land condition, and keep livestock away from sensitive areas.

Weed and Brush Management

Weeds can reduce forage production and lower forage quality (Johnson et al., 1997). Well-managed pastures present fewer weed problems as grasses can outcompete most weeds. Weed management on rangeland may involve prescribed burning or the use of herbicides (McGinty, 1996). The grazing of cattle, sheep, and goats can also be used as a weed management tool.

Grazing Systems

There is a wide range of grazing systems for rangeland and pastures that managers may select from (Table 4e-1). Specific terms and definitions used may vary considerably across the nation. In all cases, however, the key management parameters are:

❑ grazing frequency

❑ livestock stocking rates

❑ livestock distribution

❑ timing and duration of each rest and grazing period

❑ livestock kind and class

❑ forage use allocation for livestock and wildlife.

Factors to consider in determining the appropriate grazing system for any individual farm or ranch include the availability of water in each pasture, the type of livestock operation, the kind and type of forage available, the relative location of pastures, the terrain, the number and size of different pasture units available (Sedivec, 1992), and producer objectives.

While many systems may be derived from combinations of the key management parameters, the basic choice is between continuous and rotational grazing. Under continuous grazing, the livestock remain on the same grazing unit for extended periods, while rotational grazing involves moving the livestock from unit to unit during the growing season (Johnson et al., 1997). A prescribed grazing schedule for rangeland is a system in which two or more grazing units are alternately deferred or rested and grazed in a planned sequence over a period of years (USDA-NRCS, 1997b). Rest periods are generally non-grazing periods of a full year or longer, while deferment typically involves a non-grazing period of less than twelve months.

Continuous, season-long grazing is typically done on larger pastures, with less fencing and less livestock management than required for rotational grazing (Johnson et al., 1997). A central problem with this approach is the difficulty of matching the stocking rate with the changing forage growth rate during the grazing season. For example, forages may grow at a rate of 90 pounds per acre per day in spring, followed by summer growth rates of as little as 5 pounds per acre per day, resulting in a mismatch of supply and demand if the stocking rate is kept constant (Cropper, 1998).

Rotational grazing generally involves smaller pastures or paddocks, more fencing, and more livestock management than required for continuous grazing (Johnson et al., 1997). If forage growth exceeds forage intake, forage from some paddocks may be harvested and stored for winter grazing. Rotational grazing provides opportunities to better manage the available forage to meet livestock needs (Johnson et al., 1997). In some cases, the additional costs for fencing and supplying water in each paddock may be prohibitive. Options exist, however, for designing paddocks such that drinking water sources can be shared by more than one paddock, thus eliminating the need for additional water development (Drake and Oltjen, 1994). In addition, affordable, portable fencing is often used in management-intensive grazing systems (SARE, 1997).

Table 4e-1. Some commonly used grazing systems (Sedivec, 1992; McGinty, 1996; Frost and Ruyle, 1993; USDA-NRCS, 1997b).1995).

Grazing System	Description	Comments
Continuous	Unrestricted livestock access to any part of the range during the entire grazing season. No rotation or resting.	Difficult to match stocking rate to forage growth rate. Severe overgrazing occurs where cattle congregate. Other areas underutilized. Long-term productivity depends upon moderate levels of stocking. Can be year-long or seasonal continuous grazing. Less fence and labor than for rotation.
Rotation	Intensive grazing followed by resting. Livestock are rotated among 2 or more pastures during grazing season.	Each pasture may be alternately grazed and rested several times during a grazing season. Cattle are moved to different grazing area after desired stubble height or forage allowance is reached.
Switchback	Livestock are rotated back and forth between 2 pastures.	Every 2-3 weeks in ND. In TX, graze 3 months on pasture 1, 3 months on pasture 2, then 6 months on pasture 1, etc.
Rest-rotation	One pasture rested for an entire grazing year or longer. Others grazed on rotation. Multiple pastures with multiple or single herd.	In ND, 4 pastures used with 1 rested, one each grazed in spring, summer, and fall. Rest periods are generally longer than grazing periods.
Deferred rotation	Grazing discontinued on different parts of range in succeeding years to allow resting and re-growth. Generally involves multiple herds and pastures.	Length of grazing period is generally longer than the deferment period.
Twice-over rotation	Variation of deferred rotation, with faster rotation. Uses 3-5 pastures.	Long period of rest between rotations. Sequence alternates from year to year.
Short-duration grazing	Grazing for 14 days or less. Large herd, many small pastures (4-8 cells), high stocking density.	Rest period is 30-90 days. Allows 4-5 grazing cycles. Requires a high level of grass and herd management skills. Similar to high intensity-low frequency, but length of grazing and rest periods are both shorter for short-duration grazing.
High intensity-low frequency	Heavy, short duration grazing of all animals on one pasture at a time. Rotate to another pasture after forage use goal is met. Multiple pastures with single herds.	Grazing period is shorter than rest period, and grazing periods for each pasture change each year. In TX, grazing period is more than 14 days, and resting period is more than 90 days. TX typically has single herd on 4 or more pastures.
Merrill	Each of 4 pastures grazed 12 months and rested 4 months.	Three herds.
Season-long Grazing	No specific number of herds or pastures.	No set movement pattern.

A number of different stocking methods are used to manage pastures, including allocation stocking methods (continuous set stocking, continuous variable stocking, set rotational stocking, variable rotational stocking), nutrition optimization stocking methods (creep grazing, strip grazing, frontal grazing), and seasonal stocking methods (deferred stocking, sequence stocking) (USDA-NRCS, 1997b). Rotational stocking, or top grazing, is an adaptation of rotational grazing that improves the efficiency with which forage is used. This approach is based upon the fact that cattle select the highest quality forage available before grazing lower quality forage (Johnson et al., 1997). In rotational stocking, for example, a lactating dairy herd might be rotated to a paddock where it can obtain

100 percent of its forage intake needs at a low forage utilization rate (see Figure 4e-2). Forage allowances for high-producing, lactating diary cattle need to be generous to maintain milk production, resulting in utilization rates of 50 percent or less (Cannon et al., 1993). Dry cows and heifers might be rotated to the same paddock after the lactating dairy herd is removed to increase the forage utilization rate (Cropper, 1998).

Potential Environmental Impacts of Grazing

The focus of the grazing management measure is on the protection of water quality and aquatic and riparian habitat. Riparian areas may need special attention to achieve water quality and habitat related goals. The entire watershed should be evaluated to determine the sources and causes of nonpoint source pollution problems and to develop solutions to those problems. Application of this management measure will reduce the physical disturbance to sensitive areas and reduce the discharge of sediment, animal waste, nutrients, pathogens, and chemicals to surface waters.

More than half the commercial operators with beef cattle herds in the West graze federal lands. According to a report by the Council for Agricultural Science and Technology (CAST) (Laycock, 1996), a leading consortium of 33 professional scientific societies, individuals are becoming increasingly concerned about the ecological effects of improper grazing on federal lands. Major concerns include diminished biodiversity, deteriorating rangeland, watershed, and streambank conditions; soil erosion and desertification; decreased wildlife population and habitat; and lost recreational opportunities.

Riparian areas constitute important sources of livestock grazing. One acre of riparian meadow has the potential grazing capacity equal to 10 to 15 acres of surrounding forested rangeland. In the Pacific Northwest, riparian meadows often cover only 1 to 2% of the summer rangeland area, but provide about 20% of the summer forage.

The loss of streambank stability, riparian vegetation, stream habitat, and modification of hydrologic regime due to poor grazing practices has a devastating effect on stream life.

Streambank stability is directly related to the species composition of the riparian vegetation and the distribution and density of these species (Figure 4e-3). During high water, riparian vegetation protects the banks from erosion, reducing water velocity along the stream edge, and causing sediments to settle out. Platts (1991) has summarized the importance of riparian vegetation in providing cover and maintaining streambank stability. Trees provide shade and streambank stability because of their large and massive root systems. Trees that fall into or across streams create high quality pools and contribute to channel stability. Brush protects the streambank from water erosion, and its low overhanging height adds cover that is used by fish. Grasses form the vegetative mats and sod banks that reduce surface erosion and erosion of streambanks. As well-sodden banks gradually erode, they create the undercuts important to salmonids as hiding cover. Root systems of grasses and other plants trap sediment to help rebuild damaged banks.

When animals repeatedly graze directly on erodible streambanks, bank structure may be weakened causing soil to move directly into the stream. Excessive grazing on riparian vegetation can result in changes in plant community composition and density and can negatively impact bank stability and the filtering capacity of the vegetation. Within the federal government, the Bureau of Land Management (BLM) and the USDA have experience in and tools for assessing riparian system function and erodibility.

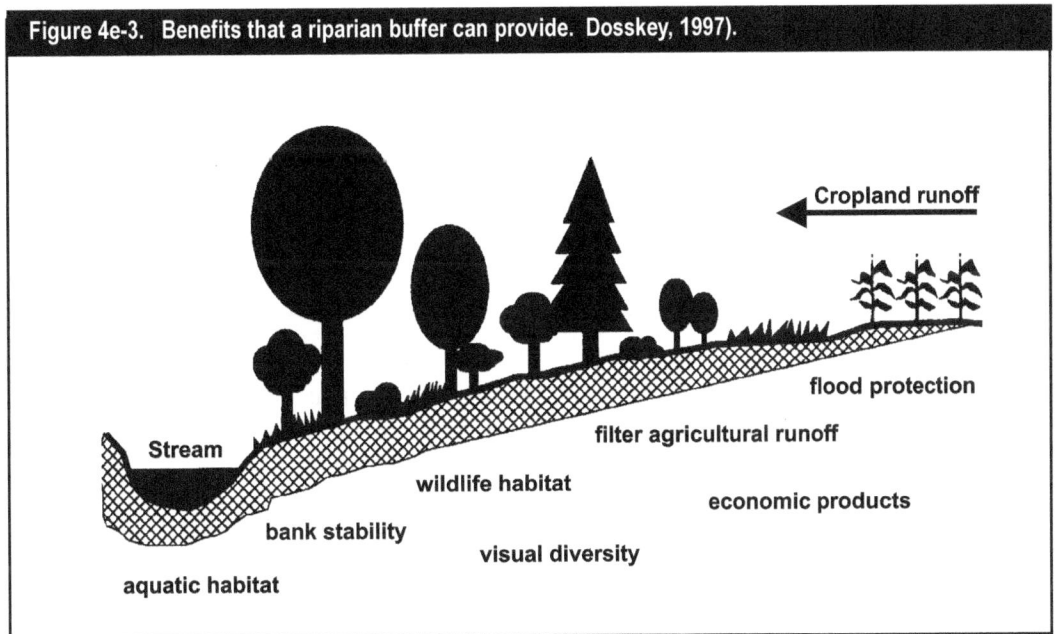

Figure 4e-3. Benefits that a riparian buffer can provide. Dosskey, 1997).

Cropland runoff

flood protection

filter agricultural runoff

economic products

wildlife habitat

visual diversity

Stream

bank stability

aquatic habitat

The loss of riparian vegetation together with collapsed streambanks increases stream width and decreases depth, which has the potential to alter stream temperature. With the loss of riparian vegetation, the stream is exposed to greater temperature fluctuations, resulting in potentially higher temperatures during the day and cooler temperatures at night. Riparian vegetation moderates stream temperatures by absorbing short-wave radiation during the day and insulating the stream from loss of long-wave radiation at night. Other reports indicate that keeping the water in the ground longer is also a major contributing factor to cooler water temperatures (Baschita, 1997).

> Compaction and vegetation loss due to improper grazing can increase runoff, erosion, and sediment delivery to streams.

Improper grazing management can contribute to the removal of most vegetative cover, soil compaction, exposure of soil, degradation of soil structure, and loss of infiltration capacity. These impacts can result in soil susceptible to wind and water erosion. Due to the steep slopes, highly erodible soils, and storm events, the sediment delivery ratio from rangeland can be very high (Carpenter et al., 1994). Improper management can also alter the plant species composition by creating a shift from desirable perennial species to undesirable annual species.

Livestock also generate microorganisms in waste deposits as they graze on pasture and rangelands. Animal wastes contain fecal coliform and fecal streptococci in numbers on the order of $10^5 - 10^8$ organisms per gram of waste, or $10^9 - 10^{10}$ excreted per animal per day (Moore et al., 1988). In addition to such indicator organisms, livestock can serve as an important reservoir of pathogens such as *E. coli* O157:H7 (Wang et al., 1996; Pell, 1997). The extent of manure and microorganism deposition on grazing land typically depends on livestock density or stocking rate (Carpenter et al., 1994; Fraser et al., 1998; Edwards et al., 2000).

> Pathogen impacts on waterways are a grazing land use issue.

Release of microbes from manure deposited on grazing land is influenced by time, temperature, moisture, and other variables. Enhanced survival of microorganisms in fecal deposits on grazing land has been documented elsewhere; the bacterial pollution potential of fecal deposits on grazing land is significant (Thelin and Gifford, 1983; Kress and Gifford, 1984). Bohn and Buckhouse (1985) reported that fecal coliforms may survive in soil only 13 days in summer and 20 days in winter, but that cow fecal deposits provide a protective medium that permit microorganisms to survive for more than a year.

Runoff from grazed land can contain high numbers of indicator microorganisms. Crane et al. (1983) cited fecal coliform counts of $10^3 - 10^5$ organisms/100 ml in pasture runoff. Edwards et al. (2000) reported that FC levels in runoff from simulated grazing plots were always higher ($2.4 \times 10^5 - 1.8 \times 10^6$ FC/100 ml) than counts from the ungrazed control plots (1.5×10^3 FC/100 ml). Microorganism counts in runoff from grazing land are, however, typically several orders of magnitude lower than numbers from land where manure is deliberately applied.

It should be noted that, because all warm-blooded animals excrete indicator bacteria in their feces, wildlife inhabiting agricultural land are likely to contribute to the pool of microorganisms available in a watershed, including both indicator organisms (Kunkle, 1970; Niemi and Niemi, 1991; Valiela et al., 1991) and pathogens such as *Giardia* (Ongerth et al., 1995).

Nutrient inputs from grazing lands to surface water come mainly in the form of nitrogen and phosphorus from manure and decaying vegetation (Carpenter et al., 1994). Nutrient impacts on water quality vary considerably in study results, and are dependent on specific site conditions such as precipitation, runoff, vegetation cover, grazing density, proximity to the stream, and period of use. The risk of nutrient enrichment is low in arid rangelands where animal wastes are distributed and runoff is comparatively light. Studies by the ARS and BLM found little evidence of nutrient enrichment from unconfined livestock grazing in Reynolds Creek, an arid watershed in southern Idaho (USDA–ARS, 1983). This risk can also be low in humid climates if grazing lands are managed correctly. In a humid site in east-central Ohio (Owens et al., 1989), nutrient concentrations did not increase significantly with summer grazing of the unimproved pasture, and were also low when continuously grazed. In another study, Schepers and Francis (1982) found increases in nutrients in a cow-calf pasture in Nebraska. Nutrient levels were correlated primarily with grazing density.

Grazing Management Practices and their Effectiveness

The Grazing Management Measure was selected based on an evaluation of available information that documents the beneficial effects of improved grazing management. Specifically, the available information shows that

❑ Riparian habitat conditions are improved with proper livestock management;

❑ The amount of time livestock spend drinking and loafing in the riparian zone is dramatically reduced through the provision of supplemental water and fencing; and

❑ Nutrient and sediment delivery is reduced through the proper use of vegetation, streambank protection, planned grazing systems, and livestock management.

For any grazing management measure to work, it must be tailored to fit the needs of the vegetation, terrain, class or kind of livestock, and particular operation involved.

For both pasture and rangeland, areas should be provided for livestock watering, supplemental minerals, and shade that are located away from streambanks and

Five Steps to a Successful Prescribed Grazing Management Plan

1. Inventory existing resources and range/pasture conditions

2. Determine management goals and objectives

3. Map out two or more grazing management units

4. Develop a grazing schedule to implement

5. Develop a monitoring and evaluation strategy

> Source: *Best Management Practices for Grazing Montana,* Montana Watershed Coordination Council's Grazing Practices Work Group, 1999.

riparian zones where necessary and practical. This will be accomplished by managing livestock grazing and providing facilities for water, minerals, and shade as needed.

The rancher may seek technical assistance from Cooperative Extension, NRCS, Soil and Water Conservation Districts, or other agencies to help identify water quality problems, develop management measures (statements of water quality goals or objectives), and select management practices. The amount or extent to which a practice is applied must be consistent with national, state, and basin water quality goals and should reflect the relative contribution of that type of land use activity toward water quality problems within the basin. This technical assistance will result in a plan, typically known as a ranch plan or conservation plan.

Contact your county Cooperative Extension agent, USDA–NRCS district conservationist, or the local Soil and Water Conservation District.

Additional information on grazing management can be found in the NRCS National Range and Pasture Handbook (USDA-NRCS, 1997b), as well as the Bureau of Land Management's (BLM) Technical Reference Series on Grazing.[1]

The Management Practices set forth below have been found by the U.S. Environmental Protection Agency (EPA) to be representative of the types of practices that can be applied successfully to achieve the management measure for grazing. The NRCS management practice number and definition are provided for each management practice, where available. Other practices may be appropriate due to site specific factors. State and local requirements may apply.

Grazing Management Practices

Appropriate grazing management systems ensure proper grazing use by adjusting grazing intensity and duration to reflect the availability of forage and feed designated for livestock uses, and by controlling animal movement through the operat-

---

[1] Four key references within the BLM's Technical Reference Series on Grazing include *Grazing Management for Riparian-Wetland Areas* (Leonard et al., 1997), *Process for Assessing Proper Functioning Condition* (Prichard et al., 1993), *A User Guide to Assessing Proper Functioning Condition and the Supporting Science for Lotic Areas* (USDOI-BLM, USDA-Forest Service, and USDA-NRCS, 1998), and *A User Guide to Assessing Proper Functioning Condition and the Supporting Science for Lentic Areas* (USDOI-BLM, USDA-Forest Service, and USDA-NRCS, 1999). Other references of similar interest include *Successful Strategies for Grazing Cattle in Riparian Zones*, Riparian Tech Bulletin #4, USDOI, Montana BLM, January 1998; and *Effective Cattle Management in Riparian Zones: A Field Survey and Literature Review*, Riparian Tech Bulletin #3, USDOI, Montana BLM, November 1997.

---

ing unit of grazing land. Grazing used as a tool for promoting vegetative vigor can help maintain live vegetation and litter cover from actively growing grasses and forbs and help reduce the soil erosion rates below the natural erosion rates for the soil type and pre-existing vegetative cover. The use of grazing management systems can help maintain riparian and other resource objectives and can help meet the specific management objectives of the desired quality, quantity, and age distribution of vegetation. Practices that accomplish this are:

- ❐ **Grazing Management Plan:** A strategy or system designed to manage the timing, intensity, frequency, and duration of grazing to protect and/or enhance environmental values while maintaining or increasing the economic viability of the grazing operation. This applies to both upland and riparian management.

- ❐ **Pasture and Hay Planting (512):** Establishing native or introduced forage species.

- ❐ **Rangeland planting (550):** Establishment of adapted perennial vegetation such as grasses, forbs, legumes, shrubs, and trees.

- ❐ **Forage Harvest Management (511):** The timely cutting and removal of forages from the field as hay, greenchop, or ensilage.

- ❐ **Prescribed Grazing (528A):** The controlled harvest of vegetation with grazing or browsing animals, managed with the intent to achieve a specified objective.

- ❐ **Use Exclusion (472):** Exclusion of animals, people, or vehicles from an area to protect, maintain, or improve the quantity and quality of the plant, animal, soil, air, water, and aesthetic resources and human health safety.

- ❐ **Nutrient Management (590):** Managing the amount, source, placement, form and timing of the application of nutrients and soil amendments.

Alternate Water Supply Practices

Providing water and mineral supplement facilities away from streams will help keep livestock away from streambanks and riparian zones. The establishment of alternate water supplies for livestock is an essential component of this measure when problems related to the distribution of livestock occur in a grazing unit. In most western states, securing water rights may be necessary. Access to a developed or natural water supply that is protective of streambank and riparian zones can be provided by using the stream crossing (interim) technology to build a watering site. In some locations, artificial shade may be constructed to encourage use of upland sites for shading and loafing. Providing water can be accomplished through the following NRCS practices and the stream crossing (interim) practice of the following section. Practices include:

- ❐ **Irrigation Water Management (449):** Irrigation water management is the process of determining and controlling the volume, frequency, and application rate of irrigation water in a planned, efficient manner.

- ❐ **Pipeline (516):** Pipeline installed for conveying water for livestock or for recreation.

- ❐ **Pond (378):** A water impoundment made by constructing a dam or an embankment or by excavation of a pit or dugout.

Practices have been developed for grazing management, alternative water supply, riparian grazing, and land stabilization.

- ❑ **Trough or Tank (614):** A trough or tank, with needed devices for water control and waste water disposal, installed to provide drinking water for livestock.

- ❑ **Well (642):** A well constructed or improved to provide water for irrigation, livestock, wildlife, or recreation.

- ❑ **Spring Development (574):** Improving springs and seeps by excavating, cleaning, capping, or providing collection and storage facilities.

Riparian Grazing Practices

When implementing a grazing management system (see table 4e-1) within a riparian area, it may at times be necessary to minimize livestock access to riparian zones, ponds or lake shores, wetlands, and streambanks to protect these areas from physical disturbance. The use of management practices for limiting access should be linked in the overall management plan to proper grazing use and other water quality goals. Practices include:

- ❑ **Fence (382):** A constructed barrier to livestock, wildlife, or people.

- ❑ **Animal Trails and Walkways (575):** A travel facility for livestock and/ or wildlife to provide movement through difficult or ecologically sensitive terrain.

- ❑ **Stream Crossing (Interim):** A stabilized area to provide controlled access across a stream for livestock and farm machinery.

Land and Streambank Stabilization Practices

It may be necessary to improve or reestablish the vegetative cover on rangeland and pastures or on streambanks to reduce erosion rates. The following practices can be used to reestablish vegetation:

- ❑ **Nutrient Management (590):** Managing the amount, source, placement, form and timing of the application of nutrients and soil amendments.

- ❑ **Channel Vegetation (322):** Establishing and maintaining adequate plants on channel banks, berms, spoil, and associated areas.

- ❑ **Pasture and Hay Planting (512):** Establishing native or introduced forage species.

- ❑ **Rangeland Planting (550):** Establishment of adapted perennial vegetation such as grasses, forbs, legumes, shrubs, and trees.

- ❑ **Critical Area Planting (342):** Planting vegetation, such as trees, shrubs, vines, grasses, or legumes, on highly erodible or critically eroding areas. (Does not include tree planting mainly for wood products.)

- ❑ **Brush Management (314):** Removal, reduction, or manipulation of non-herbaceous plants.

- ❑ **Grazing Land Mechanical Treatment (548):** Modifying physical soil and/or plant conditions with mechanical tools by treatments such as; pitting, contour furrowing, and ripping or subsoiling.

- ❑ **Grade Stabilization Structure (410):** A structure used to stabilize the grade and control erosion in natural or artificial channels, to prevent the formation and advance of gullies, and to enhance environmental quality and reduce pollution hazards.

- ❏ **Prescribed Burning (338):** Applying controlled fire to predetermined area.

- ❏ **Stream Corridor Improvement (interim):** Restoration of a modified or damaged stream to a more natural state using bioengineering techniques to protect the banks and reestablish the riparian vegetation.

- ❏ **Land Reclamation Landslide Treatment (453):** Treating inplace materials, mine spoil, mine waste, or overburden to reduce downslope movement.

- ❏ **Sediment Basin (350):** A basin constructed to collect and store debris or sediment. Stock water ponds often act as sediment basins.

- ❏ **Wetland Wildlife Habitat Management (644):** Retaining, creating or managing habitat for wetland wildlife. The construction or restoration of wetlands.

- ❏ **Stream Channel Stabilization (584):** Using vegetation and structures to stabilize and prevent scouring and erosion of stream channels.

- ❏ **Wetland Restoration (657):** A rehabilitation of a drained or degraded wetland where the soils, hydrology, vegetative community, and biological habitat are returned to the natural condition to the extent practicable.

- ❏ **Streambank and Shoreline Protection (580):** Using vegetation or structures to stabilize and protect banks of streams, lakes, or estuaries, against scour and erosion.

- ❏ **Riparian Forest Buffer/Herbaceous Cover (391A/390):** Establish an area of trees, shrubs, grasses, or forbs adjacent to and up-gradient from water bodies.

Monitoring Grazing Land Condition

Monitoring is essential to determining whether grazing management objectives are being achieved (Chaney et al., 1993). An integrated approach to monitoring will evaluate nutrient cycling, soil and water quality, and plant community dynamics. To evaluate and adjust management strategies, monitoring should be conducted on both a site specific or allotment level and at the watershed or subwatershed level to determine rangeland condition status and trends. A wide array of monitoring options exist, including the use of photo points, vegetation sampling, soil assessments, water quality and quantity analyses and assessments of watershed, riparian and stream condition. A number of methods are available for monitoring vegetation and for measuring forage utilization and residuals to determine the effects of grazing and browsing on rangelands (Interagency Technical Team, 1996 a, 1996 b; Ruyle and Forst, 1993). To assess vegetative consumption and assist in the nutritional management of livestock and wildlife, other methods, such as clipping procedures, have been developed (Brence and Sheley, 1997).

Numerous publications aid the rangeland manager in determining the status and trends of rangeland resources. Recommended publications on rangeland monitoring include:

- ❑ Monitoring the Vegetation Resources in Riparian Areas (Winward, 2000).

- ❑ Interpreting Indicators of Rangeland Health (USDOI-BLM and USGS, and USDA-NRCS and ARS, 2000).

- ❑ Monitoring Rangelands: Interpreting What You See (Rasmussen et al., 2001)

- ❑ Repeat Photography, Monitoring Made Easy (Rasmussen and Voth, 2001)

See page 143 for additional references on rangeland management.

Decisions regarding changes to stocking rates and preservation of an adequate amount of forage to ensure good rangeland health and minimize water quality impacts are dependent upon good information. Grazing land should be checked frequently to ensure that the plants and animals are meeting management objectives, depending on the management techniques being used.

Spreadsheet applications are available to make tracking and management of grazing cells much easier (Gum and Ruyle, 1993). These spreadsheets address both growing and dormant seasons, and incorporate such factors as the number and size of paddocks, the number of days each paddock is to be rested, and the relative quality of forage in each paddock. Some studies also recommend monitoring plan implementation (i.e., how well the grazing management plan is followed) and effectiveness (i.e., have objectives for vegetation condition been met) (Clary and Leininger, 2000).

Recognizing that the pattern of grazing use varies across an enclosed grazing area, or management unit, USDA recommends the identification of key grazing areas and key plant species to aid in grazing land management (USDA-NRCS, 1997b). By protecting and monitoring the key grazing areas and key plant species, it is believed that the management unit as a whole will be protected.

Practice Effectiveness

Eckert and Spencer (1987) studied the effects of a three-pasture, rest-rotation management plan on the growth and reproduction of heavily grazed native bunch grasses in Wyoming. The results indicated that rangeland improvement under this otherwise appropriate rotation grazing system is hindered by heavy grazing. Stocking rates on the study plots exceeded the carrying capacity of the land and would decrease native grasses and increase potential erosion and sedimentation.

Van Poollen and Lacey (1979) showed that herbage production was greater for managed grazing versus continuous grazing, greater for moderate versus heavy intensity grazing, and greater for light- versus moderate-intensity grazing.

Tiedemann et al. (1988) studied the effects of four grazing strategies on bacteria levels in 13 Oregon watersheds in the summer of 1984. Although wildlife were believed to be significant sources of bacteria in each of the study watersheds, results indicate that lower fecal coliform levels can be achieved at stocking rates of about 20 ac/AUM (acres per animal unit month) if management for livestock distribution, fencing, and water developments are used (Table 4e-2). The study also indicates that, even with various management practices, the highest fecal coliform levels were associated with the higher stocking rates (6.9 ac/AUM) employed in strategy D.

Owens et al. (1982) measured nitrogen losses from an Ohio pasture under a medium-fertility, 12-month pasture program from 1974 to 1979. The results included no measurable soil loss from three watersheds under summer grazing only, and increased average TN concentrations and total soluble N loads from watersheds under summer grazing and winter feeding versus watersheds under summer grazing only (Table 4e-3).

Data from a comparison of the expected effectiveness of various grazing and streambank practices in controlling sedimentation in the Molar Flats Pilot Study Area in Fresno County, California indicate that planned grazing systems are the most effective single practice for reducing sheet and rill erosion (Fresno Field Office, 1979).

By switching grazing allotments from continuous, season-long grazing to a three-pasture, rest-rotation system, the U.S. Forest Service was able to achieve

Table 4e-2. Bacterial water quality responses to four grazing strategies (Tiedemann et al., 1988).

Practice		Geometric Mean Fecal Coliform Count
Strategy A:	Ungrazed	40/L
Strategy B:	Grazing without management for livestock distribution; 20.3 ac/AUM.	150/L
Strategy C:	Grazing with management for livestock distribution: fencing and water developments; 19.0 ac/AUM.	90/L
Strategy D:	Intensive grazing management, including practices to attain uniform livestock distribution and improve forage production with cultural practices such as seeding, fertilizing, and forest thinning; 6.9 ac/AUM.	920/L

Table 4e-3. Nitrogen losses from medium-fertility, 12-month pasture program (Owens et al., 1982).

Practice	Soil Loss (kg/ha)	Total Sediment N Transport (kg/ha)	Total N Concentration (mg/l[a])	Total Soluble N Transport (kg/ha)[a]
Summer Grazing Only				
Growing season	—	—	3.7	0.4
Dormant season	—	—	1.8	0.1
Year	—	—	3.0	0.5
Summer Grazing – Winter Feeding				
Growing season	251	1.4	4.9	2.5
Dormant season	1,104	6.6	14.6	11.3
Year	1,355	8.0	10.7	13.8

[a]Five-year average (1974-1979)

major improvements in the vegetation in the Tonto National Forest in Arizona (Chaney et al., 1990). For example, cottonwood populations increased from 20 per 100 acres to more than 2,000 per 100 acres in six years, while at the same time the amount of livestock forage grazed increased by 27 percent. Similar improvements from improved grazing management were documented through case studies in Idaho, Nevada, Oregon, South Dakota, Texas, Utah, and Wyoming.

Hubert et al. (1985) showed in plot studies in Wyoming that livestock exclusion and reductions in stocking rates can result in improved habitat conditions for brook trout. In this study, the primary vegetation was willows, Pete Creek stocking density was 7.88 ac/AUM (acres per animal unit month), and Cherry Creek stocking density was 10 cows per acre (Table 4e-4).

Platts and Nelson (1989) used plot studies in Utah to evaluate the effects of livestock exclusion on riparian plant communities and streambanks. Several streambank characteristics that are related to the quality of fish habitat were measured, including bank stability, stream shore depth, streambank angle, undercut, overhang, and streambank alteration. The results clearly show better fish habitat in the areas where livestock were excluded (Table 4e-5).

Kauffman et al. (1983a) showed that fall cattle grazing decreases the standing crop of some riparian plant communities by as much as 21% versus areas where cattle are excluded, while causing increases for other plant communities. This study, conducted in Oregon from 1978 to 1980, incorporated stocking rates of 3.2 to 4.2 ac/AUM.

Buckhouse (1993) did an extensive review of livestock impacts on riparian systems. Researchers documented many factors interrelated with grazing effects, primarily dealing with instream ecology, terrestrial wildlife, and riparian vegeta-

Table 4e-4. Grazing management influences on two brook trout streams in Wyoming (Hubert et al., 1985).

Stream Parameter	Pete Creek (n=3)		Cherry Creek (n=4)	
	Heavily Grazed (mean)	Lightly Grazed (mean)	Outside Exclosure (mean)	Inside Exclosure (mean)
Width	2.9	2.2[a]	2.9	2.5[a]
Depth	0.07	0.11[a]	0.08	0.09[a]
Width/depth ratio	43	21	37	28[a]
Coefficient of variation in depth	47.3	66.6[a]	57	71
Percent greater than 22 cm deep	9.0	22.3[b]	6.7	21.0[a]
Percent overhanging bank cover	2.7	30.0[a]	24.0	15.3
Percent overhanging vegetation	0	11.7[a]	8.5	18.0
Percent shaded area	0.7	18.3[a]	23.5	28.0
Percent silt substrate	35	52	22	13[a]
Percent bare soil along banks	19.7	13.3	22.8	12.3[a]
Percent litter along banks	7.0	6.0	10.0	6.8[a]

[a] Indicates statistical significance at p<=0.05.
[b] Indicates statistical significance at p<=0.1.

Table 4e-5. Streambank characteristics for grazed versus rested riparian areas (Platts and Nelson, 1989).		
Streambank Characteristic (unit)	Grazed	Rested
Extent (m)	4.1	2.5
Bank stability (%)	32.0	88.5
Stream-short depth (cm)	6.4	14.9
Bank angle (°)	127.0	81.0
Undercut (cm)	6.4	16.5
Overhang (cm)	1.8	18.3
Streambank alteration (%)	72.0	19.0

Grazing management research indicates that local practices designed for area soils, vegetation, and stocking rates are more likely to succeed than applying one system of BMPs across the entire region.

tion. Permanent removal of grazing will not guarantee maximum herbaceous plant production. Researches found that a protected Kentucky bluegrass meadow reached peak production in six years and then declined until production was similar to the adjacent area grazed season-long. The accumulation of litter over a period of years seems to retard forage production in wetlands. Thus, some grazing of riparian areas could have beneficial effects. Stoltzfus and Lanyon (1992) also identified that fencing a riparian zone protects herd health from infectious bacteria, hoof diseases, poor quality drinking water, and provides a wildlife habitat.

The effect of grazing on streambanks depends on site conditions, management practices, timing, and other factors. Kauffman et al. (1983b) found that late-season grazing increased bank erosion relative to ungrazed areas in Oregon. If late season grazing is permitted, adequate time for regrowth should be allowed prior to the next major runoff event. Hallock (1996) found that delaying grazing in riparian pastures until the soil dries in the late spring did not degrade the streambanks or change stream morphology significantly in a Coastal California Watershed.

Lugbill (1990) estimates that stream protection in the Potomac River Basin will reduce total nitrogen (TN) and total phosphorus (TP) loads by 15%, while grazing land protection and permanent vegetation improvement will reduce TN and TP loads by 60%.

Nutrient loss is minimal where the riparian pasture remains in good condition. Vegetation buffers the stream from direct waste input and assimilates the nutrients into plant tissue. Gary et al. (1983) evaluated the effects on a small stream in central Colorado of spring cattle grazing on pastures. Nitrate nitrogen did not increase significantly and ammonia increased significantly only once.

Meals (2001) reported significant water quality improvements in Vermont streams following livestock exclusion and riparian restoration on dairy pastureland. Mean total phosphorus concentrations were reduced by 15%, and total P load was reduced by 49% over a three-year period following riparian restoration. Indicator bacteria counts in treated streams fell by 29% - 46%.

Photos have been used to document improvements in riparian condition due to such practices as rest rotations and exclusion (Chaney et al., 1993). The authors emphasize the importance, however, of looking beyond the vegetation and examining whether water quality benefits also accrue. Vegetative response usually happens in one to five years, however, stream channel changes may take decades.

Miner et al. (1991) showed that the provision of supplemental water facilities reduced the time each cow spent in the stream within 4 hours of feeding from 14.5 minutes to 0.17 minutes (8-day average). This pasture study in Oregon showed that the 90 cows without supplemental water spent a daily average of 25.6 minutes per cow in the stream. For the 60 cows that were provided a supplemental water tank, the average daily time in the stream was 1.6 minutes per cow, while 11.6 minutes were spent at the water tank. Based on this study, the authors expect that a 90% decrease in time spent in the stream will substantially decrease bacterial loading from the cows.

McDougald et al. (1989) tested the effects of moving supplemental feeding locations on riparian areas of hardwood rangeland in California. With stocking rates of approximately 1 ac/AUM, they found that moving supplemental feeding locations away from water sources into areas with high amounts of forage greatly reduces the impacts of cattle on riparian areas (Table 4e-6).

> Plant species production management is central to effective grazing BMPs. Consider ecosystem productivity, harvest rates by stock and wildlife, and regenerative capacity.

Table 4e-6. The effects of supplemental feeding location on riparian area vegetation (McDougald et al., 1989).			
	Percentage of riparian area with the following levels of residual dry matter in early October		
Practice	Low	Moderate	High
Supplemental feeding located close to riparian areas:			
1982-85 Range Unit 1	48	38	13
1982-85 Range Unit 8	59	29	12
1986-87 Range Unit 8	54	33	13
Supplemental feeding moved away from riparian area:			
1986-87 Range Unit 1	1	27	72

Factors in the Selection of Management Practices

The selection of grazing management practices for this measure should be based on an evaluation of current conditions, problems identified, quality criteria, and management goals. Successful resource management on grazing lands includes appropriate application of a combination of practices that will meet the needs of the rangeland and pasture ecosystem (i.e., the soil, water, air, plant, and animal (including fish and shellfish) resources) and the objectives of the land user.

For a sound grazing land management system to function properly and to provide for a sustained level of productivity, the following should be considered:

- ❑ Know the key factors of plant species management, their growth habits, and their response to different seasons and degrees of use by various kinds and classes of livestock.

- ❑ Know the demand for, and seasons of use of, forage and browse by wildlife species.

- ❑ Know the amount of plant residue or grazing height that should be left to protect grazing land soils from wind and water erosion, provide for plant health and regrowth, and provide the riparian vegetation height desired to trap sediment or other pollutants.

- ❑ Know the ecological site production capabilities for rangeland and the forage suitability group capabilities for pasture so an initial stocking rate can be established.

- ❑ Know how to use livestock as a tool (i.e., control timing and duration of grazing) in the management of the rangeland ecosystems and pastures to ensure the health and vigor of the plants, soil tilth, proper nutrient cycling, erosion control, and riparian area management, while at the same time meeting livestock nutritional requirements.

- ❑ Establish grazing unit sizes, watering, shade (where possible) and mineral locations, etc. to secure optimum livestock distribution and proper vegetation use.

- ❑ Provide for livestock herding, as needed, to protect sensitive areas from excessive use at critical times.

- ❑ Work with state game management agencies to agree on proper stocking numbers prior to wildlife harvest. Encourage proper wildlife harvesting to ensure proper population densities and forage balances.

- ❑ Know the livestock diet requirements in terms of quantity and quality to ensure that there are enough grazing units to provide adequate livestock nutrition for the season and the kind and classes of animals on the farm/ranch.

- ❑ Maintain a flexible grazing system to adjust for unexpected environmentally and economically generated problems.

- ❑ Follow special requirements to protect threatened or endangered species.

To speed up the rehabilitation process of riparian zones, seeding can be used as a proper management practice. This strategy, however, can be very expensive and risky. Riparian zones can be rehabilitated positively and at a lower cost through improving livestock distribution, better watering systems, fencing, or reducing stock rates. In areas where the desirable native perennial forage plants are nearly extinct, seeding is essential. Such areas will have a poor to very poor rating of forage condition and are difficult to restore.

Cost of Practices

Costs

Much of the cost associated with implementing grazing management practices is due to fencing installation, water development, and seeding. Costs vary according to region and type of practice. Generally, the more components or structures a practice requires, the more expensive it is. However, cost-share is usually available from the USDA and other federal agencies for most of these practices.

The principal direct costs of providing grazing practices vary from relatively low variable costs of dispersed salt blocks to higher capital and maintenance costs of supplementary water supply improvements. Improving the distribution of grazing pressure by developing a planned grazing system or strategically locating water troughs, salt, or feeding areas to draw cattle away from riparian zones can result in improved utilization of existing forage, better water quality, and improved riparian habitat.

Principal direct costs of excluding livestock from the riparian zone for a period of time are the capital and maintenance costs for fencing to restrict access to streamside areas and/or the cost of herders to achieve the same results. In addition, there may be an indirect cost of the forage that is removed from grazing by the exclusion.

Principal direct costs of improving or reestablishing grazing land include the costs of seed, fertilizer, and herbicides needed to establish the new forage stand and the labor and machinery costs required for preparation, planting, cultivation, and weed control (Table 4e-7). An indirect cost may be the forage that is removed from grazing during the reestablishment work and rest for seeding establishment.

Table 4e-7. Cost of forage improvement/reestablishment for grazing management (EPA, 1993a).

Location	Year	Type	Unit	Reported Capital Costs $/Unit	Constant Dollar[a] Capital Costs 1991 $/Unit	Constant Dollar[a] Annualized Costs 1991 $/Unit
Alabama[b]	1990	planting (seed, lime & fertilizer)	acre	84 - 197	83 - 195	12.37 - 29.00
Nebraska[c]	1991	establishment seeding	acre acre	47 45	47 45	7.00 6.71
Oregon[d]	1991	establishment	acre	27	27	4.02

[a] Reported costs inflated to 1991 constant dollars by the ratio of indices of prices paid by farmers for seed, 1997=100. Capital costs are annualized at 8% interest for 10 years.
[b] Alabama Soil Conservation Service, 1990.
[c] Hermsmeyer, 1991.
[d] USDA–ASCS, 1991b.

Water Development

The availability and feasibility of supplementary water development varies considerably between arid western areas and humid eastern areas, but costs for water development, including spring development and pipeline watering, are similar (Table 4e-8).

Table 4e-8. Cost of water development for grazing management (EPA, 1993a).

Location	Year	Type	Unit	Reported Capital Costs $/Unit	Constant Dollar[a] Capital Costs 1991 $/Unit	Constant Dollar[a] Annualized Costs 1991 $/Unit
California[b]	1979	pipeline	foot	0.28	0.35	0.05
Kansas[c]	1989	spring	each	1,239.00	1,282.94	191.20
		spring	each	1,389.00	1,438.26	214.34
Maine[d]	1988	pipeline	each	831.00	879.17	131.02
Alabama[e]	1990	spring	each	1,500.00	1,520.83	226.65
		pipeline	foot	1.60	1.62	0.24
		trough	each	1,000.00	1,013.89	151.10
Nebraska[f]	1991	pipeline	foot	1.31	1.31	0.20
		tank	each	370.00	370.00	55.14
Utah[g]	1968	spring	each	200.00	389.33	58.02
Oregon[h]	1991	pipeline	foot	0.20	0.20	0.03
		tank	each	183.00	183.00	27.27

[a] Reported costs inflated to 1991 constant dollars by the ratio of indices of prices paid by farmers for building and fencing, 1977=100. Capital costs are annualized at 8% interest for 10 years.
[b] Fresno Field Office, 1979.
[c] Northup et al., 1989.
[d] Cumberland County Soil and Water Conservation District, undated.
[e] Alabama Soil Conservation Service, 1990.
[f] Hermsmeyer, 1991.
[g] Workman and Hooper, 1968.
[h] USDA–ASCS, 1991b.

Use Exclusion

There is considerable difference between multistrand barbed wire, chiefly used for perimeter fencing and permanent stream exclusion and diversions, and single- or double-strand smoothwire electrified fencing used for stream exclusion and temporary divisions within permanent pastures. The latter may be all that is needed to accomplish most livestock exclusion in a smaller, managed, riparian pasture (Table 4e-9). In some cases, exclusion of livestock from waterways and riparian areas can be accomplished through the use of hedgerows, intensive herding/grazing management, or provision of feed, water, and shade at alternative sites.

Overall Costs of the Grazing Management Measure

Since the combination of practices needed to implement the management measure depends on site-specific conditions that are highly variable, the overall cost of the measure is best estimated from similar combinations of practices applied under the Agricultural Conservation Program (ACP), Rural Clean Water Program (RCWP), and similar activities.

Table 4e-9. Cost of livestock exclusion for grazing management (EPA, 1993a).

Location	Year	Type	Unit	Reported Capital Costs $/Unit	Constant Dollar[a] Capital Costs 1991 $/Unit	Constant Dollar[a] Annualized Costs 1991 $/Unit
California[b]	1979	permanent	mile	2,000	2,474.58	368.78
Alabama[c]	1990	permanent	mile	3,960	4,015.00	598.35
		net wire	mile	5,808	5,888.67	877.58
		electric	mile	2,640	2,676.67	398.90
Nebraska[d]	1991	permanent	mile	2,478	2,478.00	369.30
Great Lakes[e]	1989	permanent	mile	2,100 - 2,400	2,174.47 - 2,485.11	324.06 - 370.35
Oregon[1]	1991	permanent	mile	2,640	2,640.00	393.44

[a] Reported costs inflated to 1991 constant dollars by the ratio of indices of prices paid by farmers for building and fencing, 1977=100. Capital costs are annualized at 8% interest for 10 years.

[b] Fresno Field Office, 1979.

[c] Alabama Soil Conservation Service, 1990.

[d] Hermsmeyer, 1991.

[e] DPRA, 1989.

[1] USDA–ASCS, 1991b.

4F: Irrigation Water Management

Management Measure for Irrigation Water

To reduce nonpoint source pollution of ground and surface waters caused by irrigation:

(1) Operate the irrigation system so that the timing and amount of irrigation water applied match crop water needs. This will require, as a minimum: (a) the accurate measurement of soil-water depletion volume and the volume of irrigation water applied, and (b) uniform application of water.

(2) When chemigation is used, include backflow preventers for wells, minimize the harmful amounts of chemigated waters that discharge from the edge of the field, and control deep percolation. In cases where chemigation is performed with furrow irrigation systems, a tailwater management system may be needed.

The following limitations and special conditions apply:

(1) In some locations, irrigation return flows are subject to other water rights or are required to maintain stream flow. In these special cases, on-site reuse could be precluded and would not be considered part of the management measure for such locations. In these locations, improvements to irrigation systems and their management should still occur.

(2) By increasing the water use efficiency, the discharge volume from the system will usually be reduced. While the total pollutant load may be reduced somewhat, there is the potential for an increase in the concentration of pollutants in the discharge. In these special cases, where living resources or human health may be adversely affected and where other management measures (nutrients and pesticides) do not reduce concentrations in the discharge, increasing water use efficiency would not be considered part of the management measure.

(3) In some irrigation districts, the time interval between the order for and the delivery of irrigation water to the farm may limit the irrigator's ability to achieve the maximum on-farm application efficiencies that are otherwise possible.

(4) In some locations, leaching is necessary to control salt in the soil profile. Leaching for salt control should be limited to the leaching requirement for the root zone.

(5) Where leakage from delivery systems or return flows supports wetlands or wildlife refuges, it may be preferable to modify the system to achieve a high level of efficiency and then divert the "saved water" to the wetland or wildlife refuge. This will improve the quality of water delivered to wetlands or wildlife refuges by preventing the introduction of pollutants from irrigated lands to such diverted water.

(6) In some locations, sprinkler irrigation is used for frost or freeze protection, or for crop cooling. In these special cases, applications should be limited to the amount necessary for crop protection, and applied water should remain on-site.

A primary concern for irrigation water management is the discharge of salts, pesticides, and nutrients to ground water and discharge of these pollutants plus sediment to surface water.

Management Measure for Irrigation Water: Description

The goal of this management measure is to reduce movement of pollutants from land into ground or surface water from the practice of irrigation. This goal is accomplished through consideration of the following aspects of an irrigation system, which will be discussed in this chapter:

1. Irrigation scheduling

2. Efficient application of irrigation water

3. Efficient transport of irrigation water

4. Use of runoff or tailwater

5. Management of drainage water

Effective irrigation management reduces runoff and leachate losses, controls deep percolation and, along with cropland sediment control, reduces erosion and sediment delivery to waterways.

A well designed and managed irrigation system reduces water loss to evaporation, deep percolation, and runoff and minimizes erosion from applied water. Application of this management measure will reduce the waste of irrigation water, improve water use efficiency, and reduce the total pollutant discharge from an irrigation system. It focuses on components to manage the timing, amount and location of water applied to match crop water needs, and special precautions (i.e., backflow preventers, prevent runoff, and control deep percolation) when chemigation is used.

Irrigation and Irrigation Systems: An Overview

Irrigation, the addition of water to lands via artificial means, is essential to profitable crop production in arid climates. Irrigation is also practiced in humid and sub-humid climates to protect crops during periods of drought. Irrigation is practiced in all environments to maximize production and, therefore, profit by applying water when the plant needs it. Figure 4f-1 shows the distribution of irrigated farmland in the U.S. (USDA-ERS, 1997).

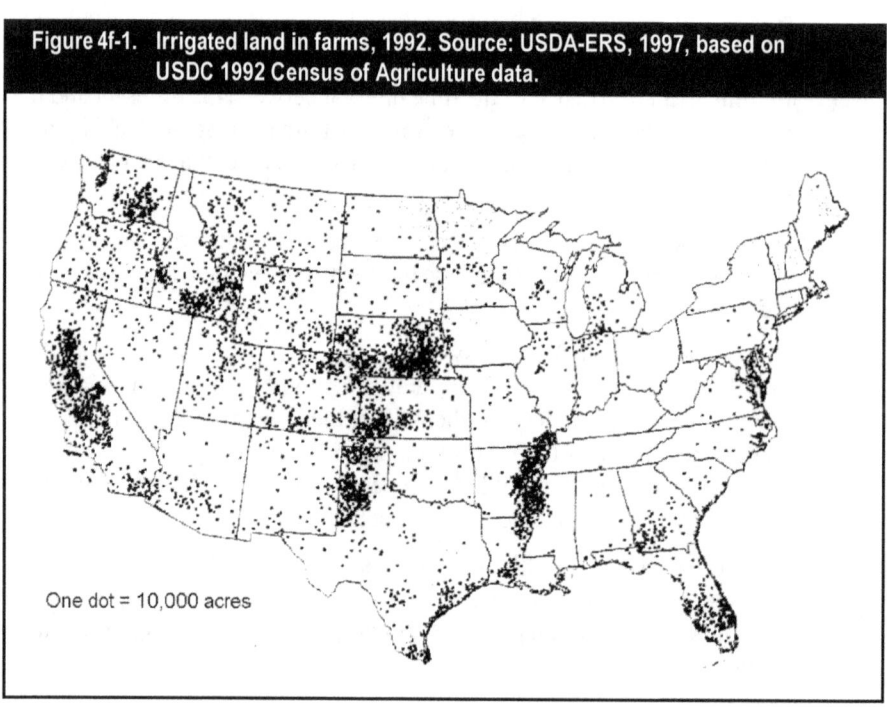

Figure 4f-1. Irrigated land in farms, 1992. Source: USDA-ERS, 1997, based on USDC 1992 Census of Agriculture data.

One dot = 10,000 acres

Soil-Water-Plant Relationships

Effective and efficient irrigation begins with a basic understanding of the relationships among soil, water, and plants. Figure 4f-2 illustrates the on-farm hydrologic cycle for irrigated lands, and Table 4f-1 provides definitions of several terms associated with irrigation. Water can be supplied to the soil through precipitation, irrigation, or from groundwater (e.g., rising water table due to drainage management). Plants take up water that is stored in the soil (soil water), and use this for growth (e.g., nutrient uptake, photosynthesis) and cooling. Transpiration is the most important component of the on-farm hydrologic cycle (Duke, 1987), with the greatest share of transpiration devoted to cooling. Water is also lost via evaporation from leaf surfaces and the soil. The combination of transpiration and evaporation is evapotranspiration, or ET. ET is influenced by several factors, including plant temperature, air temperature, solar radiation, wind speed, relative humidity, and soil water availability (USDA-NRCS, 1997a). The amount of water the plant needs, its consumptive use, is equal to the quantity of water lost through ET. Due to inefficiencies in the delivery of irrigated water (e.g., evaporation, runoff, wind drift, and deep percolation losses), the amount of water needed for irrigation is greater than the consumptive use. In arid and semi-arid regions, salinity control may be a consideration, and additional water or "leaching requirement" may be needed.

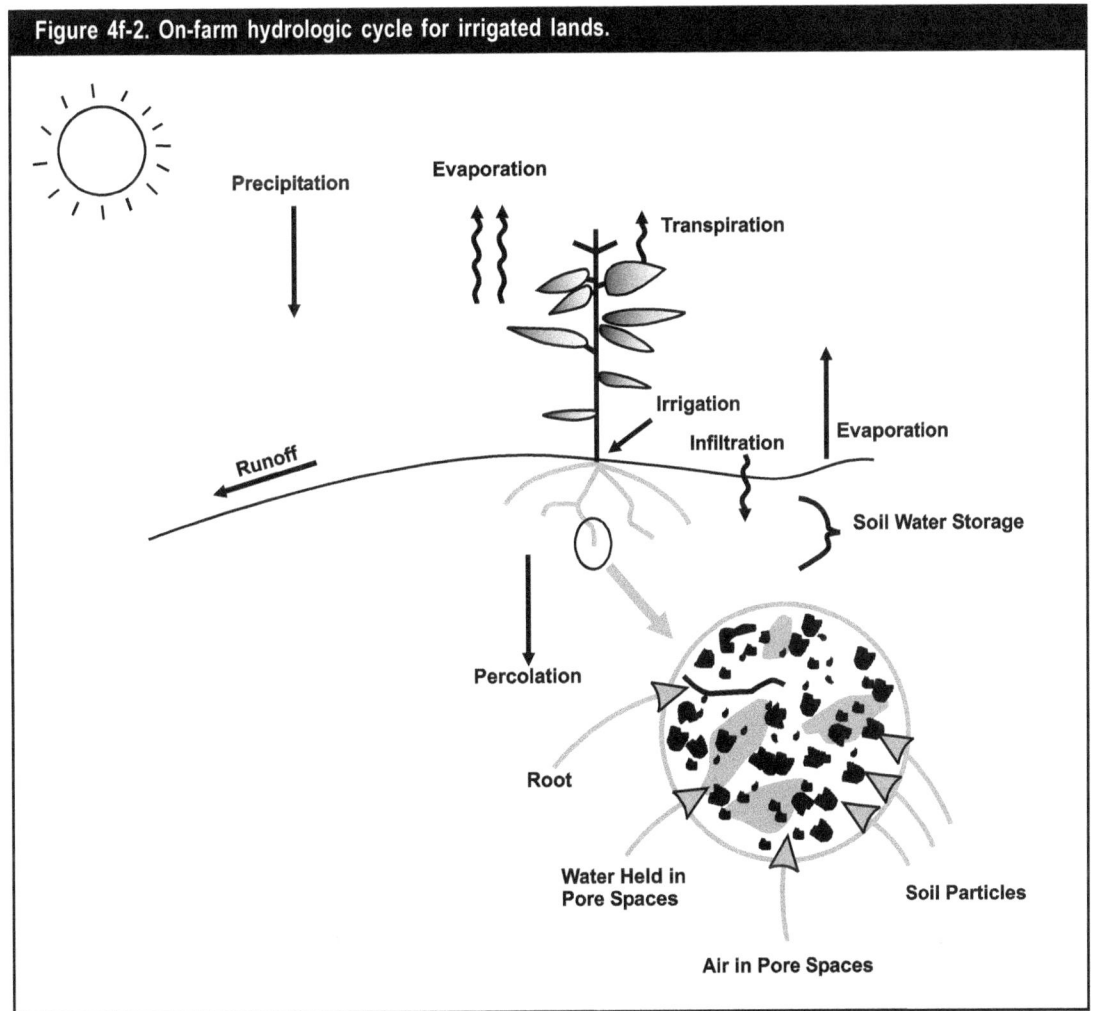

Figure 4f-2. On-farm hydrologic cycle for irrigated lands.

Table 4f-1. Soil-water-plant relationship terms.

Term	Definition
Evaporation	The transformation of water to vapor without passing through the plant.
Transpiration	The movement of water into plant roots, through the plant, and out the stomata as water vapor.
Evapotranspiration (ET)	Evaporation + Transpiration
Soil water	Water stored in the soil.
Soil-water potential Soil-water tension Soil moisture tension	A measure of the strength with which the soil holds the water. Soil water potential is the amount of work required per unit quantity of water to transport water in soil, and is measured in units of bars and atmospheres or cm. A tension is a negative potential. Water moves from high to low potential.
Gravitational water Free water	Water that moves downward freely in soils under the force of gravity.
Capillary water	Water that moves slowly through smaller pores in soils, due to surface tension forces in unsaturated conditions.
Field capacity	The amount of soil water stored in the soil after free water (gravitational water) passes through the soil profile. Sometimes referred to as 2-3 day drainage or a soil water potential of about -1/3 bar. For a sandy soil, this might occur in less than one day.
Available water capacity	The amount of stored soil water that is available to the plant.
Water holding capacity	The amount of water that can be stored in the soil at field capacity.
Permanent wilting point	The soil-water content at which most plants cannot obtain sufficient water to prevent permanent tissue damage, about -15 bars.
Management allowable depletion (MAD)	The greatest amount of water that can be removed by plants before irrigation is needed to avoid undesirable crop water stress.
Consumptive use	The amount of water that is used by the plant. Is equal to ET.
Soil texture	The proportion of the various sizes of soil particles (sand, silt, and clay). Defines coarseness or fineness of soil, along with structure, and controls the hydraulic characteristics of the soil.
Soil structure	The arrangement and organization of soil particles into natural units of aggregation.
Bulk density	The weight of a unit volume of dry soil.

Build up of salts typically occurs in regions where evapotranspiration exceeds precipitation. Salts contained in precipitation or dissolved in the soil are left behind as evaporation and capillary action transports and deposits these salts near the surface. Salinity is not normally a problem in humid regions, where natural leaching of salts from rainfall occurs.

Excess salts in the soil have an adverse impact on plant growth. The total concentration of salts in the soil solution exerts an osmotic force, and therefore makes it more difficult for plants to uptake water. In addition, specific ions, such as chloride, sodium, boron and others may have a toxic effect on plants at certain levels. Crops respond differently to both total and specific salts, some being more sensitive than others.

Plant growth depends upon a renewable supply of soil water, which is governed by the movement of water in the soil, the soil-water holding capacity, the amount of soil water that is readily available to plants, and the rate at which soil water can be replenished (Duke, 1987). Efficient irrigation provides plants with this renewable supply of soil water with a minimum of wasted time, energy, and water. Knowledge and understanding of the factors that affect water movement in the soil, storage of water in the soil, and the availability of water to plants are essential to achieving maximum irrigation efficiencies.

Movement of soil water

When water is applied to soils it moves via such pathways as infiltration, runoff, and evaporation (Figure 4f-2). The ultimate fate and transport of applied water is determined by various forces, including gravity and capillary force. Gravity pulls water downward freely in soils with large pores, causing it to move through the root zone quickly if not taken up by the crop (Duke, 1987). As the water passes through the soil, the pores are filled again with air, preventing crop damage that could arise due to excess water. In soils with smaller pores, water moves via capillary forces. This "capillary water" moves more slowly than gravitational water, and tends to move from wetter areas to drier areas. The lateral distribution of capillary water makes it more important to the irrigated crop since it provides greater wetting of the soil (Duke, 1987). In saturated conditions, gravity is the primary force causing downward water movement (Watson, et al. 1995), while capillary action is the primary force in unsaturated soil.

The above discussion uses subjective terms such as "capillary water" and "gravitational water" (see Table 4f-1) to simplify the description of how water moves in soils. USDA describes this movement in the more technically correct terms of soil-water potential, measured in units of bars and atmospheres (USDA-NRCS, 1997a). Soil-water potential is the sum of matric, solute, gravitational, and pressure potential, detailed discussions of which are beyond the scope of this document. In simple terms, however, water in the soil moves toward decreasing potential energy, or commonly from higher water content to lower water content (USDA-NRCS, 1997a).

Storage and availability of soil water

The amount of water that soil can hold, its water holding capacity, is a key factor in irrigation planning and management since the soil provides the reservoir of water that the plant draws upon for growth. Water is stored in the soil as a film around each soil particle, and in the pore spaces between soil particles (Risinger and Carver, 1987). The magnified area in Figure 4f-2 illustrates how soil water and air are held in the pore spaces of soils.

All soil water is not equally available for extraction and use by plants. The ability of plants to take water from the soil depends upon a number of factors, including soil texture, soil structure, and the layering of soils (Duke, 1987). Texture is classified based upon the proportion of sand, silt, and clay particles in the soil (Figure 4f-3). Structure refers to how the soil particles are arranged in groups or

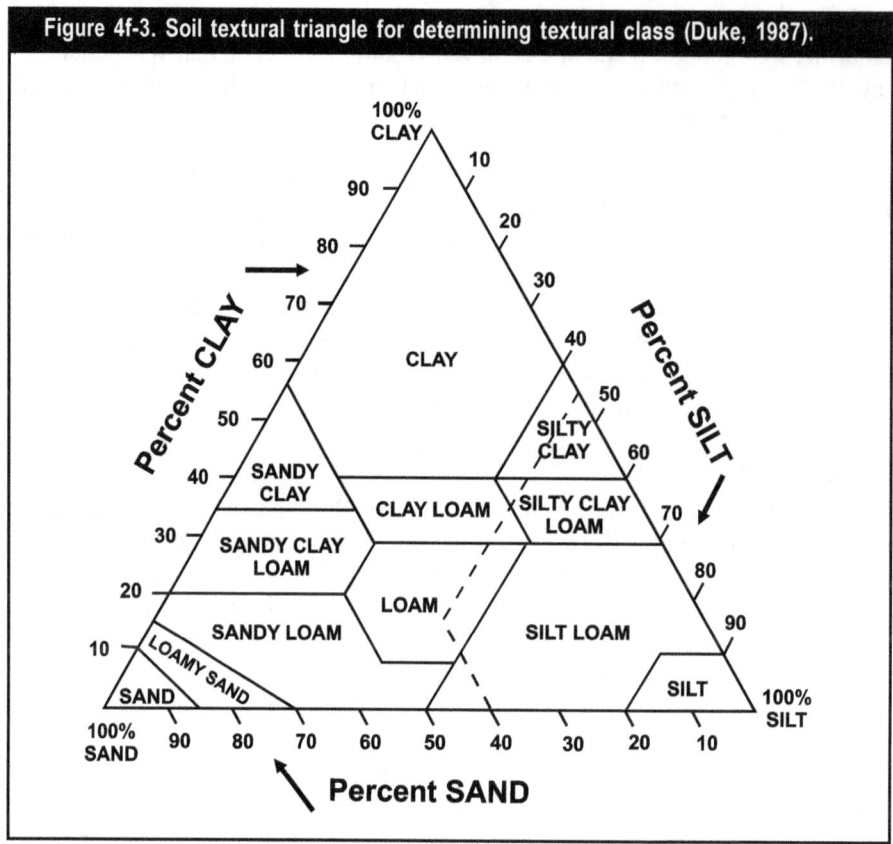

Figure 4f-3. Soil textural triangle for determining textural class (Duke, 1987).

aggregates, while layering refers to the vertical distribution of soils in the soil profile (e.g., clay soils underlying a sandy loam layer). The type and extent of layering can influence the percolation and lateral distribution of applied water.

Soil texture and structure affect the size, shape, and quantity of pores in the soil, and therefore the space available to hold air or water. For example, the available water capacity of coarse sand ranges from 0.1 to 0.4 inches of water per foot of soil depth (in/ft), while silt holds 1.9–2.2 in/ft, and clay holds 1.7–1.9 in/ft (USDA-NRCS, 1997a). The structure of some volcanic ash soils allows them to carry very high water content at field capacity levels, but pumice and cinder fragments may contain some trapped water that is not available to plants (USDA-NRCS, 1997a). In fine-textured soils and soils affected by salinity, sodicity, or other chemicals, a considerable volume of soil water may not be available for plant use due to greater soil water tension (USDA-NRCS, 1997a).

Field capacity is the amount of water a soil holds after "free" water has drained because of gravity (USDA-NRCS, 1997a). "Free" water, which is conceptually similar to "gravitational" water, can drain from coarse-textured (e.g., sandy) soils in a few hours from the time of rainfall or irrigation, from medium-textured (e.g., loamy) soils in about 24 hours, and from fine-textured (e.g., clay) soils in several days. Soil properties that affect field capacity are texture, structure, bulk density, and strata within the soil profile that restrict water movement. Available water capacity is the difference between the amount of water held at field capacity and the amount held at the permanent wilting point (Burt, 1995).

Uptake of soil water by plants

Water stored in soil pore spaces is the easiest for the plant to extract, while water stored in the film around soil particles is much more difficult for the plant to withdraw (Risinger and Carver, 1987). As evapotranspiration draws water from the soil, the remaining water is held more closely and tightly by the soil. Soil moisture tension increases as soils become drier, making it more difficult for the plant to extract the soil water. Figure 4f-4 is a soil moisture release curve that shows how greater energy (tension measured in bars, or potential measured in negative (-) bars) is needed to extract water from the soil as soil-water content decreases (USDA-NRCS, 1997a). This figure also illustrates the greater soil-water tension (or lesser soil-water potential) in clays versus loam and sand for any given soil-water content. Because clay holds water at greater tension than medium-textured soils (e.g., loam) at similar water contents, it has less *available* water capacity despite its greater water holding capacity (USDA-NRCS, 1997a).

Wilting occurs when the plant cannot overcome the forces holding the water to the soil particles (i.e., the soil-water tension). Irrigation is needed at this point to save the plant. The permanent wilting point (represented as -15 bars in Figure 4f-4) is the soil-water content at which most plants cannot obtain sufficient water to prevent permanent tissue damage (USDA-NRCS, 1997a). Based upon yield and

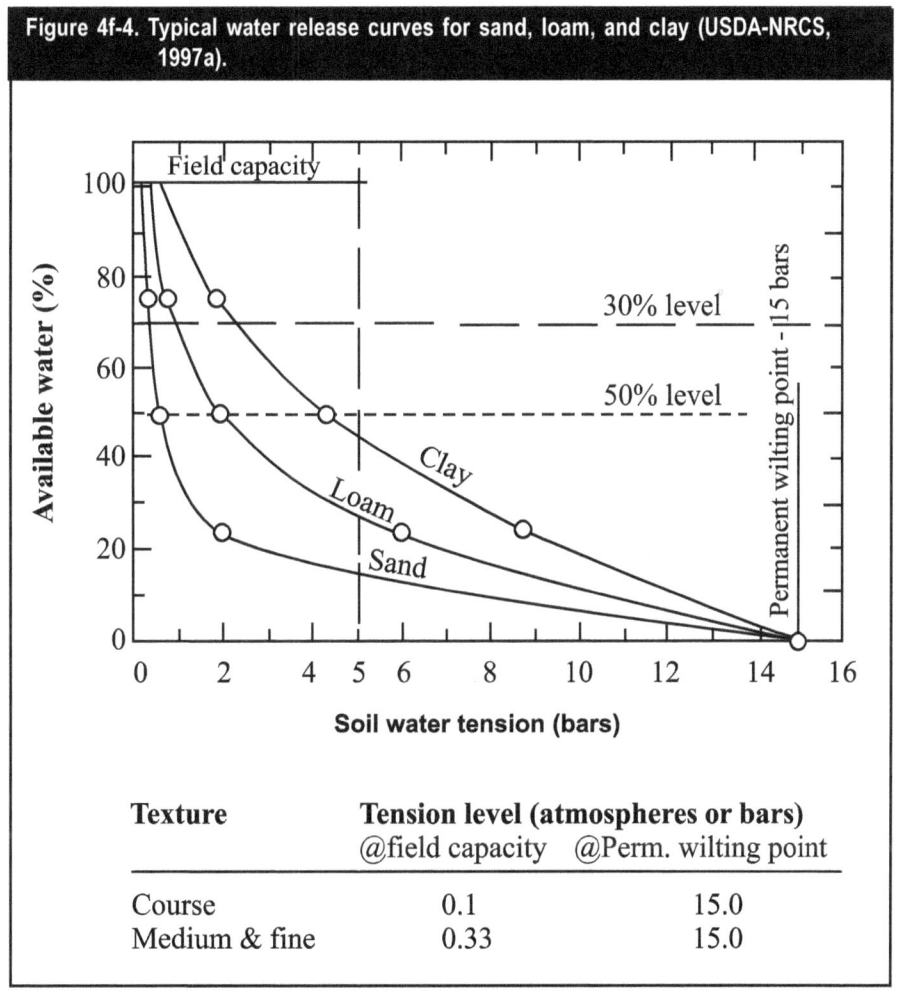

Figure 4f-4. Typical water release curves for sand, loam, and clay (USDA-NRCS, 1997a).

Texture	Tension level (atmospheres or bars)	
	@field capacity	@Perm. wilting point
Course	0.1	15.0
Medium & fine	0.33	15.0

product quality objectives, growers decide how much water to allow plants to remove from the soil before irrigation. This amount, the Management Allowable Depletion (MAD), is expressed as a percentage of the available water-holding capacity and varies for different crops and irrigation methods. As a general rule of thumb, MAD is 50%. Smaller MAD values, which result in more frequent irrigations, may be desirable where micro-irrigation is practiced, when saline water is used, for shallow root zones, and in cases where the water supply is uncertain (Burt, 1995). Large MAD values might be desirable when hand-move and hose-pull sprinklers are used, where furrows are long and soils are sandy, or for crops such as some varieties of cotton that need to be stressed on heavy soil to develop a sufficient number of cotton bolls (Burt, 1995).

Irrigation Methods and System Designs

Irrigation systems consist of two basic elements: (1) the transport of water from its source to the field, and (2) the distribution of transported water to the crops in the field. A number of soil properties and qualities are important to the design, operation, and management of irrigation systems, including water holding capacity, soil intake characteristics, permeability, soil condition, organic matter, slope, water table depth, soil erodibility, chemical properties, salinity, sodicity, and pH (USDA-NRCS, 1997a). Some soils cannot be irrigated due to various physical problems, such as low infiltration rates and poor internal drainage which may cause salt buildup. The chemical characteristics of the soil and the quantity and quality of the irrigation water will determine whether irrigation is a suitable management practice that can be sustained without degrading the soil or water resources (Franzen et al., 1996; Scherer et al., 1996; and Seelig and Richardson, 1991).

Water supply and demand

Producers need to factor the availability of good quality water (in terms of amount, timing, and rate) into their irrigation management decisions. Both surface water and ground water can be used to supply irrigation water. An assessment of the total amount of water available during an irrigation season is essential to determining the types and amounts of irrigated crops that can be grown on the farm.

The quality of some water is not suitable for irrigating crops. Irrigation water must be compatible with both the crops and soils to which it will be applied (Scherer and Weigel, 1993; Seelig and Richardson, 1991). The quality of water for irrigation purposes is generally determined by its salt content, bicarbonate concentration, and the presence of potentially toxic elements. Irrigation water can also contain appreciable amounts of nutrients that should be factored into the overall nutrient management plan.

Efficient irrigation scheduling depends upon knowledge of when water will be available to the producer. In some areas, particularly west of the Mississippi River, irrigation districts or some other outside entities may manage the distribution of water to farms, while farmers in other areas have direct access to and control over their water supplies. An irrigation district is defined as blocks of irrigated land within a defined boundary, developed or administered by a group or agency (USDA-NRCS, 1997a). Water is delivered from a source to individual turnouts via a system of canals, laterals, or pipelines. Figure 4f-5 depicts the Ainsworth Unit in northern Nebraska within which water from the Merritt Reservoir is distributed to the Ainsworth Irrigation District via the 53-mile long, concrete-lined Ainsworth Canal (Hermsmeyer, 1991). A system of laterals and drains serves approximately 35,000 acres of cropland in the irrigation district. Irrigation districts that deliver water to farms on a rotational basis control when the farmer can irrigate, leaving the farmer to choose only the rate and methods of irrigation. In cases where farmers are able to control the availability of irrigation water it is possible, however, to develop a predetermined irrigation schedule.

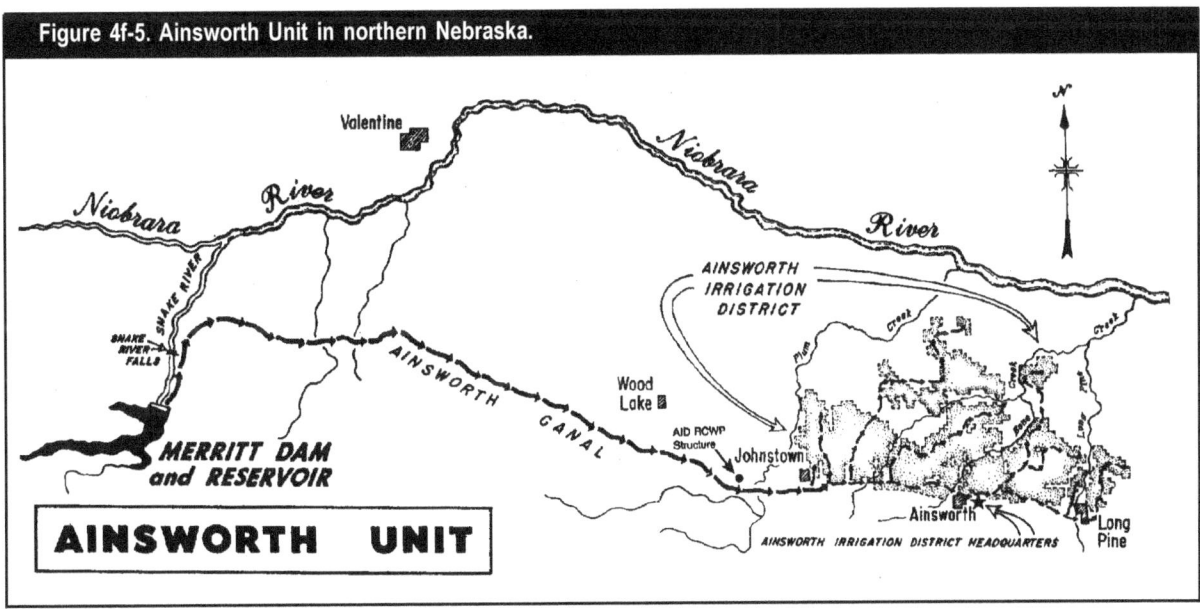

Figure 4f-5. Ainsworth Unit in northern Nebraska.

The amount of water that is needed for adequate irrigation depends upon climate and crop growth stage. Different crops require different amounts of water, and the water demand for any particular crop varies throughout the growing season. Producers need to factor the peak-use rates, the amount of water used by a crop during its period of greatest water demand (usually during period of peak growth), into both the initial design of an irrigation system and annual irrigation planning.

Irrigation methods

There are four basic *methods* of applying irrigation water: (1) surface (or flood), (2) sprinkler, (3) trickle, and (4) subsurface. Types of surface irrigation are furrow, basin, border, contour levee or contour ditch. Factors that are typically considered in selecting the appropriate irrigation method include land slope, water intake rate of the soil (i.e., how fast the soil can absorb applied water), water tolerance of the crops, and wind. For example, sprinkler, surface, or trickle methods may be used on soils (e.g., fine soils) with low water intake rates, but surface irrigation may not be appropriate for soils (e.g., coarse soils) with high water intake rates. Key factors that determine water intake rates are soil texture, surface sealing due to compaction and sodium content of the soil and/or irrigation water, and electrical conductivity of the irrigation water.

Water available to the farm from either on-site or off-site sources can be transported to fields via gravity (e.g. canals and ditches) or under pressure (pipeline). Pressure for sprinkler systems is usually provided by pumping, but gravity can be used to create pressure where sufficient elevation drops are available.

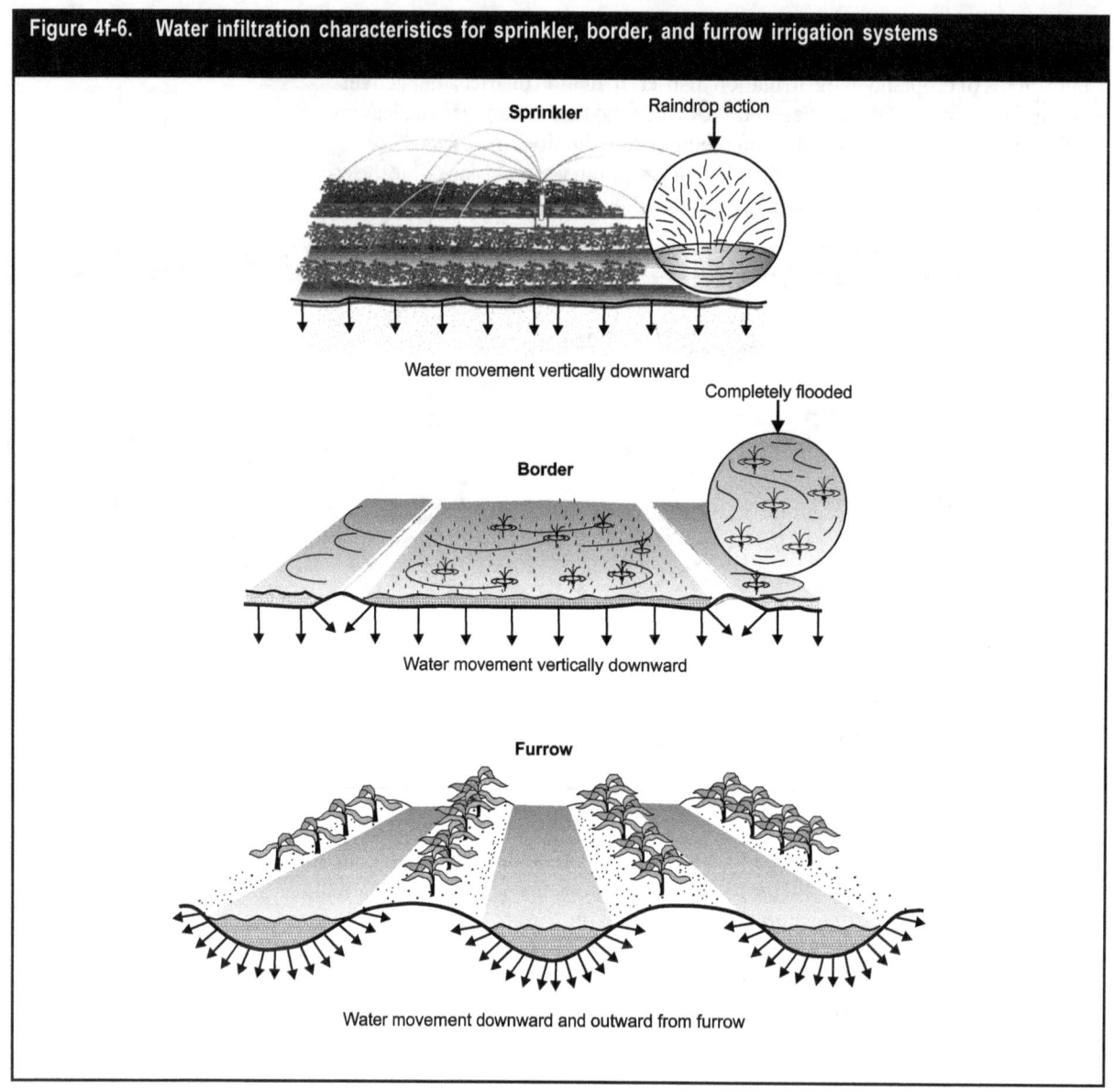

Figure 4f-6. Water infiltration characteristics for sprinkler, border, and furrow irrigation systems

Gravity-based, or surface irrigation systems, rely on the ponding of water on the surface for delivery through the soil profile (Figure 4f-6), whereas pressure-based sprinkler systems are generally operated to avoid ponding for all but very short time periods (USDA-NRCS, 1997a).

Irrigation systems

There are several irrigation *system* options for each irrigation *method* selected for the farm. The options for irrigation by gravity include level basins or borders, contour levees, level furrows, graded borders, graded furrows, and contour ditches (Figure 4f-7) (USDA-NRCS, 1997a). Pressure-based irrigation systems include periodic move, fixed or solid-set, continuous (self) move, traveling gun, and traveling boom sprinkler systems, as well as micro-irrigation and subirrigation systems. Operational modifications to center pivot and linear move systems, including Low Energy Precision Application (LEPA) and Low Pressure In Canopy (LPIC), increase the range of pressure-based options to select from (USDA-

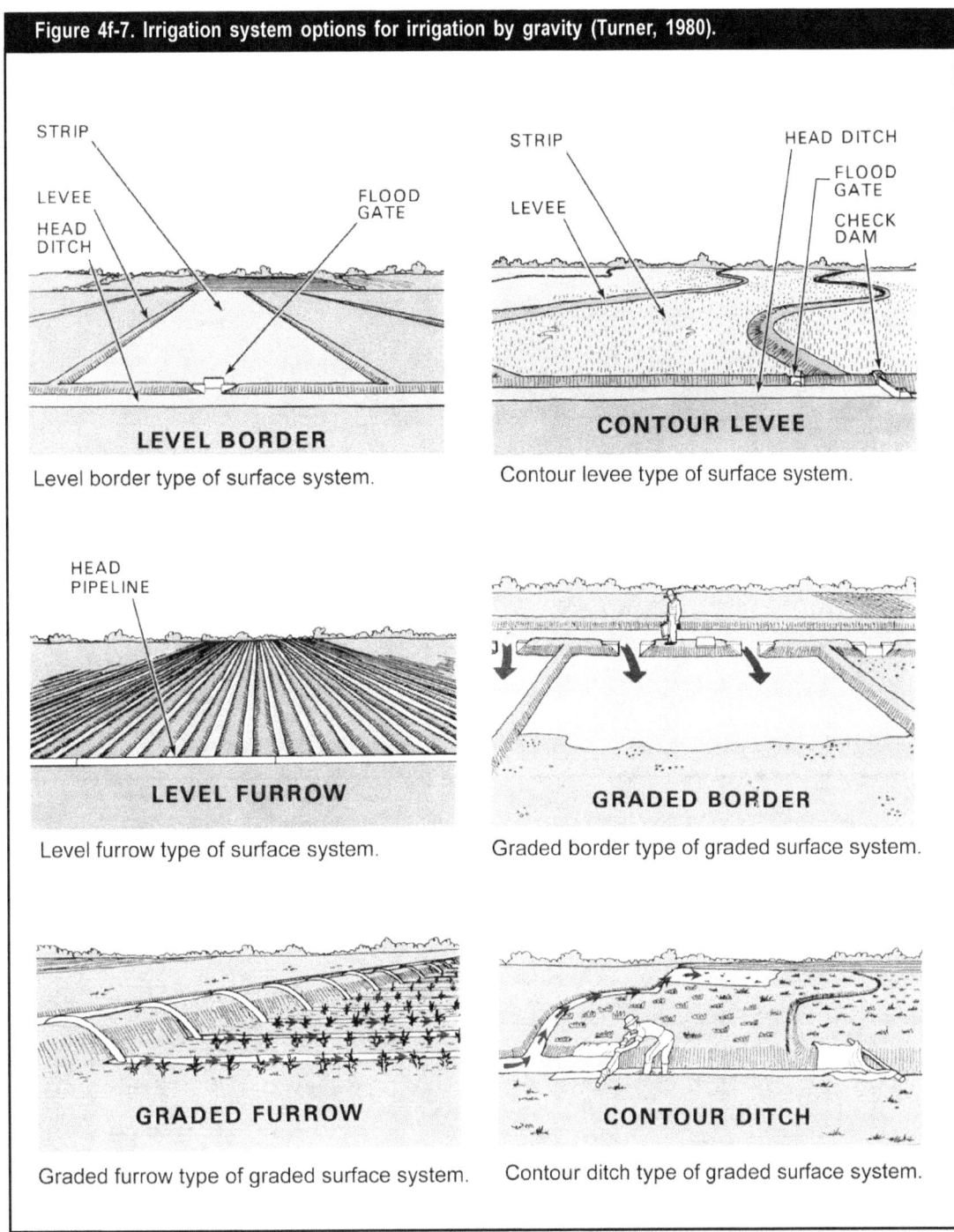

Figure 4f-7. Irrigation system options for irrigation by gravity (Turner, 1980).

LEVEL BORDER
Level border type of surface system.

CONTOUR LEVEE
Contour levee type of surface system.

LEVEL FURROW
Level furrow type of surface system.

GRADED BORDER
Graded border type of graded surface system.

GRADED FURROW
Graded furrow type of graded surface system.

CONTOUR DITCH
Contour ditch type of graded surface system.

NRCS, 1997a). Figure 4f-8 illustrates a range of sprinkler systems. Micro-irrigation systems (Figure 4f-9) include point-source emitters (drip, trickle, or bubbler emitters), surface or subsurface line-source emitters (e.g., porous tubing), basin bubblers (Figure 4f-10), and spray or mini-sprinklers. Table 4f-2 summarizes the basic features of each type of irrigation system (USDA-NRCS, 1997a), and Figure 4f-11 shows typical layouts of graded-furrow with tailwater recovery and reuse, solid-set, center pivot, traveling gun, and micro-irrigation systems (USDA-NRCS, 1997a; Turner, 1980).

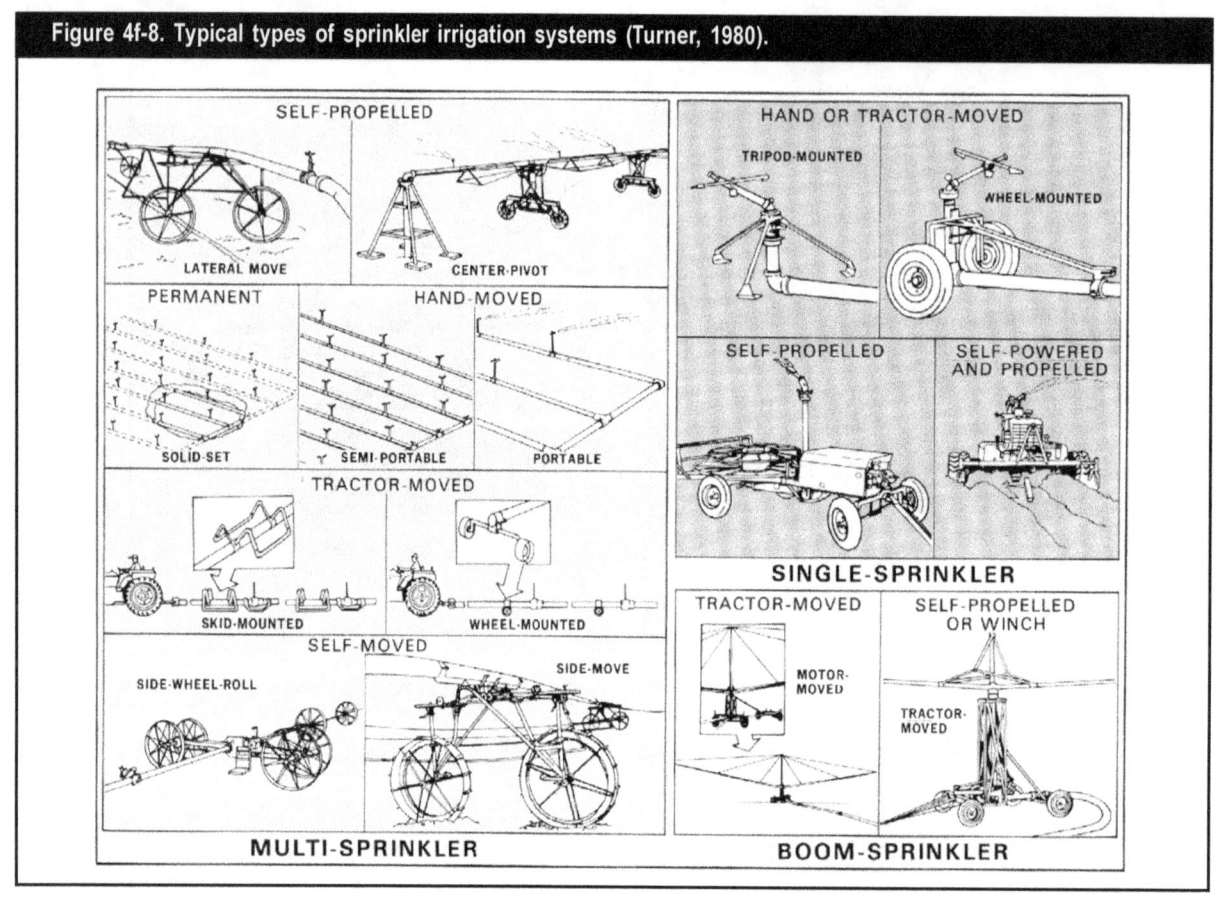

Figure 4f-8. Typical types of sprinkler irrigation systems (Turner, 1980).

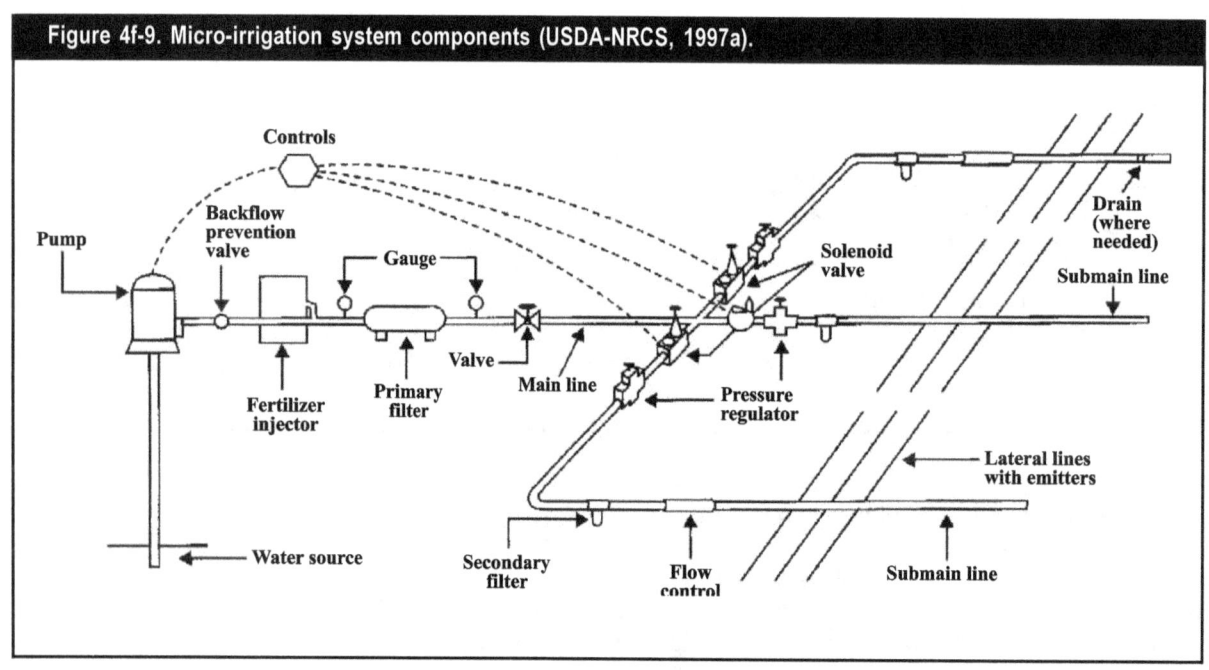

Figure 4f-9. Micro-irrigation system components (USDA-NRCS, 1997a).

The advantages and disadvantages of the various types of irrigation systems are described in a number of existing documents and manuals (USDA-NRCS, 1997a; EduSelf Multimedia Publishers Ltd., 1994).

A comprehensive set of publications, videos, interactive software, and slides on irrigation has been assembled by the U.S. Department of Agriculture to train its employees (USDA-NRCS, 1996a). This irrigation "toolbox" covers soil-water-plant relationships, irrigation systems planning and design, water measurement, irrigation scheduling, soil moisture measurement, irrigation water management planning, and irrigation system evaluation. Updated material is provided periodically as it becomes available. Other sources of material may be found in USDA-NRCS, 1997a, Sec. 652-1502.

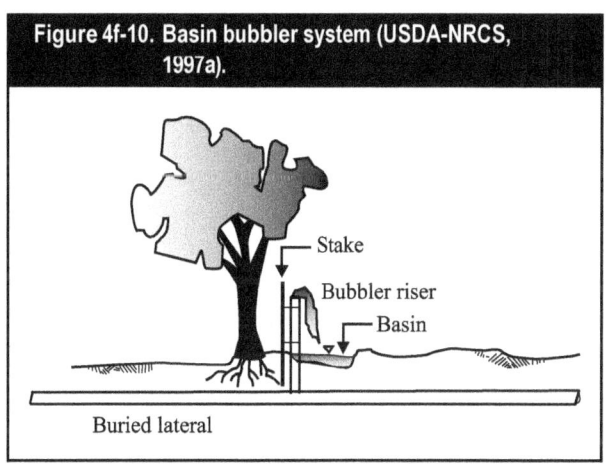

Figure 4f-10. Basin bubbler system (USDA-NRCS, 1997a).

Pollutant Transport from Irrigated Lands

Return flows, runoff, and leachate from irrigated lands may transport the following types of pollutants to surface or ground waters:

❏ Sediment and particulate organic solids;

❏ Particulate-bound nutrients, chemicals, and metals, such as phosphorus, organic nitrogen, a portion of applied pesticides, and a portion of the metals applied with some organic wastes;

❏ Soluble nutrients, such as nitrogen, soluble phosphorus, a portion of the applied pesticides, soluble metals, salts, and many other major and minor nutrients; and

❏ Bacteria, viruses, and other pathogens.

❏ If soils or drainage in the irrigated area contain toxic substances that may concentrate in the drainage or reuse system, this factor must be considered in any decisions about use of the water and design of the reuse system. Discharge of drainage water containing selenium into wetlands is an example of where this type of problem can occur.

The movement of pollutants from irrigated lands is affected by the timing and amount of applied water and precipitation; the physical, chemical, and biological characteristics of the irrigated land; the type and efficiency of the irrigation system used; crop type; the degree to which erosion and sediment control, nutrient management, and pesticide management are employed; and the management of the irrigation system.

Transport of irrigation water from the source of supply to the irrigated field via open canals and laterals can be a source of water loss if the canals and laterals are not lined. Water is also transported through the lower ends of canals and laterals as part of flow-through requirements to maintain water levels. In many soils, unlined canals and laterals lose water via evaporation and seepage in bottom and side walls. Seepage water either moves into the ground water through percolation or forms wet areas near the canal or lateral. This water will carry with it any soluble pollutants in the soil, thereby creating the potential for pollution of ground or surface water (Figure 4f-12).

Figure 4f-11. Typical irrigation system layouts (USDA-NRCS, 1997a; Turner, 1980).

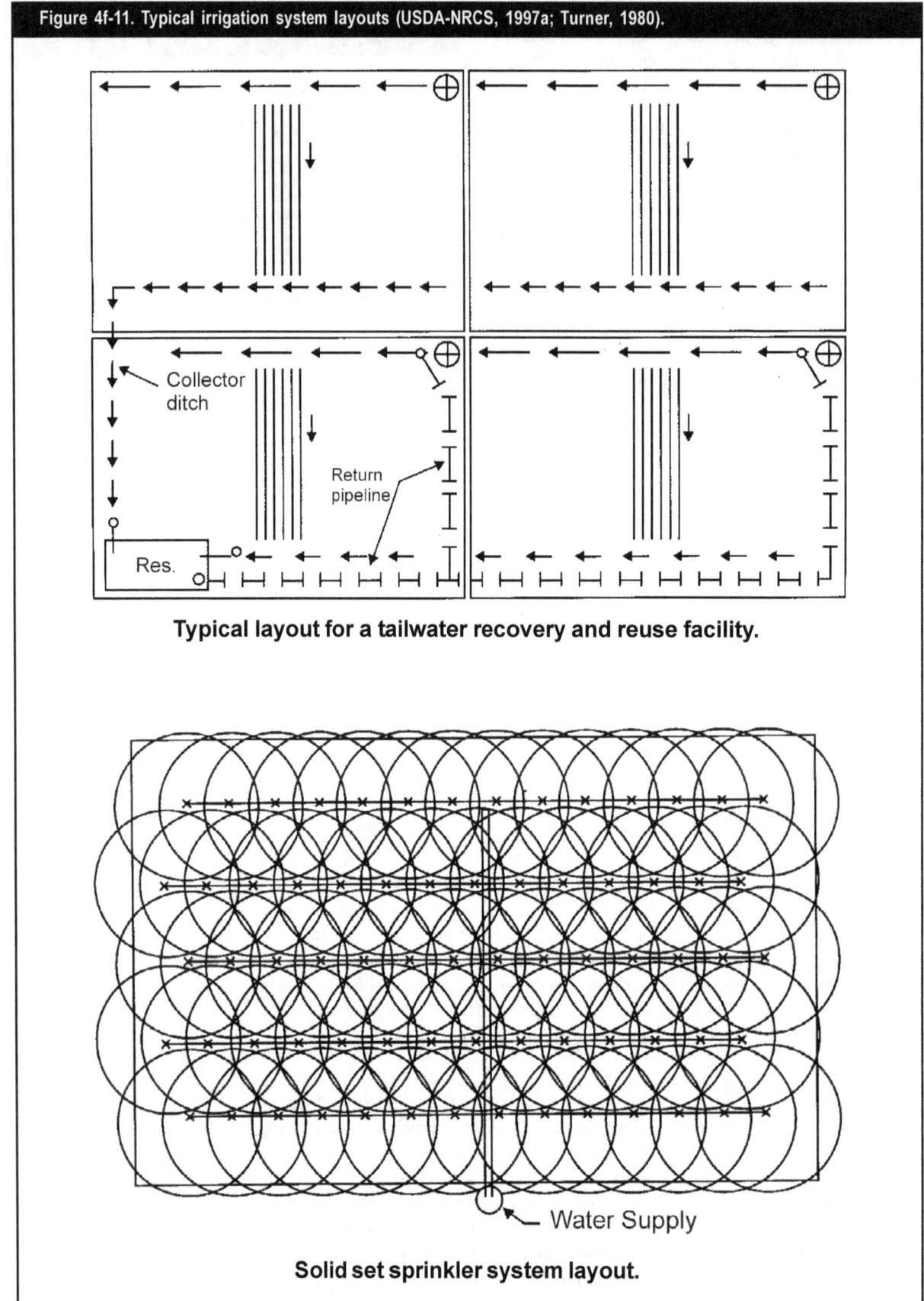

Typical layout for a tailwater recovery and reuse facility.

Solid set sprinkler system layout.

Figure 4f-11. Typical irrigation system layouts (USDA-NRCS, 1997a; Turner, 1980). Continued

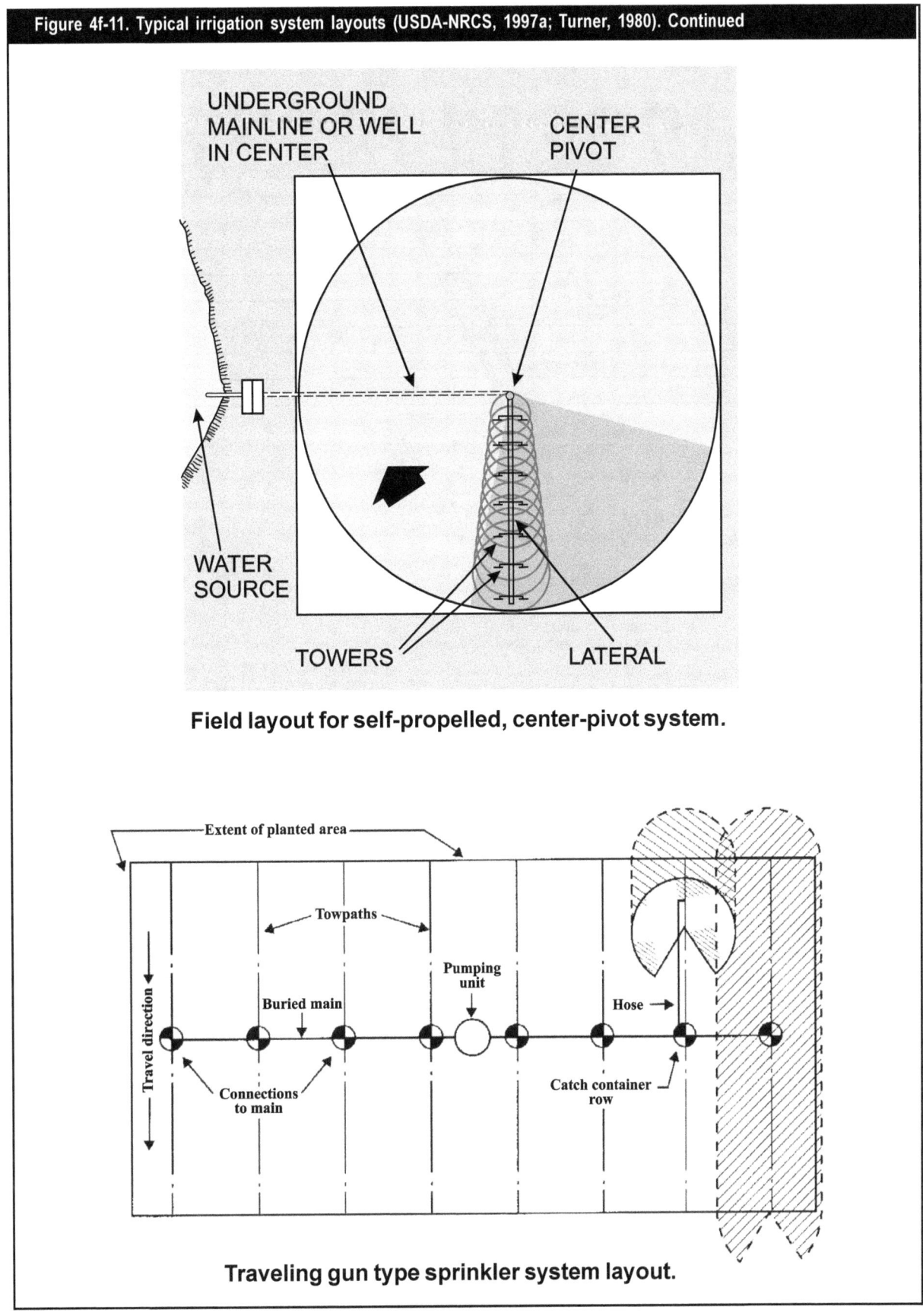

Field layout for self-propelled, center-pivot system.

Traveling gun type sprinkler system layout.

Figure 4f-11. Typical irrigation system layouts (USDA-NRCS, 1997a; Turner, 1980). Continued

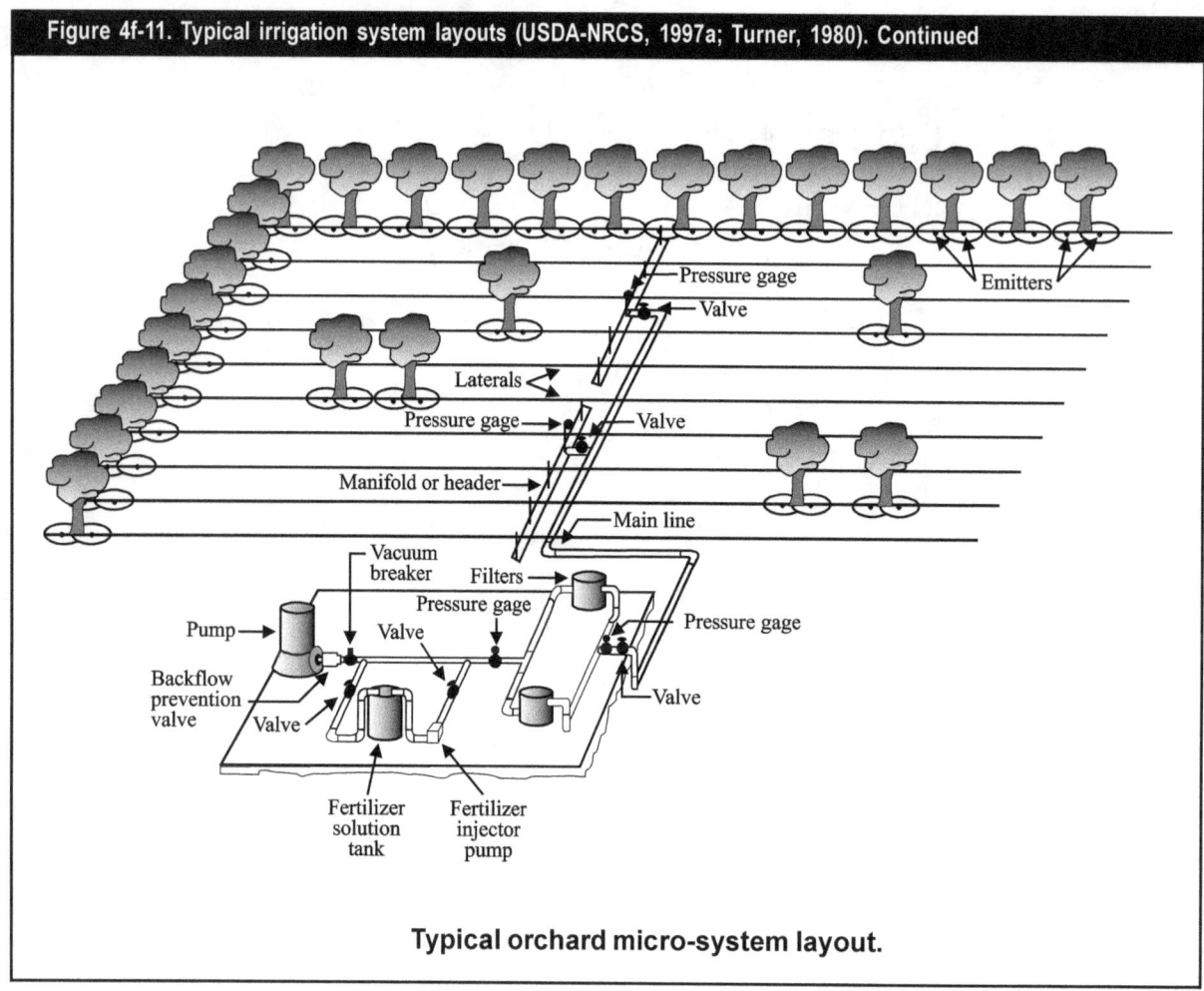

Typical orchard micro-system layout.

Table 4f-2. Types of Irrigation Systems.

Irrigation System Type	Major Features of System
Gravity-Level Basins	Large flow rates over short periods to flood entire field or basin. Level fields surrounded by low diko or levee. Best for soils with low to medium water intake rate.
Gravity-Contour Levees	Similar to level basins except for rice. Small dikes or levees constructed on contour. For rice, ponding is maintained. Best for soils with very low intake rate.
Gravity-Level Furrows	Large flow rates over short periods. Level fields. End of furrow or field is blocked to contain water. Best for soils with moderate to low water intake rate and moderate to high available water capacity.
Gravity-Graded Borders	Controlled surface flooding. Field divided into strips bordered by parallel dikes or border ridges. Water introduced at upper end.
Gravity-Graded Furrows	Like graded borders, but only furrows are covered with water. Water distribution via vertical and lateral infiltration. Water application amount is a function of intake rate of soil, spacing of furrows, and length of field. Heavy soils (small pores sizes) provide slower infiltration and greater lateral movement.
Gravity-Contour Ditches	Controlled surface flooding. Water discharged with siphon tubes, over ditch banks, or from gated pipes located upgradient and positioned across the slope on contour. Sheet flow is goal.
Pressure-Periodic Move Sprinkler	Sprinkler is operated in a fixed location for a specified period of time, then moved to the next location. Many design options including hand-moved laterals, side-roll laterals, end-tow laterals, hose-fed (pull) laterals, guns, booms, and perforated pipe.
Pressure- Fixed or Solid-Set Sprinkler	Laterals are not moved, but one or more sections of sprinklers are cycled on and off to provide coverage of entire field over time.
Pressure-Continous Move Sprinkler	Center pivot (irrigates in circular patterns, or rectangular with end guns or swing lines) or linear (straight lateral irrigates in rectangular patterns) move continuously to irrigated field. Multiple sprinklers located along the laterals.
Pressure-Traveling Gun Sprinkler	High-capacity, single-nozzle sprinkler fed by flexible hose. Hose is dragged or on a reel. Gun is guided by cable, and moved from field to field. Best for soils with high water intake rates.
Pressure-Traveling Boom Sprinkler	Similar to traveling gun, except a boom with several nozzles is used.
Micro/Pressure-Point Source Emitters	Frequent, low-volume, low-pressure applications through small tubes and drop, trickle, or bubbler emitters. Water must be filtered. Used for orchards, vineyards, ornamental landscaping. Emitters discharge from 0.5 to 30 gallons per hour.
Micro/Pressure-Line Source Emitters	Frequent, low-volume, low-pressure applications through surface or buried tubing that is porous or has uniformly spaced emitter points. For permanent crops, but also vegetables, cotton, melons.
Micro/Pressure-Basin Bubblers	Water applied via risers into small basins adjacent to plant. Bubblers discharge less than 60 gallons per hour. Water filtration not required. Orchards and vineyards. Best for medium to fine textured soils.
Micro/Pressure-Spray or Mini-Sprinklers	Water applied as spray droplets from small, low-pressure heads. Wets a greater area (2 to 7 feet in diameter) than drop emitters. Discharges less than 30 gallons per hour.
Subirrigation	Manage water table by providing subsurface drainage, providing controlled drainage, and irrigating via buried laterals.

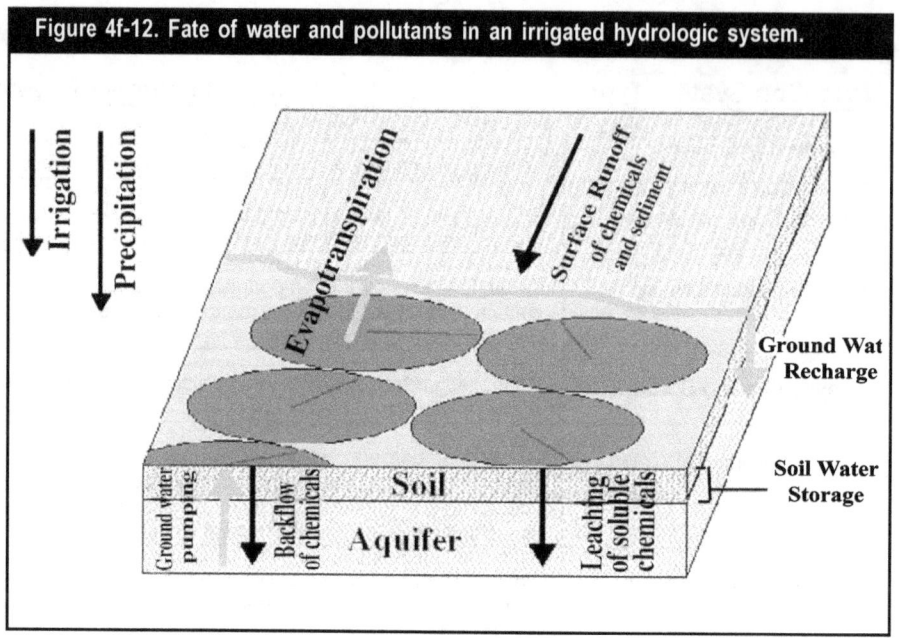

Figure 4f-12. Fate of water and pollutants in an irrigated hydrologic system.

Since irrigation is a consumptive use of water, any pollutants in the source waters that are not consumed by the crop (e.g., salts, pesticides, nutrients) can be concentrated in the soil, concentrated in the leachate or seepage, or concentrated in the runoff or return flow from the system. Salts that concentrate in the soil profile must be managed in order to sustain crop production. In such cases, a carefully calculated additional amount of water may be applied to leach the salts below the root zone. The application of this "leaching requirement" should be timed to prevent the leaching of other potential pollutants when possible (e.g., after the growing season when nutrients are low, or after a cover crop that has used excess nutrients).

Irrigation Scheduling

Both long-term and short-term irrigation decisions must be made by the producer. Long-term decisions, which are associated with system design and the allocation of limited seasonal water supplies among crops, rely on average water use determined from historical data (Duke, 1987) and average water availability. Particularly in arid areas, long-term irrigation decisions are needed to determine seasonal water requirements of different possible crops, determine which crops to grow based upon crop adaptability and water availability, and in some cases to determine when and how much to stress the various crops to maximize economic return. Short-term decisions determine when and how much to irrigate, and are based upon daily water use. In areas where rainfall is either insignificant or falls predictably during the growing season, long-term decisions can be used to construct an irrigation schedule at the beginning of the growing season (Duke, 1987), although better water management is obtained by constant updating of information. In semi-arid and humid areas where weather varies significantly on a daily basis, short-term irrigation decisions are used in lieu of pre-determined irrigation schedules. The emphasis of this guidance is placed on short-term irrigation decisions.

Irrigation scheduling is the use of water management strategies to prevent over-application of water while minimizing yield loss from water shortage or drought stress (Evans et al., 1991c). Irrigation scheduling will ensure that water is applied to the crop when needed and in the amount needed (USDA-NRCS, 1997a). Effective scheduling requires knowledge of the following factors (Evans et al., 1991b; Evans et al., 1991c):

- ❏ Soil properties
- ❏ Soil variability within the field
- ❏ Soil-water relationships and status
- ❏ Type of crop and its sensitivity to drought stress
- ❏ The stage of crop development and associated water use
- ❏ The status of crop stress
- ❏ The potential yield reduction if the crop remains in a stressed condition
- ❏ Availability of a water supply
- ❏ Climatic factors such as rainfall and temperature

Much of the above information can be found in Natural Resources Conservation Service soil surveys and Extension literature. However, all information should be site-specific and verified in the field.

In environments where salts tend to concentrate in the soil profile, additional information is needed to sustain crop production, including:

- ❏ Salt tolerance of the crop
- ❏ Salinity of the soil
- ❏ Salinity of the irrigation water
- ❏ Leaching requirement of the soil

Deciding when to irrigate

There are three ways to determine when irrigation is needed (Evans et al., 1991c):

- ❏ Measuring soil water
- ❏ Estimating soil water using an accounting approach
- ❏ Measuring crop stress

Research in irrigation scheduling indicates the need for specific site-dependent data for plan development.

Soil water can be measured directly by sampling the soil and determining the water content through gravimetric analysis. The distribution of plant roots and their pattern of development during the growing season are very important considerations in deciding where and at what depth to take soil samples to determine soil water content (USDA-NRCS, 1997a). For example, all plants have very shallow roots early in their development, and the concentration of moisture-absorbing roots of most plants is usually greatest in the upper quarter of the root zone. Further, since roots will not grow into a dry soil, it may be important to measure soil moisture beyond the current root zone to determine irrigation needs associated with full root development. Figure 4f-13 illustrates the typical water extraction pattern in a uniform soil, again pointing out the need to relate soil sampling decisions to crop development.

Soil moisture can also be determined indirectly using a range of devices (Evans et al., 1991a; Werner, 1992), including tensiometers (Figure 4f-14), electrical resistance blocks (Figure 4f-14), neutron probes, heat dissipation sensors, time domain reflectometers, and carbide soil moisture testers (USDA-NRCS, 1997a). Table 4f-3

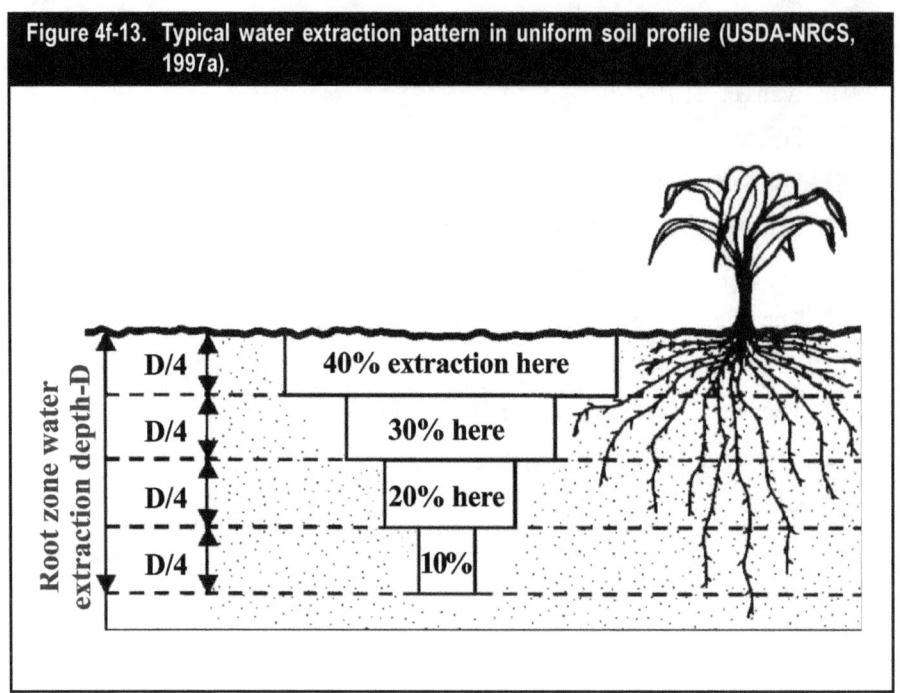

Figure 4f-13. Typical water extraction pattern in uniform soil profile (USDA-NRCS, 1997a).

Root zone water extraction depth-D

D/4   40% extraction here

D/4   30% here

D/4   20% here

D/4   10%

Figure 4f-14. Soil moisture measurement devices: (a) tensiometer and (b) electrical resistance block.

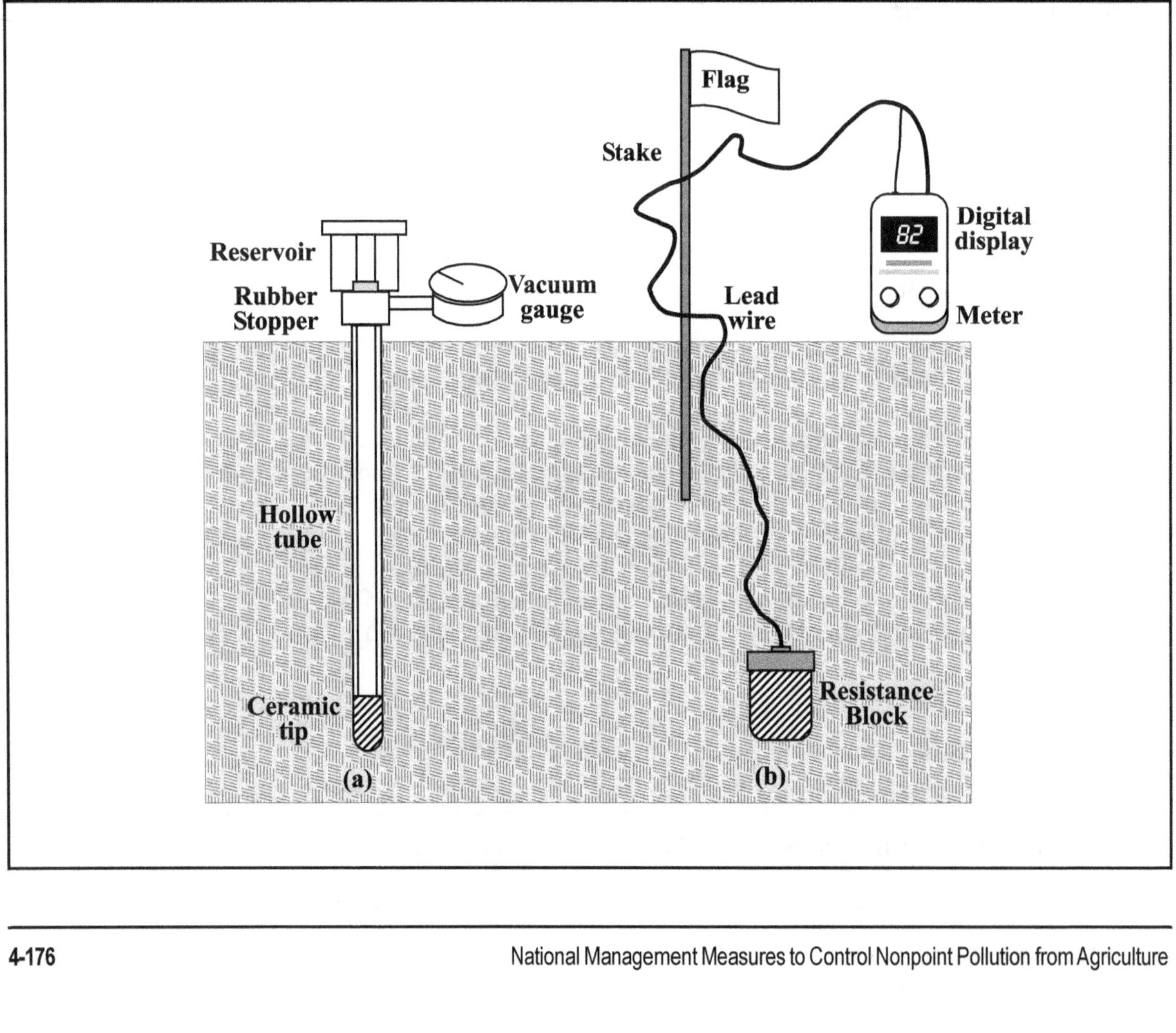

Flag

Stake

Reservoir

Rubber Stopper

Vacuum gauge

Lead wire

Digital display

82

Meter

Hollow tube

Ceramic tip

(a)

Resistance Block

(b)

Table 4f-3. Devices and methods to measure soil moisture.		
Device (Other Names)	**How It Works**	**Comments**
Tensiometer	Measures soil suction which is related to soil water content.	Available in lengths from 6 to 72 inches. Requires careful installation and field maintenance. Most applicable when soil moisture is between 50-75 percent of field capacity, and on medium to fine-textured soils with frequent irrigation.
Electrical Resistance Block (Gypsum or Moisture or Porous Block)	Measures electrical resistance which is related to soil water content via a calibration curve.	Inexpensive. Simple to use. Gives accurate readings over wider moisture range than tensiometers, but limited to medium to coarse-textured soils. Most accurate when soil moisture is below field capacity. Sodic soils problematic. Gypsum blocks need replacement each growing season; nylon, plastic, fiberglass more durable.
Neutron Probe (Neutron Scattering)	Measures thermalized neutrons (fast neutrons that are slowed by collisions with hydrogen molecules in water) which are related to volumetric soil water content by a calibration curve.	Can be most accurate and precise method. Requires calibration using gravimetric procedures, especially if used for top 6 inches of soil profile, in clay soils, soils with high organic matter content, and soils with boron ions. Requires licensed operator since radioactive. Expensive.
Thermal Dissipation Block (Heat Dissipation Sensor)	Estimates soil water based upon the relationship between heat conductance and soil water content.	Requires calibration. Work across wide range of soil-water content.
Time Domain Reflectometer (TDR) & Frequency Domain Reflectometer (FDR) (Dialectric Constant Method)	Senses the dielectric property of soil which is related to water content.	Requires careful installation. TDR works across wide range of soil texture, bulk density, and salinity. FDR results may be skewed as salinity increases.
Carbide Soil Moisture Tester (Speedy Moisture Tester)	Measures gas pressure from reaction of calcium carbide with water in soil sample.	Provides percent water content of soil. Works in field. Practice necessary for reliable results.
Feel and Appearance Method	Soil samples are compared to tables or pictures that give moisture characteristics of different soil textures.	Experienced individuals can estimate soil moisture within 10 percent of true value, but tables and pictures use ranges of 25 percent.
Gravimetric Method (Oven Dry)	Soil samples from field are weighed, dried, and weighed again in the lab.	Accurate measure of water content. Requires sensitive scales, drying method, and known or estimated bulk density value to calculate % volume of water.

provides an overview of these devices. The appropriate device for any given situation is a function of the acreage of irrigated land, soils, cost, available trained labor, and other site-specific factors.

Direct measurement of soil water status or crop status is always more accurate than estimating its magnitude, but because of the cost associated with obtaining representative samples in some situations, it may be more appropriate to use estimation techniques (Duke, 1987). Accounting approaches estimate the quantity of plant-available water remaining in the effective root zone. A variety of methods can be used to estimate and predict the root zone water balance, including a simple checkbook method (USDA-NRCS, 1997a), computer-assisted methods (Hill, 1997 and Allen, 1991), graphical methods (Figure 4f-15), and tabular methods. In essence, these methods begin with an estimate of initial soil-water depletion and use measurements or estimates of daily water inputs (rain, irrigation) and outputs (evapotranspiration) to determine the current soil-water depletion volume (Equation 4f-1).

Net irrigation depth is the depth of water applied multiplied by the irrigation efficiency, which ranges from 75-100% for drip systems to 20-60% for furrow irrigation on sandy soils (Duke, 1987). Effective precipitation is the amount of precipitation minus losses due to runoff or unnecessary deep percolation. At some pre-determined moisture deficit (e.g., the MAD value), irrigation must be started (Figure 4f-15). The water balance must be updated at least weekly, including field checks on estimated parameters, to be useful for scheduling irrigations (Duke, 1987).

Potential sources of data for Equation 4f-1 include field measurements to determine the initial soil-water content, field measurements to determine effective rooting depth as the plant matures, ET measurements or estimates based upon data from weather stations, irrigation depth measurements, measured precipitation,

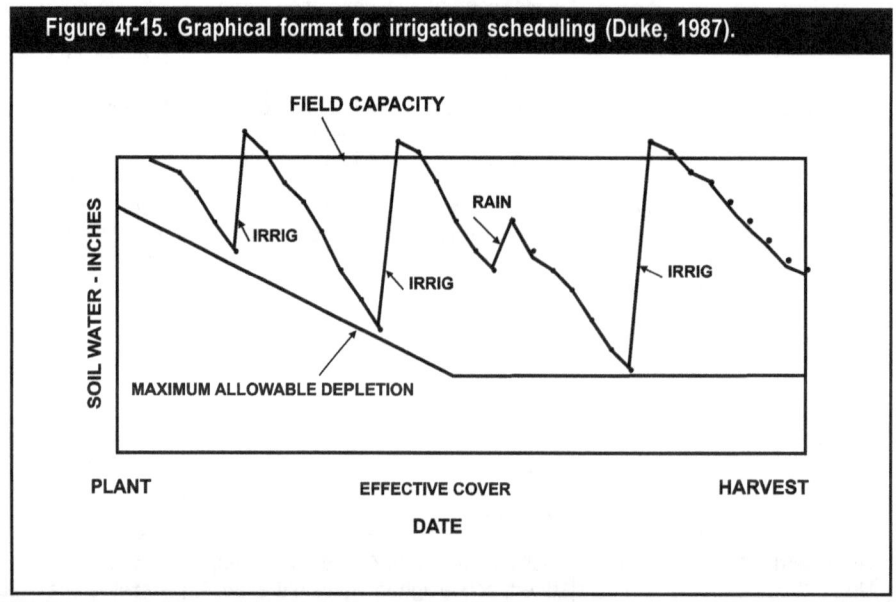

Figure 4f-15. Graphical format for irrigation scheduling (Duke, 1987).

Equation 4f-1. Soil-water depletion volume (Duke, 1987).

$$D = D_0 + ET - IR - R - WT$$

where D = soil-water depletion at end of day (D=0 at field capacity)

D_0 = soil-water depletion for previous day

ET = ET for the day

IR = net irrigation depth (depth of applied water which is stored in soil root zone) for the day

R = effective precipitation during the day

WT = upward movement of water during the day from water table close to bottom of root zone

If the water table is not near the root zone, the last term (WT) may be dropped.

and estimates of water table contributions. Clearly, good estimates or measurements of ET are essential to successful accounting approaches since crop water use can vary considerably with crop type, stage of growth, temperature, sunshine, wind speed, relative humidity, and soil moisture content (Figure 4f-16). Direct measurement of ET with lysimeters may not be practical for most farms, but evaporation pans and atmometers can be used effectively. There is also, however, a wide range of computational techniques for estimating ET from weather data (Doorenbos and Pruitt, 1975; Jensen et al., 1990; USDA-SCS, 1993). Crop ET data are often available in newspapers, through telephone dial-up service, or on television, and some farms have on-site weather stations that provide the necessary ET data (USDA-NRCS, 1997a). There is also a growing number of computer programs that aid the irrigation decisionmaker, including the NRCS Scheduler (Figure 4f-17) and others (Smith, 1992; Allen, 1991; and Hill, 1991).

Figure 4f-16. Crop water use for corn, wheat, soybean, and potato based on average climatic conditions for North Dakota (Lundstrom and Stegman, 1991).

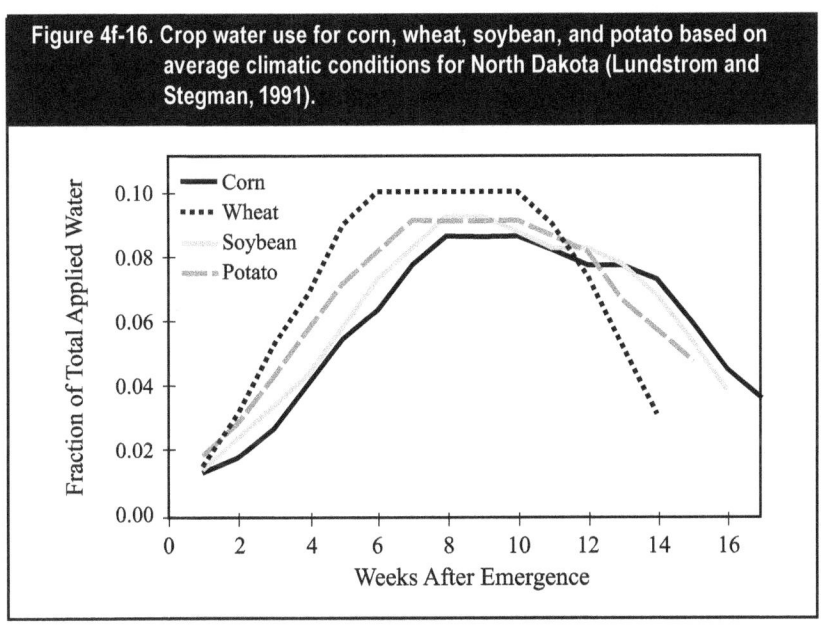

Figure 4f-17. NRCS (SCS) Scheduler – seasonal crop ET (USDA-NRCS, 1997a).

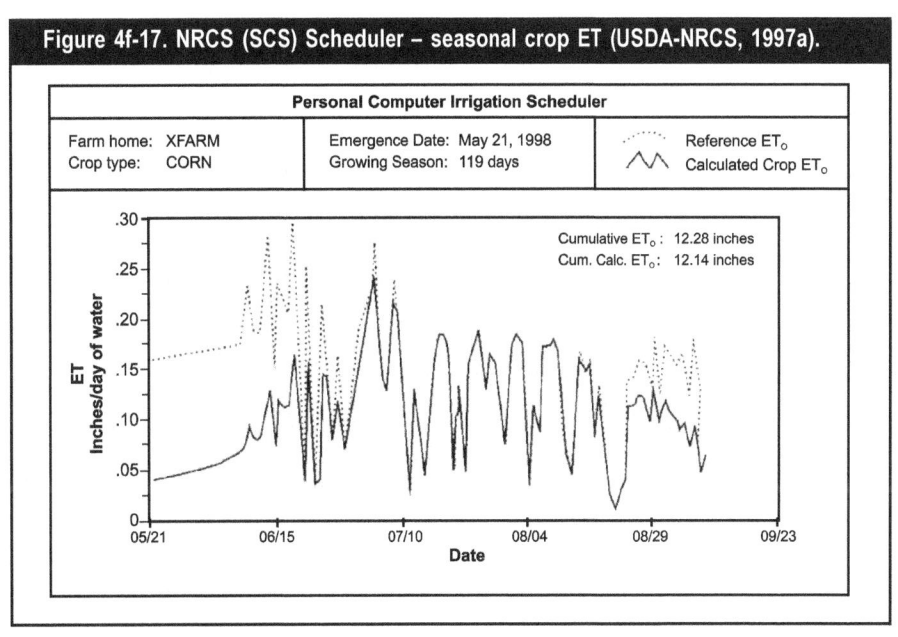

Measuring crop stress is another way to determine when irrigation is needed. Unavailability of water during crop stress periods could result in crop failure or reduced yields that leave unused nutrients vulnerable to runoff and deep percolation. Devices and methods used to measure crop stress include the crop water stress gun, leaf moisture stress as measured in a pressure chamber, and infrared photography (USDA-NRCS, 1997a). However, infrared photography is typically not an option for "real time" water management due to slow turnaround times. The crop water stress gun calculates plant water stress and expresses it as an index value based on measurements of plant canopy temperature, ambient air temperature, relative humidity, and a range of solar radiation. Using a crop water stress index, irrigation can be scheduled depending on the severity of moisture stress. Threshold values must be developed for each crop.

Deciding how much water to apply

Once the decision to irrigate has been made, the amount of water to apply must be determined. A decision rule should be established to determine how much water to apply, with the basic choices being full irrigation to replenish the root zone to field capacity or partial irrigation. Partial irrigation, which is more easily achieved via sprinkler systems, may be preferred if there is opportunity for rainfall to provide some of the water needed to reach field capacity.

Factors in determining the amount of irrigation water to apply include the soil-water depletion volume in the effective root zone and local weather forecasts for rain. The application rate should not exceed the water intake rate of the soil when using sprinkler systems, and the application depth should not exceed the soil-water depletion volume, except as necessary for leaching of salts (Duke, 1987). Local weather forecasts for rain should be considered before irrigating to avoid over-application.

The relationship between irrigation system capacity, irrigated area, and time of irrigation may be expressed as

$$Q = \frac{453\,Ad}{fT}$$

where Q is system discharge capacity (gpm), A is irrigated area (acres), d is gross application depth (in), f is time allowed for completion of one irrigation (days), and T is actual operating time (hr/day) (USDA, 1983). Normally A, T, and d are fixed in a design process. The time allowed for completion of one irrigation should be set to insure that the area initially irrigated does not become stressed before the next irrigation is applied. Note that a system design that just meets the peak crop water demand may be determined as illustrated in Table 4f-4. Partial irrigations may facilitate covering a larger area to prevent immediate crop damage, but they increase the frequency of irrigation necessary, and could impede root growth or harm a crop that will be stressed if the soil is not adequately saturated.

Deep percolation of irrigation water can be greatly reduced by limiting the amount of applied water to the amount that can be stored in the plant root zone. The deep percolation that is necessary for salt management can be accomplished with a sprinkler system by using longer sets or very slow pivot speeds or by applying water during the non-growing season. Salt management by surface irrigation methods is much less efficient than other irrigation methods, and water used to leach salts should be applied when nutrients or pesticides are least vulnerable to

Table 4f-4.	System capacity needed in gal/min-acre for different soil textures and crops to supply sufficient water in 9 out of 10 years (Scherer, 1994).						

Crop	Root Zone Depth (ft)	Coarse Sand and Gravel	Sand	Loamy Sand	Sandy Loam	Fine Sandy Loam	Loam and Silt Loam
Potatoes[a]	2.0[b]	8.2	7.5	7.0	6.4	6.1	5.7
Dry beans	2.0	7.9	7.1	6.4	6.1	5.7	5.4
Soybeans	2.0	7.9	7.1	6.4	6.1	5.7	5.4
Corn	3.0	7.3	6.6	5.9	5.5	5.3	4.9
Sugarbeets	3.0	7.3	6.6	5.9	5.5	5.3	4.9
Small grains	3.0	7.3	6.6	5.9	5.5	5.3	4.9
Alfalfa	4.0	6.8	5.9	5.6	5.1	5.0	4.5

[a] Adjusted for 40% depletion of available water.
[b] An application efficiency of 80% and a 50% depletion of available soil water were used for calculations.

leaching, such as when maximum uptake or dissipation of the chemical has occurred.

Accurate measurements of the amount of water applied are essential to maximizing irrigation efficiency. A wide range of water measurement devices is available (USDA-NRCS, 1997a). For example, the quantity of water applied can be measured by such devices as a totalizing flow meter that is installed in the delivery pipe or calibrated canal gates. If water is supplied by ditch or canal, weirs or flumes in the ditch can be used to measure the rate of flow. Rain gauges should also be used in the field to determine the quantity of water added through rainfall. Such gauges are also a valuable tool for checking uniformity of application of sprinkler systems.

Efficient Transport and Application of Irrigation Water

There are several measures of irrigation efficiency, including conveyance efficiency (Table 4f-5), irrigation efficiency, application efficiency, project application efficiency, potential or design application efficiency, uniformity of application, distribution uniformity, and Christiansen's uniformity (USDA-NRCS, 1997a). Project water conveyance and control facility losses can be as high as 50% or more in long, unlined, open channels in alluvial soils (USDA-NRCS, 1997a). Seepage losses associated with canals and laterals can be reduced by lining them, or can be eliminated by conversion from open canals and laterals to pipelines. Flow-through losses or spill, however, will not be changed by lining canals and laterals, but can be eliminated or greatly reduced by conversion to pipelines or through changes in operation and management. Flow-through water constitutes over 30% of canal capacity in some water districts, but simple automatic gate/valve control devices can limit flow-through water to less than 5% (USDA-NRCS, 1997a). Conversion to pipelines may in some cases cause impacts to wildlife due to loss of beneficial wet areas, and an environmental assessment or environmental impact statement may be needed before the conversion is made (USDA-NRCS, 1997a).

Table 4f-5. Measures of irrigation efficiency.

Measure of Irrigation Efficiency	Definition
Conveyance Efficiency (to farm)	$\dfrac{W_{Delivered}}{W_{Diverted}} *100$
Irrigation Efficiency (on farm)	$\dfrac{W_{Beneficial}}{W_{Applied}} *100$
Application Efficiency (on farm)	$\dfrac{W_{Stored}}{W_{Applied}} *100$
Project Application Efficiency (to and on farm)	$\dfrac{W_{Stored}}{W_{Diverted}} *100$

Where
$W_{delivered}$ = Water delivered
$W_{diverted}$ = Total water diverted or pumped into an open channel or pipeline at upstream end
$W_{beneficial}$ = Avg. depth of water beneficially used
$W_{applied}$ = Avg. depth of applied water
W_{stored} = Avg. depth of water infiltrated and stored in the plant root zone

Water application efficiency can vary considerably by method of application. Increased application efficiency reduces erosion, deep percolation, and return flows. In general, trickle and sprinkler application methods are more efficient than surface and subsurface methods. Two major hydraulic distinctions between surface irrigation methods and sprinkler and micro irrigation are key to this difference in efficiencies (Burt, 1995):

1. The soil surface conveys the water along border strips or furrows in surface irrigation, whereas the water infiltrates into the soil very near to the point of delivery from sprinkler and micro irrigation systems.

2. Water application rate exceeds soil water infiltration rate in surface irrigation, and the soil controls the amount of water that will infiltrate. In properly designed and managed sprinkler and micro irrigation systems, the application rate is equal to the soil water infiltration rate.

The type of irrigation system used will dictate which practices can be employed to improve water use efficiency and to obtain the most benefit from scheduling. Flood systems will generally infiltrate more water at the upper end of the field than at the lower end because water is applied to the upper end of the field first and remains on that portion of the field longer. This will cause the upper end of the field to have greater deep percolation losses than the lower end. This situation can sometimes be improved by changing slope throughout the length of the field or shortening the length of run. For example, furrow length can be reduced by cutting the field in half and applying water in the middle of the field. This will require more pipe or ditches to distribute the water across the middle of the field. Other methods used to improve application efficiency in surface systems are surge and cut-back irrigation. In surge irrigation, flow is pulsed into the furrow allowing for wet and dry cycles, while in cut-back irrigation, the furrow inflow rate is reduced after a period of time. Both of these methods improve irrigation efficiency by allowing for a more uniform time of infiltration. A wide range of options exist for manipulating field lengths, slopes, flow rate, irrigation time, and other management variables to increase surface irrigation efficiency (Burt, 1995; USDA-NRCS, 1997a).

A properly designed, operated, and maintained sprinkler irrigation system should have a uniform distribution pattern. The volume of water applied can be changed by altering the total time the sprinkler runs; by altering the pressure at which the sprinkler operates; or, in the case of a center pivot, by adjusting the speed of travel of the system. There should be no irrigation runoff or tailwater from most well-designed and well-operated sprinkler systems (USDA-NRCS, 1997a). Operating outside of design pressures and using worn equipment can greatly affect irrigation uniformity.

Use of Runoff or Tailwater

Surface irrigation systems are usually designed to have a percentage (up to 30%) of the applied water lost as tailwater. The volume and peak runoff rate of tailwater will depend upon both the irrigation method and its management. Tailwater recovery and reuse facilities collect irrigation runoff and return it to the same, adjacent, or lower fields for irrigation use (USDA-NRCS, 1997a). If the water is pumped to a field at higher elevation, the facility is a return-flow or pumpback facility. Sequence-use facilities deliver the water to adjacent or lower-elevation fields. Those facilities that store runoff and precipitation for later use are reservoir systems, while cycling-sump facilities have limited storage and pump the water automatically to irrigate fields.

The components of a tailwater reuse or pumpback facility include tailwater collection ditches to collect the runoff; drainageways, waterways, or pipelines to convey the water to a central collection area; a sump (cycling-sump facilities) or reservoir (reservoir systems); a pump and power unit for pumpback facilities; and pipelines or ditches to deliver the recovered water (USDA-NRCS, 1997a). A typical pumpback facility plan is illustrated in Figure 4f-18. For new facilities, runoff flows must be measured or estimated to properly size tailwater reuse sumps, reservoirs, and pumping facilities. Capacity should be provided to handle concurrent peak runoff events from both precipitation and tailwater, unexpected interruption of power, and other uncertainties.

Tailwater management is needed to reduce the discharge of pollutants such as suspended sediment and farm chemicals which can be found in the runoff. In reservoir systems, tailwater is typically stored until it can be either pumped back to the head of the field and reused or delivered to additional irrigated land. The quality of tailwater, including nutrient concentrations, should be considered in reuse systems. Water quality testing may be necessary. In some locations, there may be downstream water rights that are dependent upon tailwater, or tailwater

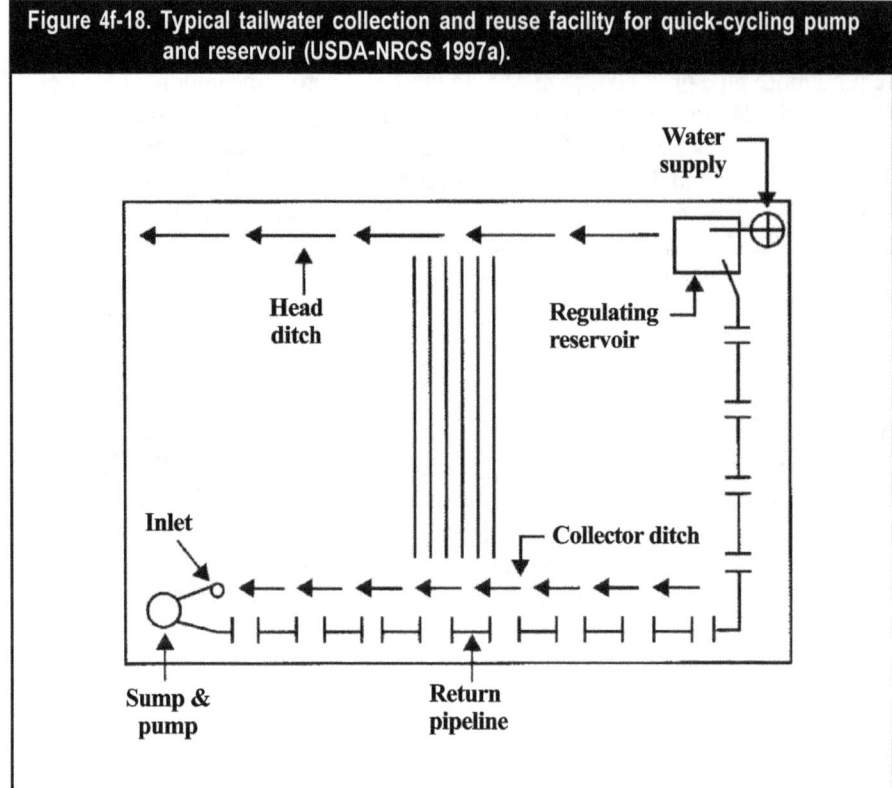

Figure 4f-18. Typical tailwater collection and reuse facility for quick-cycling pump and reservoir (USDA-NRCS 1997a).

may be used to maintain flow in streams. These requirements may take legal precedence over the reuse of tailwater.

If a tailwater recovery system is used, it should be designed to allow storm runoff to flow through the system without damage. Where reservoir systems are used, storm runoff containing a large sediment volume should bypass or be trapped before entering the storage reservoir to prevent rapid loss of storage capacity (USDA-NRCS, 1997a). Additional surface drainage structures such as filter strips, field drainage ditches, subsurface drains, and water table control may also be used to control runoff and leachate if site conditions warrant their use.

Management of Drainage Water

Drainage of agricultural lands is intended to control and manage soil moisture in the crop root zone, provide for improved soil conditions, and improve plant root development (USDA-NRCS, 1997a). In cases where the water table impinges upon the root zone, water table control is an essential element of irrigation water management. However, installation of subsurface drainage should only be considered when good irrigation water management, good nutrient management, and good pesticide management are being conducted. Further, impacts to wetlands, wildlife habitat, and water quality must be thoroughly investigated, and relevant federal, state, and local laws fully considered prior to installation of drainage practices.

Drainage increases water infiltration, which reduces soil erosion and also allows application of excess water to keep salts leached below the root zone. Drainage also provides more available soil moisture and plant food by increasing the depth of the root zone. Subsurface drainage may concentrate soluble nutrients in irrigation return flows. Properly installed subsurface drainage systems can be used successfully as a supplemental source of irrigation water if the water is of good quality (USDA-NRCS, 1997a).

Irrigation Water Management Practices and Their Effectiveness

The practices that can be used to implement this management measure on a given site are commonly used and are recommended by NRCS for general use on irrigated lands. Many of the practices that can be used to implement this measure (e.g., water-measuring devices, tailwater recovery systems, and backflow preventers) may already be required by State or local rules or may otherwise be in use on irrigated fields.

The NRCS practice number and definition are provided for each management practice, where available. Additional information about the purpose and function of individual practices is presented in Appendix A. Another useful reference is "Irrigation Management Practices to Protect Ground Water and Surface Water Quality–State of Washington" (WSU Cooperative Extension, 1995).

Irrigation Scheduling Practices

Proper irrigation scheduling is a key element in irrigation water management. Irrigation scheduling should be based on knowing the daily water use of the crop, the water-holding capacity of the soil, and the lower limit of soil moisture for each crop and soil, and measuring the amount of water applied to the field. Also, natural precipitation should be considered and adjustments made in the scheduled irrigations.

Daily accounting for the cropland field water budget helps determine irrigation scheduling.

Whether the irrigation source is surface or ground water, water availability during the growing season should be adequate to support the most water sensitive crop in the rotation. The design capacity of the irrigation system depends on regional climate, irrigation efficiency, crop, and soil (USDA-SCS, 1993; USDA-SCS, 1970). See Table 4f-4 for typical required system capacities for various crops and soils.

A practice that may be used to accomplish proper irrigation scheduling is:

❑ **Irrigation Water Management (449)**: Determining and controlling the rate, amount, and timing of irrigation water in a planned and efficient manner.

Tools to assist in achieving proper irrigation scheduling include:

❑ **Water-Measuring Device**: An irrigation water meter, flume, weir, or other water-measuring device installed in a pipeline or ditch.

❑ **Soil and Crop Water Use Data**: From soils information the available water-holding capacity of the soil can be determined along with the amount of water that the plant can extract from the soil before additional irrigation is needed (MAD). Water use information for various crops can be obtained from various United States Department of Agriculture (USDA) publications. Crop water use for some selected irrigated crops is shown in Figure 4f-16.

Drainage Systems: An Overview

Drainage is as old as agriculture and dates back to the Roman Empire and probably earlier. Modern drainage practices began in the 1800s. The purpose of drainage is to provide a root environment suitable for plant growth, thereby increasing production and yield of crops. Artificial drainage is essential on poorly drained agricultural fields to provide optimum air and salt environments in the root zone (Ritzema, 1996). Artificial drainage provides for more management control in areas where the water table is in or near the root zone (USDA-NRCS, 1997a). By controlling soil moisture, drainage can also provide for easier farm operations and lessen compaction by animal and equipment traffic (Luthin, 1973).

In 1985, about 107 million acres of land had been drained in the U.S., of which 72 percent was crop land (Zucker and Brown, 1998). Illinois, Iowa, Indiana, and Ohio are the states with the highest total acreage of drained crop land. Together, these states account for 28.6 million acres of drained crop land. In Ohio and Indiana approximately 50 percent of all crop land is drained. In Illinois and Iowa respectively about 35 and 25 percent of all crop land is drained (USDA, 1987).

Arid Lands

In arid lands, drainage may be required to prevent salts from accumulating in the root zone, and to prevent a water table from building up. Drainage has also been used to bring saline soils into production by leaching salts through the soil profile. In many arid regions, it is not uncommon to apply water via irrigation in excess of crop water requirements to keep salts from building up in the soil profile. The amount of water applied in excess of crop water needs is called a "leaching requirement."

Humid Lands

Drainage in humid lands is required for reasons different from those in arid lands. High water tables are caused by water that builds up over impermeable soil layers due either to clay or compaction. Land may also be subjected to periodic inundation due to topography. Drainage systems are installed to allow for cultural operations (seedbed preparation, planting, harvesting, tillage) and to prevent extended periods of saturated soil conditions (Zucker and Brown, 1998).

Drainage Systems

Subsurface drainage can be achieved through the use of either open ditches or by buried pipe.

Open Ditches

Open ditches are used for collector drains which receive drainage from the buried drains in the field or are sometimes used as field drains. Controlled drainage is oftentimes used with open field drains. Typically, field drains are 3-5 feet deep and spaced between 500 and 600 feet. In a controlled drainage system, the water level is controlled by a water control structure and is used also to irrigate. Irrigation with this method is called "sub-irrigation" or "seepage irrigation." This method is practiced in humid regions on drought-prone soils in order to reduce drought stress on high value crops.

Buried drainage systems

Historically, buried pipe was made of clay, but today drain pipe is made of plastic. In some cases, mole drains are used. Mole drains are open channels formed beneath the ground by pulling a cylindrical bullet shaped object through the soil. Drain depth and spacing are designed to keep the water table below the root zone. Drain depths may range from 2.5 – 8 feet and drain spacing can range from 50 to over 1,000 feet. The downstream end of the drains are connected to a collector drain. (Figure 1 depicts a buried field drainage system.)

Outlets

There are generally two types of outlets for a drainage system: gravity outlets and pump outlets. As the name implies, in a gravity outlet water flows by gravity into an open ditch or natural channel. If the topography is limiting, pumped outlets may be required. With pumped outlets, a sump normally collects the drainage water from the field drains, and the pump lifts the water to a gravity outlet.

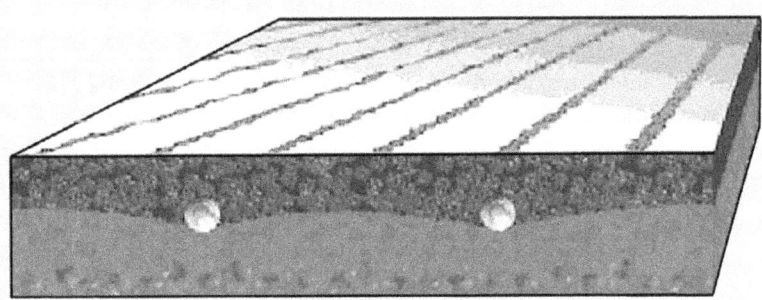

Figure 1. Subsurface field drains showing water table (Zucker and Brown, 1998)

Water quality issues of drainage systems

The installation of drainage systems can result in changes to the ecosystem. These changes can be positive or negative. When compared to agricultural land without subsurface drainage, drained agricultural land can actually have a positive impact on some nonpoint source pollution problems (Zucker and Brown, 1998). The NRCS has listed the subsurface drain as a conservation practice with purposes of reducing erosion and improving water quality (USDA-NRCS, 1997c). However, subsurface drainage water from irrigated agriculture is normally degraded compared with the quality of the original water supply (FAO, 1997). Loss of habitat is also an issue as more than half of the original wetlands in the United States have been lost to drainage practices. Approximately 80 percent of this loss is due to agricultural production (NRC, 1992).

Some of the potential adverse impacts of subsurface drainage systems are:

- **Increased nutrient discharge**
  The two major nutrients in subsurface drainage water are nitrogen and phosphorus. At elevated levels these nutrients contribute to the eutrophication of surface waters which can result in depressed levels of oxygen in receiving waters. The form of nitrogen most prevalent in subsurface drainage is nitrate. Due to strong sorption in the soil, little phosphorus is normally found in subsurface drainage water (Johnson et al., 1965; Mackenzie and Viets, 1974; Madramootoo et al., 1992). The exception to this may be in soils with a highly developed macropore systems (Simard et al., 2000).

- **Pesticide discharge**
  Pesticides may also be of concern, although they are more typically transported with soil particles in surface water drainage (Munster et al., 1995). Although typically low in export loads, pesticide transport may be increased by preferential flow paths resulting in concentrations exceeding drinking water standards (Gentry et al., 2000). Kladivko et al. (1999) found that closer drain tile spacing resulted in more pesticide transport although the total amounts leached were small.

- **Trace elements in effluent**
  Trace elements are commonly present in low levels in nature and may be concentrated in drainage water. Trace elements will depend on geology and, therefore, be different in arid and humid regions. Many of these elements can become toxic a low levels. Mercury (Hg) and selenium (Se) are of particular concern for aquatic life, but arsenic (As), boron (B), molybdenum (Mo), and uranium (U) are also potentially harmful.

- **Sediment**
  Sediment is not normally a problem in subsurface drainage systems since the effluent is primarily ground water. If the system is poorly constructed, sediment can become an issue. More likely, the sediment free water discharging from the subsurface drains might erode the banks of unlined surface drains, thereby increasing the sediment load of the drainage water.

- **Bacteria**

Contamination from bacteria is normally assessed by the presence of coliform and fecal coliform. Irrigated crop land would not be expected to produce adverse bacteriological levels in surface or subsurface drainage water. The presence of coliform or fecal coliform would indicate that wastewater or animal manure has been applied. Since soil is a biological filter, it is not normally expected that micro-organisms will move through the soil from surface water to a subsurface drainage system (FAO, 1997). However, some researchers have implicated subsurface drainage systems in bacteria transport. Geohring, et al. (1998) found that manure applied at nominal rates and followed by a precipitation event can result in bacterial contamination of subsurface drainage in soils exhibiting preferential flow.

- **Salinity**

Salinity of agricultural drainage water is a problem in arid regions. Salts are concentrated in the drainage water. The major cations are sodium (Na), calcium (Ca$_2$), and potassium (K). Major anions are chloride (Cl), sulfate (SO$_4$), bicarbonate (HCO$_3$), nitrate (NO$_3$), and carbonate (CO$_3$). Salinity is generally a problem in agricultural reuse of water, as salinity in general can be detrimental to yield and some crops are sensitive to specific ions such as chloride, boron and sodium.

Management Practices for Drainage Water

There are several management practices which may used for effective drainage water management. A few of them are described below. The applicability of drainage practices to a particular site should be determined on a case-by-case basis. When planning to implement a drainage water management program, a producer should contact state and local authorities regarding any specific requirements or limitations. The assistance of NRCS, Cooperative Extension, or another entity familiar with the design and operation of drainage systems should also be sought.

Water Table Management

Water table management or controlled drainage has the potential to significantly reduce NO$_3$-N. Nitrogen reduction is accomplished by reducing drainage outflow and by providing a denitrifying environment via a higher field water table level. Controlled drainage has been shown to reduce the annual transport of total nitrogen at the field edge by 9 lbs/ac/yr or 45% on the average (Gilliam et al., 1997). Phosphorus transport has also been documented to be reduced by controlled drainage (Gilliam et al., 1997). Water table management has been practiced in the humid environments of the mid-western and eastern parts of the United States in relatively flat landscapes.

Treatment of Drainage Water

Constructed wetlands may be used to treat drainage water. Wetlands are effective in removing sediment, nitrogen and phosphorus. Other physical and chemical treatment processes may be used to treat drainage water (e.g., flocculation, chemical precipitation, or membrane microfiltration), but these are normally only applied where the value of the crop justifies the treatment costs or regulatory requirements exist.

Re-Use of Drainage Water

Drainage water reuse may be appropriate in regions where water is in short supply. The benefit of drainage water reuse is to reduce chemical and nutrient loads to receiving waters. Water quality of re-use water may be of concern, especially in arid regions where salt content of drainage water may be high. Where soils, geologic and hydrologic conditions do not permit constructed wetlands, agricultural drainage water may be re-used on successively salt tolerant crops. Drainage water may also be applied to forested systems. The reduced volume of final drainage water can be discharge to an evaporation pond. With such reuse, care must be taken to insure that concentrations of chemicals do not exceed toxic levels.

The purpose of collecting these data is to allow the manager to estimate the amount of available water remaining in the root zone at any time, thereby indicating when the next irrigation should be scheduled and the amount of water needed. Methods to measure or estimate the soil moisture should be employed, especially for high-value crops or where the water-holding capacity of the soil is low.

Practices for Efficient Irrigation Water Application

Irrigation water should be applied in a manner that ensures efficient use and distribution, minimizes runoff or deep percolation, and minimizes soil erosion.

The method of irrigation employed will vary with the type of crop grown, the topography, and soils. There are several systems that, when properly designed and operated, can be used as follows:

☐ **Irrigation System, Drip or Trickle (441)**: A planned irrigation system in which all necessary facilities are installed for efficiently applying water directly to the root zone of plants by means of applicators (orifices, emitters, porous tubing, or perforated pipe) operated under low pressure (Figure 4f-19).

☐ **Irrigation System, Sprinkler (442):** A planned irrigation system in which all necessary facilities are installed for efficiently applying water by means of perforated pipes or nozzles operated under pressure.

☐ **Irrigation System, Surface and Subsurface (443)**: A planned irrigation system in which all necessary water control structures have been installed for efficient distribution of irrigation water by surface means, such as furrows, borders, contour levees, or contour ditches, or by subsurface means.

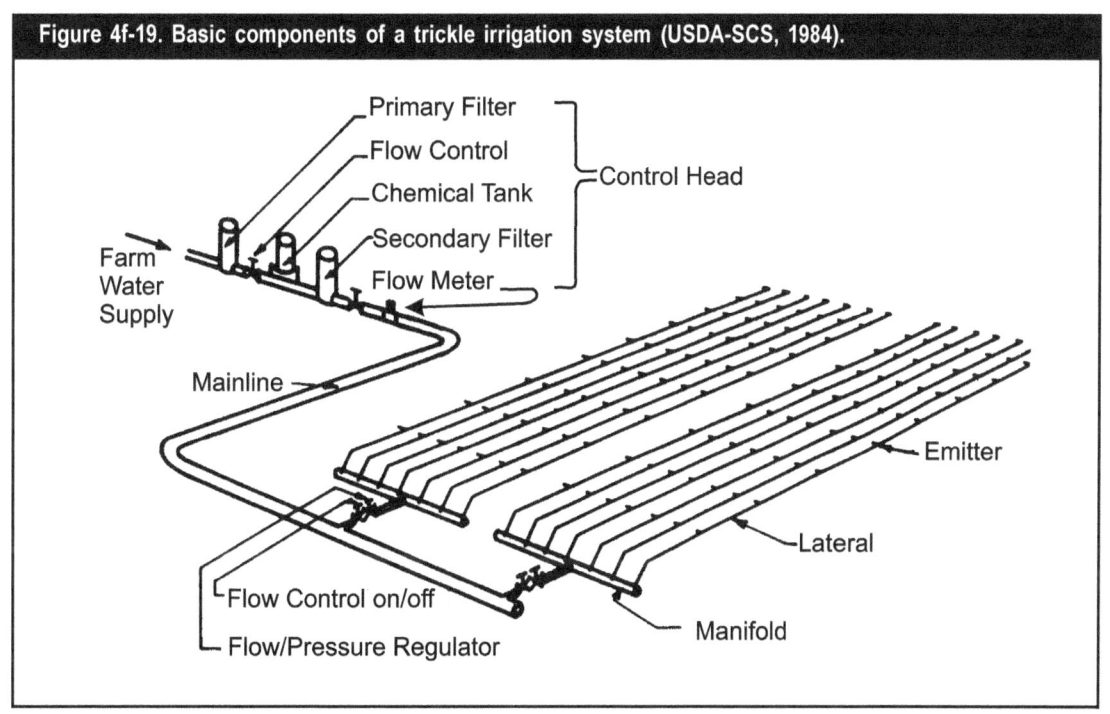

Figure 4f-19. Basic components of a trickle irrigation system (USDA-SCS, 1984).

❑ **Irrigation Field Ditch (388)**: A permanent irrigation ditch constructed to convey water from the source of supply to a field or fields in a farm distribution system.

❑ **Irrigation Land Leveling (464)**: Reshaping the surface of land to be irrigated to planned grades.

Practices for Efficient Irrigation Water Transport

Irrigation water transportation systems that move water from the source of supply to the irrigation system should be designed and managed in a manner that minimizes evaporation, seepage, flow-through water losses from canals and ditches, and leakage from pipes. Delivery and timing need to be flexible enough to meet varying plant water needs throughout the growing season.

Transporting irrigation water from the source of supply to the field irrigation system can be a significant source of water loss and cause of degradation of both surface water and ground water. Losses during transmission include seepage and evaporation from canals and ditches. The primary water quality concern is the development of saline seeps below the canals and ditches and the discharge of saline waters. Another water quality concern is the potential for erosion within canals and at their turnouts. Practices that are used to ensure proper transportation of irrigation water from the source of supply to the field irrigation system can be found in the USDA-NRCS *Handbook of Practices* (USDA-NRCS, 1977) and include:

❑ **Irrigation Water Conveyance, Ditch and Canal Lining (428)**;

❑ **Irrigation Water Conveyance, Pipeline (430)**; and

❑ **Structure for Water Control (587)**.

Practices for Irrigation Erosion Control

The design of farm irrigation systems must provide for conveying and distributing irrigation water without causing damaging soil erosion. All unlined ditches should be located on nonerosive gradients. If water must be conveyed down slopes that are steep enough to cause excessive flow velocities, the irrigation system design should provide for the installation of such erosion-control structures as drops, chutes, buried pipelines, or erosion-resistant ditch linings. Conservation treatments such as land leveling, irrigation water management, reduced tillage, and crop rotations should be used to control irrigation-induced erosion.

On surface irrigated lands susceptible to irrigation-induced erosion, the addition of polyacrylamide (PAM) to surface irrigation water may be appropriate to minimize or control soil erosion. However, PAM cannot make up for failure to implement effective overall conservation practices, or replace environmentally responsible farm management. PAM can provide erosion protection in situations where other solutions have proven uneconomical or ineffective. Further description of the use of PAM in irrigation water is found on page 194. This summary reports that application by irrigators is relatively new and requires current information on effective application rates. Research and associated outreach should continue to provide this type of information. Research on the environmental fate and potential ecological effects of PAM use should continue as well.

On sprinkler irrigated land, the design rate of application should be within a range established by the minimum practical application rate under local climatic conditions and the maximum rate consistent with the intake rate of the soil and the conservation practices used on the land. Sprinkler systems should be designed for zero runoff so no water leaves the point of application. The effects on erosion and the movement of sediment, and soluble and sediment-attached substances carried by runoff should be considered whether surface or sprinkler irrigation systems are employed.

Practices for Use of Runoff Water or Tailwater

The use of runoff water to provide additional irrigation or to reduce the amount of water diverted increases the efficiency of use of irrigation water. For surface irrigation systems that require runoff or tailwater as part of the design and operation, a tailwater management practice is needed. The practice is described as follows:

> ❑ **Irrigation System, Tailwater Recovery (447)**: A facility to collect, store, and transport irrigation tailwater for reuse in the farm irrigation distribution system.

Practices for Drainage Water Management

Drainage water from an irrigation system should be managed to reduce deep percolation, move tailwater to the reuse system, reduce erosion, and help control adverse impacts on surface water and ground water. A total drainage system should be an integral part of the planning and design of an efficient irrigation system.

There are several practices to accomplish this:

> ❑ **Filter Strip (393)**: A strip or area of vegetation for removing sediment, organic matter, and other pollutants from runoff and waste water.

> ❑ **Surface Drainage Field Ditch (607)**: A graded ditch for collecting excess water in a field.

> ❑ **Subsurface Drain (606)**: A conduit, such as corrugated plastic tile, or pipe, installed beneath the ground surface to collect and/or convey drainage water.

> ❑ **Water Table Control (641)**: Water table control through proper use of subsurface drains, water control structures, and water conveyance facilities for the efficient removal of drainage water and distribution of irrigation water.

> ❑ **Controlled Drainage (335)**: Control of surface and subsurface water through use of drainage facilities and water control structures.

Practices for Backflow Prevention

The American Society of Agricultural Engineers recommends, in standard EP409, safety devices to prevent backflow when injecting liquid chemicals into pressurized irrigation systems (ASAE, 1989).

The process of supplying fertilizers, herbicides, insecticides, fungicides, nematicides, and other chemicals through irrigation systems is known as chemigation. A backflow prevention system will "prevent chemical backflow to the water source" in cases when the irrigation pump shuts down (ASAE, 1989).

Three factors an operator must take into account when selecting a backflow prevention system are the characteristics of the chemical that can backflow, the water source, and the geometry of the irrigation system. Areas of concern include whether injected material is toxic and whether there can be backpressure or backsiphonage (ASAE, 1989; EPA, 1991b). Several different systems used as backflow preventers are:

- ❑ **Air gap.** A physical separation in the pipeline resulting in a loss of water pressure. Effective at end of line service where reservoirs or storage tanks are desired.

- ❑ **Check valve with vacuum relief and low pressure drain.** Primarily used as an antisiphon device (Figure 4f-20).

- ❑ **Double check valve.** Consists of two single check valves coupled within one body and can handle both backsiphonage and backpressure.

- ❑ **Reduced pressure principle backflow preventer.** This device can be used for both backsiphonage and backpressure. It consists of a pressure differential relief valve located between two independently acting check valves.

- ❑ **Atmospheric vacuum breaker.** Used mainly in lawn and turf irrigation systems that are connected to potable water supplies. This system cannot be installed where backpressure persists and can be used only to prevent backsiphonage.

- ❑ **Pump interlocking.** Application of chemicals in sprinkler systems require an injection pump. By interlocking the injection pump with the water pump, the injection pump is only powered when the water pump is operating.

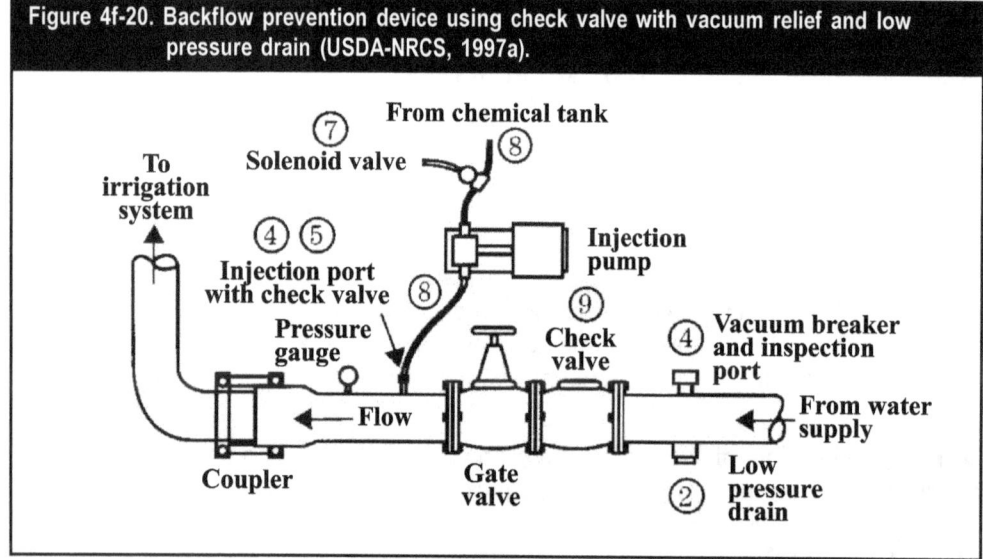

Figure 4f-20. Backflow prevention device using check valve with vacuum relief and low pressure drain (USDA-NRCS, 1997a).

Practice Effectiveness

The following is information on pollution reductions that can be expected from installation of the management practices outlined within this management measure. However, it should be noted that practice effectiveness is determined through experience and evaluations based on system limitations, topography, climate, etc., and cannot merely be selected from a chart. The efficiency and effectiveness figures given below are for illustrative purposes.

In a review of a wide range of agricultural control practices, EPA (1982a) determined that increased use of call periods, on-demand water ordering, irrigation scheduling, and flow measurement and control would all result in decreased losses of salts, sediment, and nutrients. Various alterations to existing furrow irrigation systems were also determined to be beneficial to water quality, as were tailwater management and seepage control.

Logan (1990) reported that chemical backsiphon devices are highly effective at preventing the introduction of pesticides and nitrogen to ground water. The American Society of Agricultural Engineers (ASAE) specifies safety devices for chemigation that will prevent the pollution of a water supply used solely for irrigation (ASAE, 1989).

Properly designed sprinkler irrigation systems will have little runoff (Boyle Engineering Corp., 1986). Furrow irrigation and border check or border strip irrigation systems typically produce tailwater, and tailwater recovery systems may be needed to manage tailwater losses (Boyle Engineering Corp., 1986). Tailwater can be managed by applying the water to additional fields, by treating and releasing the tailwater, or by reapplying the tailwater to upslope cropland.

> Irrigation management practice systems can reduce suspended sediment loading to streams.

The Rock Creek Rural Clean Water Program (RCWP) project in Idaho is the source of much information regarding the benefits of irrigation water management (USDA, 1991). Crops in the Rock Creek watershed are irrigated with water diverted from the Snake River and delivered through a network of canals and laterals. The combined implementation of irrigation management practices, sediment control practices, and conservation tillage resulted in measured reductions in suspended sediment loadings ranging from 61% to 95% at six stations in Rock Creek (1981-1988). Similarly, 8 of 10 sub-basins showed reductions in suspended sediment loadings over the same time period. The sediment removal efficiencies of selected practices used in the project are given in Table 4f-6.

Normally, drip irrigation will have the greatest irrigation efficiency and contour ditch irrigation will have the lowest irrigation efficiency. See Table 4f-7 for application efficiencies of various systems and Table 4f-8 for a range of deep percolation and runoff losses from surface and sprinkler methods. Tailwater recovery irrigation systems are expected to have the greatest percolation rate. USDA projects significant increases in overall irrigation efficiencies when tailwater recovery facilities are used (Table 4f-9).

Plot studies in California have shown that in-season irrigation efficiencies for drip irrigation and Low Energy Precision Application (LEPA) are greater than those for improved furrow and conventional furrow systems (Table 4f-10). LEPA is a linear move sprinkler system in which the sprinkler heads have been removed and replaced with tubes that supply water to individual furrows (Univ. Calif., 1988). Dikes are placed in the furrows to prevent water flow and reduce soil effects on infiltrated water uniformity.

Mielke and Leavitt (1981) studied the effects of tillage practice and type of center pivot irrigation on herbicide (atrazine and alachlor) losses in runoff and sediment. Study results clearly show that, for each of three tillage practices studied, low-pressure spray nozzles result in much greater herbicide loss in runoff than either high-pressure or low-pressure impact heads.

Table 4f-6. Sediment removal efficiencies and comments on BMPs evaluated (USDA, 1991).

Practice	Sediment Removal Efficiency (%)		Comment
	Average	Range	
Sediment basins: field, farm, subbasin	87	75-95	Cleaning costly.
Mini-basins	86[a]	0-95	Controlled outlets essential. Many failed. Careful management required.
Buried pipe systems (incorporating mini-basins with individual outlets into a buried drain)	83	75-95	High installation cost. Potential for increased production to offset costs. Eliminates tailwater ditch. Good control of tailwater.
Vegetative filters	50[a]	35-70	Simple. Proper installation and management needed.
Placing straw in furrows	50	40-80	Labor-intensive without special equipment. Careful management required.

[a] Mean of those that did not fail.

Table 4f-7. Ranges of irrigation application efficiencies from various sources.

Irrigation System	Application Efficiency, %		
	Duke, 1987[1]	USDA-NRCS, 1997a	Hill, 1994[2]
Center Pivot	70-90	75-85	80
Linear Move		80-87	80
LEPA		90-95	
Solid Set Sprinklers		60-75	70-80
Periodic Move Lateral		60-75	70-80
Drip	75-100		80-90
Level Basin	70-90		80
Border			60-75
Furrow			60-70
Furrow – sandy soil	20-60		40-50
Furrow – clay soil	50-90		65
Contour Ditch		35-60	45-55

[1] Typical single event efficiencies

[2] Possible values for various systems with good design and above average management practices

Table 4f-8.   Ranges of Application Efficiency E_a and runoff, deep percolation, and evaporation losses (Hill, 1994).[1]

Method	%		
	Hi	Low	Typical
Surface Irrigation			
E_a	72	24	50
Runoff Losses	55	5	20
Deep Percolation Losses	65	20	30
Sprinkler Irrigation			
E_a	84	52	70
Evaporation Losses	45	8	12
Deep Percolation Losses	37	8	18

[1] determined from field evaluations in Utah

Table 4f-9. Overall efficiencies obtainable by using tailwater recovery and reuse facility (USDA-NRCS, 1997a).

Original applic effic %	% of water reused	-----First reuse-----			-----Second reuse-----			-----Third reuse-----			----- Fourth reuse-----		
		% of orig water used	Effect use - %of orig	Accum effect %	% of orig water used	Effect use - %of orig	Accum effect %	% of orig water used	Effect use - %of orig	Accum effect %	% of orig water used	Effect use - %of orig	Accum effect %
60	40	16	9.6	69.6	2.6	1.5	71.1	1.1	0.7	71.8	0.2	0.1	71.9
	60	24	14.4	74.4	5.8	3.5	77.9	1.4	0.8	78.7	0.4	0.2	78.9
	80	32	19.2	79.2	10.2	6.1	85.3	3.3	2.0	87.3	1.0	0.6	87.9
50	40	20	10.0	60.0	4.0	2.0	62.0	0.8	0.4	62.4	0.2	0.1	62.5
	60	30	15.0	65.0	9.0	4.5	69.5	2.7	1.4	70.9	0.8	0.4	71.3
	80	40	20.0	70.0	16.0	8.0	78.0	6.4	3.2	81.2	2.6	1.3	82.5
40	40	24	9.6	49.6	5.8	2.3	52.9	1.4	0.6	53.5	0.3	0.1	53.6
	60	36	14.4	54.4	13.0	5.2	59.6	4.7	1.9	61.5	1.7	0.7	62.2
	80	48	19.2	59.2	23.0	9.2	68.4	11.0	4.4	72.8	5.3	2.1	74.9
30	40	28	8.4	38.4	7.8	2.4	40.8	2.2	0.7	41.5	0.6	0.2	41.7
	60	42	12.6	42.6	17.8	5.3	49.9	7.5	2.3	52.2	3.1	0.9	53.1
	80	56	16.8	46.8	31.4	9.4	56.2	17.6	5.3	61.5	9.8	3.0	64.5
20	40	32	6.4	26.4	10.2	2.1	28.5	3.2	0.7	29.2	1.0	0.2	29.4
	60	48	9.6	29.6	23.0	4.6	34.2	11.0	2.2	36.4	5.3	1.1	37.5
	80	64	12.8	32.8	41.0	8.2	41.0	26.2	5.3	46.3	17.5	3.5	49.8

Table 4f-10. Irrigation efficiencies of selected irrigation systems for cotton (California SWRCB, 1992).

System	Year	Seasonal Irrigation (in.)	Distribution Uniformity (%)	Irrigation Efficiency (%)	Deep Percolation (in.)
Subsurface Drip Irrigation	1989[1]	23.54	79	86	2.43
	1990[1]	24.04	76	81	3.98
LEPA (Low Energy Precision Application)	1989	19.89	80	82	2.88
	1990	26.55	92	74	6.13
Improved Furrow	1988	29.77	60	35	18.9
	1990	20.19	82	66	6.06
Conventional Furrow	1989	30.75	61	35	19.39
	1990	28.76	72	62	9.85

[1] includes one preirrigation with hand move sprinklers

Factors in Selection of Management Practices

Irrigation Scheduling

Selecting a water scheduling method will depend on the availability of climatic data. Crop water use depends on the type of crop, stage of growth, temperature, sunshine, wind speed, relative humidity and soil moisture content. Water use can be estimated based on maximum daily temperatures and the growth stage of the crop. If climatic data cannot be measured on site or is not available nearby, it may be more appropriate to schedule irrigation from representative field soil water measurements.

Determining water holding capacity for the field is critical in water scheduling. Where large differences in soil texture are found in an irrigated field, particular attention should be paid to the coarsest textures. Coarse textures will hold less available water than finer textured soils and will reach depletion sooner. Knowledge of soil texture and soil moisture status will help determine the appropriate application rate and depth, so runoff and deep percolation are minimized. Variable rate application of water should be considered if water holding capacities range significantly.

Efficient Irrigation Water Application

The selection of an appropriate irrigation system should be based on having sufficient capacity to adequately meet peak crop water demands for the crop with the highest peak water demand in the rotation. The system capacity is dependent on the peak period evapotranspiration rate, crop rooting depth, available water holding capacity of the soil, and irrigation efficiency. Other potentially limiting factors are water delivery capacity and permitted water allocation (Table 4f-4).

Other factors that should be considered when selecting an irrigation system are the shape and size (acres) of the field and the topography. Field slope and steepness will determine whether surface or sprinkler irrigation can be used. If surface application of water is chosen, land leveling may be required to more efficiently spread water over the field.

A sprinkler system can and should be designed to apply water uniformly without runoff or erosion. The application rate of the sprinkler system should be matched to the intake rate of the most restrictive soil in the field. If the application rate exceeds the soil intake rate, the water will run off the field or relocate within the field resulting in areas of over application that could percolate soluble chemicals to ground water. Care should be taken in a pivot system to match endguns with soil water intake rates.

If secondary salinization from irrigation is a problem, an application method must be chosen to keep salts leached below the root zone.

The selected water application method will also depend on whether chemigation is to be used. Coverage, timing, and type of chemical application will determine which application method will be most efficient. Chemigation with surface irrigation should be avoided when alternative methods are available for the application of fertilizers and pesticides. Additional costs for pollution prevention may be incurred when chemigating.

Tailwater recovery may be required if surface chemigation is practiced, and backflow prevention is needed if sprinkler chemigation is used.

Cost and Savings of Practices

Costs

Costs to install, operate and maintain an irrigation system will depend on the type of irrigation system used. In order to efficiently irrigate and prevent pollution of surface and ground waters, the irrigation system must be properly maintained and water measuring devices used to estimate water use.

A cost of $10 per irrigated acre is estimated to cover investments in flow meters, tensiometers, and soil moisture probes (EPA, 1992a; Evans, 1992). The cost of devices to measure soil water ranges from $3 to $4,900 (Table 4f-11). Gypsum blocks and tensiometers are the two most commonly used devices. A more expensive and instantaneous device is a neutron probe. It uses a radioactive source of neutrons and a probe to measure the amount of moisture in the soil. The probe is inserted into the soil through a tube and the energy, produced by neutons colliding with hydrogen and oxygen atoms that make up water, is measured in the probe indicating the soil moisture content.

For quarter-section center pivot systems, backflow prevention devices cost about $416 per well (Stolzenburg, 1992). This cost (1992 dollars) is for: (1) an 8-inch, 2-foot-long unit with a check valve inside ($386); and (2) a one-way injection point valve ($30). Assuming that each well will provide about 800-1,000 gallons per minute, approximately 130 acres will be served by each well. The cost for backflow prevention for center pivot systems then becomes approximately $3.20 per acre. In South Dakota, the cost for an 8-inch standard check valve is about $300, while an 8-inch check valve with inspection points and vacuum release costs about $800 (Goodman, 1992). The latter are required by State law. For quarter-section center pivot systems, the cost for standard check valves ranges from about $1.88 per acre (corners irrigated, covering 160 acres) to $2.31 per acre (circular pattern, covering about 130 acres). To maintain existing equipment so that water delivery is efficient, annual maintenance costs can be figured at 1.5% of the new equipment cost (Scherer, 1994).

Table 4f-11. Cost of soil water measuring devices.

Device	Approximate Cost
Tensiometers[a]	$50 and up, depending on size
Gypsum blocks[b]	$3-4, $200-400 for meter
Neutron Probe[c]	$4,900
Phene Cell[a]	$4,000-4,500
Tensiometers and soil moisture probes[d]	$10 per irrigated acre

[a] Hydratec, 1998.
[b] Sneed, 1992.
[c] Cambell Pacific Nuclear, 1998.
[d] Evans, 1992.

Polyacrylamide Application for Erosion and Infiltration Management

Polyacrylamide (PAM) is a water soluble polymer produced for agricultural use to control erosion and promote infiltration on irrigated lands. When applied to soils, erosion-prevention PAM binds fine-grained soil particles within the top 1/16 inch (1-2 mm) of soil. It is not only used for erosion control, but it is also employed in municipal water treatment, paper manufacturing, food and animal feed processing, cosmetics, friction reduction, mineral and coal processing, and textile production.

PAM comes in many formulations which should not be confused. The super water-absorbent PAM used to increase soil water holding capacity is not the PAM used for erosion control. Most states require environmental, safety, and efficacy evaluation for registration, labeling, and sale of soil amendments. Erosion control PAM formulations have been registered and labeled by individual states where sales and use occur, and farmers should purchase only registered and properly labeled PAM from reputable agrichemical dealers. A compendium of PAM-related research and user information is available at the website http://kimberly.ars.usda.gov/pamPage.shtml.

Availability and Application

Erosion-prevention PAM is available in blocks or cubes, or as a powder, aqueous concentrate or emulsified concentrate. Each form has benefits and drawbacks that would alter efficacy in different settings and with different application methods. Additional factors that affect PAM's effectiveness include irrigation inflow rate, duration of furrow exposure, and soil salinity. Erosion prevention PAM costs range from $3-$8 per pound, depending on the application form purchased, and is typically effective at applications of 1 lb. per crop-acre with each treated irrigation (Sojka, 1999). Amounts applied per crop-acre can be reduced with repeat irrigations.

Application rates of PAM recommended by the Natural Resources Conservation Service (NRCS) and Agricultural Research Service (ARS) are 10 ppm in the irrigation inflow during the furrow-advance period (only). ARS has reported results using the following application methods:

- ☐ adding dry granules to the irrigation water in a gated irrigation pipe;
- ☐ adding a stock solution to furrow heads; and
- ☐ placing 1/2 to 1 oz. powder patches directly on the soil immediately below furrow inlets.

Environmental Pros and Cons

Studies using erosion-prevention PAM have shown a 94% reduction of sediment loss in irrigation runoff, although there is some variability in results due to differing application techniques and management practices. At the same time, PAM has resulted in some cases in higher crop yields, improved crop emergence, and decreased soil crusting. In addition to sediment removal, PAM-based erosion control has been shown to improve off-site water quality through reduction of N, P, BOD, herbicides, pesticides, microorganisms and weed seeds in irrigated runoff contributing to return flows to riparian surface waters (see Table 1).

PAM, like conservation tillage, no-till, and various other infiltration and runoff management systems, increases infiltration. As with any soil management system that reduces return flow pollution through improved infiltration and runoff prevention, greater attention should be paid to irrigation water volume application, inflow control, and crop irrigation scheduling. The NRCS and ARS encourage increasing the furrow irrigation inflow rate, resulting in shortened advance times and preventing leaching of surface applied nutrients or agrichemicals from over-irrigation of the near end of the field when using PAM for erosion control.

Most of the concern regarding PAM has arisen because of acrylamide (AMD), the monomer associated with PAM and a contaminant of the PAM manufacturing process. AMD has been shown to be both a neurotoxin and a carcinogen in laboratory experiments. Current regulations require that AMD not exceed 0.05% in PAM products. At the application rates prescribed by the NRCS, the concentration of AMD in outflow waters is several orders of magnitude less than what is considered toxic. According to the ARS, AMD decomposes in 18 to 45 hours in biologically active environments (Barvenik et al., 1996). Although there seems to be little risk from AMD as a result of prescribed application of PAM, care should be taken to avoid spills, over-application, or other unforeseen accidents as their effects are uncertain (See Table 2).

Table 1. PAM 's beneficial effects on the environment and crop production (Sojka and Lentz, 1996).

What PAM Does	Environmental Benefit
Decrease sediment loading	Decrease turbidity Improve clarity Decrease P, N, pesticides, salts, pathogens Decrease BOD, eutrophication Decrease weed seed in runoff
Improve soil tilth	Increase infiltration Decrease runoff
Binds fine soil particles	Decrease wind erosion Accelerates clarification of turbid water bodies Prevents erosion
Increase soil water storage	Improves irrigation efficiency Decrease plant stress Improve plant vigor

Table 2. PAM 's potential detrimental effects on the environment and crop production (Dawson et al., 1996 in Sojka and Lentz, 1996; Sojka, personal communication, 2000).

What PAM Does	Potential Detrimental Effect	Preventative Measures
Increased infiltration	At prescribed rates on fine or medium textured soil, PAM can increase infiltration comparable to no-till, risking drainage and leaching of nutrient or chemicals.	Increase irrigation flow rate to prevent over-irrigation of the near end of the field.
Reduce infiltration	Over-application of PAM, or use on coarse textured soil, can reduce infiltration.	Careful application suited to site-specific needs.
Unknown effects on fish and wildlife	While safe at prescribed rates, large spills or excessive application may affect habitat.	Take care to avoid spills; use as directed.

Anionic PAM (containing less than 0.05% AMD), the form registered by states for use in erosion control products, is not toxic to aquatic, soil, or crop species when used as directed at specified rates. The molecule is too large to cross membranes, so it is not absorbed by the gastrointestinal tract, is not metabolized, and does not bioaccumulate in living tissue. PAM effects on aquatic biota are buffered if the water contains sediments, humic acids, or other impurities (Barvenik et al., 1996). While assessments of PAM effects directly on wildlife have not been conducted, the fact that PAM is applied in very dilute form to land via irrigation water, and largely stays on targeted fields, coupled with highly positive effects on several important runoff water quality components, suggests little danger if label directions and cautions are followed. This perception is strengthened by the fact that PAM has been used in a variety of industrial water treatment uses and land disposed for decades, with no reported adverse effects on wildlife. Published soil microbial studies have shown no negative impact on soil microflora or microfauna in treated fields. Furthermore, erosion control PAMs are restricted to anionic forms that are also used in human food processing and cosmetic and pharmaceutical preparations.

Conclusion

Anionic PAM has proven an effective erosion control technology since research began in 1991. Continued USDA research and extension efforts since 1995 have resulted in a million acres of PAM use annually since 1998, with no reports of adverse environmental consequences. PAM has been shown to prevent the entry of sediment, nutrients, and pesticides into riparian waters via irrigation runoff and return flows. However, the learning curve for effective PAM use is steep and sometimes counter intuitive. Farmers need to be well informed of PAM properties and application requirements. While PAM is an important additional erosion-combating conservation tool that can often be effective where other approaches fail, it should not be used as a substitute for good overall farm management and a balanced and effective conservation plan. PAM cannot make up for failure to implement effective overall conservation practices and environmentally responsible farm management, but can provide essential erosion protection in many situations where other solutions have proven uneconomical or ineffective.

Tailwater can be prevented in sprinkler irrigation systems through effective irrigation scheduling, but may need to be managed in furrow systems. The reuse of tailwater downslope on adjacent fields is a low-cost alternative to tailwater recovery and upslope reuse (Boyle Engineering Corp., 1986). Tailwater recovery systems require a suitable drainage water receiving facility such as a sump or a holding pond, and a pump and pipelines to return the tailwater for reapplication (Boyle Engineering Corp., 1986). The cost to install a tailwater recovery system was about $125/acre in California (California SWRCB, 1987) and $97.00/acre in the Long Pine Creek, Nebraska, RCWP (Hermsmeyer, 1991). Additional costs may be incurred to maintain the tailwater recovery system.

The cost associated with surface and subsurface drains is largely dependent upon the design of the drainage system. In finer textured soils, subsurface drains may need to be placed at close intervals to adequately lower the water table. To convey water to a distant outlet, land area must be taken out of production for surface drains to remove seeping ground water and for collection of subsurface drainage.

The Agricultural Conservation Program (ACP) has been phased out and replaced by the Environmental Quality Incentive Program (EQIP) in the 1996 Farm Bill. However, the Statistical Summaries (USDA-FSA, 1996) from the ACP contain reliable cost-share estimates. The following cost information is taken from these summaries and assumes a 50% cost-share to obtain capital cost estimates. The ACP program has a unique set of practice codes that are linked to a conservation practice. The cost to install irrigation water conservation systems (FSA practice WC4) for the primary purpose of water conservation in the 33 States that used the practice was about $73.00 per acre served in 1995. Practice WC4 increased the average irrigation system efficiency from 47% to 64% at an amortized cost of $10.41 per acre foot of water conserved. The components of practice WC4 are critical area planting, canal or lateral, structure for water control, field ditch, sediment basin, grassed waterway or outlet, land leveling, water conveyance ditch and canal lining, water conveyance pipeline, trickle (drip) system, sprinkler system, surface and subsurface system, tailwater recovery, land smoothing, pit or regulation reservoir, subsurface drainage for salinity, and toxic salt reduction. When installed for the primary purpose of water quality, the average installation cost for WC4 was about $67 per acre served. For erosion control, practice WC4 averaged approximately $82 per acre served. Specific cost data for each component of WC4 are not available.

Water management systems for pollution control, practice SP35, cost about $94 per acre served when installed for the primary purpose of water quality. When installed for erosion control, SP35 costs about $72 per acre served. The components of SP35 are grass and legumes in rotation, underground outlets, land smoothing, structures for water control, subsurface drains, field ditches, mains or laterals, and toxic salt reduction.

The design lifetimes for a range of salt load reduction measures are presented in Table 4f-12 (USDA-ASCS, 1988).

Savings

Savings associated with irrigation water management generally come from reduced water and fertilizer use.

Steele et al. (1996) found that improved methods of irrigation scheduling can produce significant savings in seasonal irrigation water totals without yield reductions. In a six-year continuous corn field study, a 31% savings in seasonal irrigation totals was realized compared to the average commercial grower in the same irrigation district. Corn grain yields were maintained at 3% above average corn grain yields in the irrigation district.

Table 4f-12. Design lifetime for selected salt load reduction measures (USDA-ASCS, 1988).

Practice/Structure	Design Life (Years)
Irrigation Land Leveling	10
Irrigation Pipelines – Aluminum Pipe	20
Irrigation Pipelines – Rigid Gated Pipe	15
Irrigation Canal and Ditch Lining	20
Irrigation Head Ditches	1
Water Control Structure	20
Trickle Irrigation System	10
Sprinkler Irrigation System	15
Surface Irrigation System	15
Irrigation Pit or Regulation Reservoir	20
Subsurface Drain	20
Toxic Salt Reduction	1
Irrigation Tailwater Recovery System	20
Irrigation Water Management	1
Underground Outlet	20
Pump Plant for Water Control	15

Using Management Measures to Prevent and Solve Nonpoint Source Problems in Watersheds

5

Watershed Approach

Watersheds are areas of land that drain to a single stream or other water resource. Watersheds are defined solely by drainage areas and not by land ownership or political boundaries. The *watershed approach* is a coordinating framework for environmental management that focuses public and private sector efforts to address priority problems within hydrologically defined geographic areas (e.g., watersheds), taking into consideration both ground and surface water flow (EPA, 1995b).

EPA supports watershed approaches that aim to prevent pollution, achieve and sustain environmental improvements and meet other goals important to the community. Although watershed approaches may vary in terms of specific objectives, priorities, elements, timing, and resources, all should be based on the following guiding principles.

❑ *Partnerships*: Those people most affected by management decisions are involved throughout and shape key decisions.

This ensures that environmental objectives are well integrated with those for economic stability and other social and cultural goals. It also provides that the people who depend upon the natural resources within the watersheds are well informed of and participate in planning and implementation activities.

❑ *Geographic Focus:* Activities are directed within specific geographic areas, typically the areas that drain to surface water bodies or that recharge or overlay ground waters or a combination of both.

❑ *Sound Management Techniques based on Strong Science and Data*: Collectively, watershed stakeholders employ sound scientific data, tools, and techniques in an iterative decision making process. This includes:

i. assessment and characterization of natural resources and the communities that depend upon them;

ii. goal setting and identification of environmental objectives based on the condition or vulnerability of resources and the needs of the aquatic ecosystem and the people within the community;

iii. identification of priority problems;

iv. development of specific management options and action plans;

v. implementation; and

vi. evaluation of effectiveness and revision of plans, as needed.

Because stakeholders work together, actions are based upon shared information and a common understanding of the roles, priorities, and responsibilities of all involved parties. Concerns about environmental justice are addressed and, when possible, pollution prevention techniques are adopted. The iterative nature of the watershed approach encourages partners to set goals and targets and to make maximum progress based on available information while continuing analysis and verification in areas where information is incomplete.

Watershed projects should have a strong monitoring and evaluation component. Using monitoring data, stakeholders identify and prioritize stressors that may pose health and ecological risk in the watershed and any related aquifers. Monitoring is also essential to determining the effectiveness of management options chosen by stakeholders to address high priority stressors. Because many watershed protection activities require longterm commitments from stakeholders, stakeholders need to know whether their efforts are achieving real improvements in water quality. Monitoring is described in greater detail in Chapter 6.

Watershed projects should also be consistent with state regulatory programs such as development and implementation of total maximum daily loads (TMDLs) and basinwide water quality assessments. In fact, a watershed may be selected for special attention because of the need for a complex TMDL involving point and nonpoint sources (see Chapter 7 for a discussion of TMDLs).

Operating and coordinating programs on a watershed basis makes good sense for environmental, financial, social, and administrative reasons. For example, by jointly reviewing the results of assessment efforts for drinking water protection, point and nonpoint source pollution control, fish and wildlife habitat protection and other resource protection programs, managers from all levels of government can better understand the cumulative impacts of various human activities and determine the most critical problems within each watershed. Using this information to set priorities for action allows public and private managers from all levels to allocate limited financial and human resources to address the most critical needs. Establishing environmental indicators helps guide activities toward solving those priority problems and measuring success.

The final result of the watershed planning process is a plan that is a clear description of resource problems, goals to be attained, and identification of sources for technical, educational, and funding assistance needed. A comprehensive plan will provide a basis for seeking support and for maximizing the benefits of that support.

Implementing Management Measures in Watersheds

Management measures can be implemented in either a preventive or restorative mode depending upon the State and local needs identified through the watershed planning process. Similarly, although management measures are generally considered to be technology-based, they can also be used as key elements of a water quality-based approach to solving identified water quality problems. Technology-based pollution control measures are identified based upon technical and economic achievability rather than on the cause-and-effect linkages between particular land use activities and particular water quality problems that drive water quality-based approaches.

Technology-based Implementation

As noted earlier, a clear assessment of the problem is essential to identifying appropriate solutions. For example, the Section 6217 management measures were specified to address water quality problems in the Nation's coastal areas. These management measures were developed as affordable technology-based controls that could be implemented broadly within coastal drainage areas to improve and protect the quality of coastal waters. The Section 6217 program also includes provisions for implementing additional control measures where water quality problems are not solved through implementation of the management measures alone (USDOC and EPA, 1993). This iterative approach to solving coastal problems is consistent with the guiding principles of the watershed approach.

Primary justification for applying management measures through a technology-based approach is that the measures are known to reduce pollution and are generally acceptable and affordable. Therefore, the measures should be applied to as much land as possible, regardless of location. This has been the approach of most USDA and state agencies for many years. For example, Vermont's *Accepted Agricultural Practices* are "basic practices that all farmers must follow as part of their normal operations" (Vermont Department of Agriculture, 1995). They "are intended to reduce, not eliminate, pollutants associated with nonpoint sources." By implementing management measures or practices in a technology-based approach, a level of water quality protection can be achieved which makes it easier to then focus on remaining sources that need additional control.

The means by which management measures are implemented in a technology-based approach can range from voluntary to regulatory. All States have some form of voluntary program for addressing agricultural nonpoint source pollution. These programs include USDA's Farm Bill programs (Chapter 1) and State and local cost-share and assistance programs. Cost-share programs are very often technology-based and can be directed to high-priority watersheds in much the same way that Section 6217 is focused within coastal drainage areas. Private sector efforts are also technology-based in many cases, including, for example, precision farming techniques.

Water Quality-based Implementation

In areas where specific water quality problems have been identified and characterized in detail, it is possible to tailor implementation to achieve well-defined goals. For example, TMDLs result in allocations of the quantity of pollution that can be discharged from point sources (wasteload allocation) and nonpoint sources (load allocation) to ensure that water quality standards are achieved within a specified margin of safety (see Chapter 7). Management measures can be applied to achieve all or part of the pollution control needed by agricultural sources to achieve the load allocation. Management measures can also be used in permits to address the portion of a wasteload allocation assigned to animal operations designated as point sources.

Understanding Hydrology

Understanding site and watershed hydrology is essential to understanding nonpoint source problems and the impacts that management measure implementation may have on water quality. Each action taken on a farm has the potential to impact hydrology (see Garen et al., 1999). For example, diversions and buffers clearly affect water movement, and even grazing management affects hydrology through its changes to grazing land quality and/or riparian condition. Nutrient management can also affect hydrology directly if the application of nutrients includes liquids, and indirectly through its effects on crop growth which control plant water and nutrient uptake. The extent to which management decisions affect hydrology needs to be understood and estimated since hydrology is so important to the detachment, transport, and delivery of pollutants.

In agricultural watersheds, hydrology can be affected by a number of factors including the use of tile drains and irrigation practices, installation of grassed waterways and diversions, field buffers and buffer strips, crop type, and tillage type. The combined effects on hydrology of all management measures and management practices implemented should be considered both at the farm level and at the watershed scale in order to estimate the impacts on receiving water quality. Field-scale and watershed-scale models can aid analysis of the impacts on hydrology, and thus decisions on appropriate selection and placement of measures and practices in the watershed. In some cases, a thoughtful discussion or simple analysis will provide the answers regarding impacts to hydrology, but some form of modeling will usually be needed to integrate the various small and large impacts that management measures and practices are likely to have on watershed hydrology. However, models often have many limitations. Therefore, a thorough understanding of the hydrology of the area gained through monitoring or experience is usually needed to properly interpret model results.

If the watershed within which agricultural management measures will be implemented includes land uses other than agriculture, then planners will need to consider agriculture's role within the watershed. In other words, the degree to which agricultural lands control watershed hydrology should be investigated and understood to enable analysis of the potential impacts that management measures and practices will have on watershed hydrology. Once again, some sort of watershed modeling capability will usually be needed to aid this analysis.

Assessing On-Site Treatment Needs

Once watershed hydrology is understood, analysis of on-site treatment needs and the impacts of management measures on pollutant sources and delivery patterns can be conducted. At a particular farm it may be simple to determine which management measures are needed. For example, if nutrients and pesticides are applied, then nutrient and pesticide management should be implemented. If runoff from a confined animal facility leaves the farm without any attenuation or treatment, then storage and treatment of runoff is probably needed. More difficult cases will be those in which some management is practiced, but not enough to fully achieve the management measures. Even more difficult may be the cases where management measures are fully achieved but water quality goals or standards are still not being met.

On-site assessments should be performed to determine the needs on any individual farm. USDA-NRCS, soil and water conservation districts, state cooperative extension, and other public and private organizations have expertise in performing on-site assessments. EPA has developed guidance for tracking and evaluating the implementation of nonpoint source control measures (EPA, 1997b). Tools such as Farm*A*Syst (Jackston et al., undated) can be helpful when performing self-assessments of on-farm conditions.

It is usually beneficial to examine the water resource (e.g., to perform a stream walk) to view the watershed from the perspective of the receiving water body. This may lead to discovery of sources that would not be found from a typical on-site assessment. USDA's *Stream Visual Assessment Protocol* (USDA-NRCS, 1998) is a potential tool for stream assessment. In some watershed projects upland erosion control and riparian protection have been implemented with the expectation that sedimentation problems would be solved. Results, however, indicated that sedimentation problems persisted. For example, in the Rock Creek, Idaho, Rural Clean Water Program project, improved irrigation, sediment retention structures, filter strips, and conservation tillage were implemented to address sediment problems impacting a cold-water fishery (EPA, 1990a). The project did achieve and measure reduced levels of suspended sediment, but it was concluded that the project should have included the contribution of sediment from streambanks and the effects of hydromodification to fully achieve water quality objectives. A thorough examination of the water resource could have helped in the initial planning stages for this project.

Targeting

Even properly designed management practice systems constitute only part of an effective land treatment strategy. In order for a land treatment strategy to be most effective, properly designed management practice systems must be placed in the correct locations in the watershed (i.e., "critical areas") and the extent of land treatment must be sufficient to achieve water quality improvements (Line and Spooner, 1995). RCWP results indicate that 75% of the critical areas (as designated in that program) need to be treated to achieve water quality goals. For livestock-related water quality problems, generally 100% of the critical area should be treated with BMP systems (Meals, 1993). "Critical areas" are generally considered to be sub-areas within a watershed or recharge area that encompass the major pollutant sources that have a direct impact on the impaired water resource (Gale et al., 1993). The discussion below and in Chapter 7 provides information related to the delineation of critical areas. Although the term "critical area" is not generally used in TMDLs, the allocation of loads to sources in the watershed is entirely consistent with the concept.

In cases where implementation of management measures is water-quality based or voluntary, the implementation should be prioritized based upon the water quality benefits to be derived. Phased implementation on a priority basis may be best if financial resources are limited.

Estimating On-Site and Off-Site Impacts

On-site benefits are highly desirable, yet unless the needed off-site benefits are derived from the collective implementation of management measures and practices across the watershed, then implementation has not been fully successful. It is important to estimate the collective impacts of all management activities in the watershed to gage whether water quality goals will be achieved. In watersheds with easily characterized problems (e.g., bacterial contamination is due to a few obviously polluting animal operations in a watershed that has no other identifiable sources of pathogens) it may be very easy to project that water quality benefits will be achieved through implementation of the management measures for nutrient management, erosion and sediment control, and facility wastewater and runoff, for example. However, in a watershed with multiple land uses where agriculture is considered to contribute about one-third or so of the pollutants, it is more complicated to estimate the combined impacts of a variety of management measures and practices on a fairly large number of diverse farming operations. Further complicating the assessment may be that historic loading of pollutants has caused the water quality impairment and several years are required for the water resource to recover or cleanse itself (i.e., current loading may be low). In this type of situation, computer modeling may be needed.

A variety of models exist to help assess the relative benefits of implementing practices at the field and watershed level. However, an understanding of the model's limitations and assumptions is necessary for appropriate interpretation of modeling results. It is also important that models be adequately validated and calibrated for a range of circumstances. The following are some models that have been evaluated for a relatively wide range of conditions and have been shown to be appropriate for the farm or field:

☐ GLEAMS (Knisel et al., 1991) simulates the effects of management practices and irrigation options on edge of field surface runoff, sediment, and dissolved and sediment attached nitrogen, phosphorus, and pesticides. The model considers the effects of crop planting date, irrigation, drainage, crop rotation, tillage, residue, commercial nitrogen and phosphorus applications, animal waste applications, and pesticides on pollutant movement. The model has been used to predict the movement of pesticides (Zacharias et al., 1992) and nutrients and sediment from various combinations of land uses and management (Knisel and Leonard, 1989; Smith et al., 1991).

☐ EPIC (Sharpley and Williams, 1990) simulates the effect of management strategies on edge of field water quality and nitrate nitrogen and pesticide leaching to the bottom of the soil profile. The model considers the effect of crop type, planting date, irrigation, drainage, rotations, tillage, residue, commercial fertilizer, animal waste, and pesticides on surface and shallow ground water quality. The EPIC model has been used to evaluate various cropland management practices (Sugiharto et al., 1994; Edwards et al., 1994).

☐ NLEAP (Follet et al., 1991) evaluates the potential of nitrate nitrogen leaching due to land use and management practices. The NLEAP model has been used to predict the potential for nitrogen leaching under various management scenarios (Wylie et al., 1994; Wylie et al., 1995).

❏ PRZM (Mullens et al. 1993) simulates the movement of pesticides in unsaturated soils within and immediately below the root zone. Several different field crops can be simulated and up to three pesticides are modeled simultaneously as separate parent compounds or metabolites. The PRZM model has been used under various conditions to assess pesticide leaching under fields (Zacharias et al., 1992; Smith et al., 1991).

❏ DRAINMOD (Skaggs, 1980) simulates the hydrology of poorly drained, high water table soils. Breve et al. (1997) developed DRAINMOD-N, a nitrogen version of the model to evaluate nitrogen dynamics in artificially drained soils. The DRAINMOD model has been used to predict pollutant losses associated with various drainage management scenarios (Deal et al., 1986). Website is http://www.bae.ncsu.edu/bae/research/soil_water/www/watmngmnt/drainmod/index.htm.

❏ REMM (Riparian Ecosystem Management Model) is a computer simulation model used to simulate hydrology, nutrient dynamics and plant growth for land areas between the edge of fields and a water body. Output from REMM allows designers to develop buffer systems to help control non-point source pollution. REMM was developed by ARS at the Southeast Watershed Research Laboratory, Coastal Plain Experiment Station, Tifton, GA. Web site is http://www.cpes.peachnet.edu/remmwww/.

❏ NTRM (Shaffer and Larson, 1985) simulates the impact of soil erosion on the short and long-term productivity of soil, and is intended to assist with evaluation of existing and proposed soil management practices in the subject areas of erosion, soil fertility, tillage, crop residues, and irrigation. The NTRM model has been applied to evaluate effects of conservation tillage, supplemental nitrogen and irrigation practices (Shaffer, 1985) and moldboard plow and chisel plow tillage (Shaffer et al., 1986) on soil erosion and productivity. This model has had limited use.

The following models can be used for either farm field or small watershed scale analysis:

❏ WEPP (Flanagan and Nearing, 1995) simulates water runoff, erosion, and sediment delivery from fields or small watersheds. Management practices including crop rotation, planting and harvest date, tillage, compaction, stripcropping, row arrangement, terraces, field borders, and windbreaks can be simulated. The WEPP model has been applied to various land use and management conditions (Tiscareno-Lopez et al., 1993; Liu et al., 1997). Web site is http://topsoil.nserl.purdue.edu/nserlweb/weppmain/wepp.html.

❏ SWAT (which incorporates SWRRBWQ) (Arnold et al., 1990) simulates the effect of agricultural management practices such as crop rotation, conservation tillage, residue, nutrient, and pesticide management; and improved animal waste application methods on water quality. The SWRRB model has been used on several watersheds to assess management practices and to test its validity (Arnold and Williams, 1987; Bingner et al., 1987). Web site is http://www.brc.tamus.edu/swat.

❏ AnnAGNPS (Cronshey and Theurer, 1998) is a spatially-distributed model for estimating pollutant runoff from agricultural watersheds.

BASINS 3.0: A Powerful and Improved Tool for Managing Watersheds

BASINS (Better Assessment Science Integrating Point and Nonpoint Sources) is a multipurpose environmental analysis system for use by regional, state, and local agencies in performing watershed and water quality-based studies. This software makes it possible to quickly assess large amounts of point source and nonpoint source data in a format that is easy to use and easy to understand. Installed on a personal computer, BASINS allows the user to assess water quality at selected stream sites or throughout an entire watershed. It is an invaluable tool that integrates environmental data, analytical tools, and modeling programs to support development of cost-effective approaches to environmental protection.

BASINS addresses three objectives: (1) to facilitate examination of environmental information, (2) to provide an integrated watershed and modeling framework, and (3) to support analysis of point and nonpoint source management alternatives. It also supports the development of total maximum daily loads, which requires a watershed-based approach that integrates both point and nonpoint sources. Basins can support a number of pollutants at a variety of scales, using tools that range from simple to sophisticated.

Originally released in 1996, with a second release in 1998, BASINS comprises a suite of interrelated components. BASINS' databases and assessment tools are directly integrated within an ArcView environment. These components work together to support the user performing various aspects of environmental analysis. The components include (1) nationally derived databases with Data Extraction and Project Builder tools; (2) assessment tools (TARGET, ASSESS, and Data Mining) that address large- and small-scale characterization needs; (3) utilities to facilitate importing local data and for organizing and evaluating data; (4) Watershed Delineation tools; (5) utilities for classifying elevation (DEM), land use, soils, and water quality data; (6) Watershed Characterization Reports that facilitate compilation and output of information on selected watersheds; (7) an in-stream water quality model; (8) two watershed loading and transport models and (9) a simplified GIS based nonpoint annual loading model.

What's New in BASINS 3.0?

This major release includes an overhaul of the system architecture that packages system components as ArcView extensions and external programs. This architecture is open and flexible. It promotes the growth of BASINS by allowing users and developers to write their own extensions to the system. BASINS 3.0 also includes many new features and improvements.

- An automatic delineation tool that allows users to delineate watershed based on a Digital Elevation Model (DEM) grid formatted data.

- An enhanced manual delineation tool that allows users additional flexibility in editing shapes and attributes of manually delineated watersheds.

- A new Windows interface for the HSPF model that fully supports interaction with the entire HSPF input sequence.

- A watershed model called Soil Water Assessment Tool (SWAT), developed by the U.S. Department of Agriculture's ARS.

- A model called PLOAD, developed by CH2M-Hill, which uses export coefficients to estimate watershed loading.

- A model postprocessor and scenario generator called GenScn. Originally developed for the U.S. Geological Survey (USGS), GenScn allows users to manage, visualize, analyze, and compare the results of several HSPF and/or SWAT simulations.

- A time series data management utility called WDMUtil.

- A grid projector that allows the user to project grid data.

- An improved Permit Compliance System point source (PCS) database with annual loadings updated through 1999.

- DEM (grid format) data on the distribution CD buffered to 8 digit HUC boundaries.

For more information on content, availability, and training, please contact:
basins@epa.gov
Exposure Assessment Branch
Standards and Applied Science Division
Office of Science and Technology
U.S. Environmental Protection Agency
1200 Pennsylvania Avenue, NW
Washington, DC 20460
www.epa.gov/ost/basins/

Within cells, the model can evaluate practices such as feedlot management, terraces, vegetative buffers, grassed waterways, and farm ponds. Simulated nutrient, sediment, and pesticide concentrations and yields are available for any cell within the watershed. The AnnAGNPS model has been applied to many field and watershed size areas to estimate pollutant runoff from various land uses and management practices (Bosch et al., 1998; Line et al., 1997; Young et al., 1994; Sugiharto et al., 1994; Bingner et al., 1987). Web site is http://www.sedlab.olemiss.edu/AGNPS.html.

❏ ANSWERS (Beasley et al., 1980) is a spatially-distributed watershed model. The model is primarily a runoff and sediment model as soil nutrient processes are not simulated. The ANSWERS model has been applied to several small field-sized areas with various management practices (Griffin et al., 1988; Bingner et al., 1987).

❏ BASINS (EPA, 2001d) is a user-friendly GIS-based program containing several models capable of simulating watershed loadings and receiving water impacts at various levels of complexity. This new version allows you to subdivide large watersheds into very small watershed segments using either an automated delineation tool or a manual delineation tool. BASINS 3.0 includes three watershed models. The HSPF model, present in earlier versions, is supported by a new Windows interface that makes it easier to run the urban and rural watershed simulations. A rural watershed model called Soil Water Assessment Tool (SWAT), developed by the U.S. Department of Agriculture's Agricultural Research Service, has been added to BASINS. It is anticipated that this model will be widely used in agricultural watersheds. A third very simple model called PLOAD has also been added. PLOAD is most applicable for screening analyses. In addition, there is a new model postprocessor and scenario generator called GenScn that allows users to manage, visualize, analyze, and compare the results of several HSPF and/or SWAT simulations. Web site is www.epa.gov/ost/basins.

A series of pollutant specific protocols has been developed by EPA to assist in the development of TMDLs and implementation plans to achieve the TMDLs (EPA, 1997d; 1999b; 1999c; 2001c). These protocols focus primarily on the application of computer models that simulate watershed conditions and the changes that could result from implementation of various land management scenarios. Some models contain default values for the quantity of pollutants that are delivered in runoff from various sources (e.g., cropland deliver X pounds of nitrogen per acre per inch of runoff). These default values can generally be replaced with better information that is available for a particular watershed. Models should have functions that are intended to simulate the implementation of management practices, enabling modelers to estimate changes due to a range of land management options. Such models can be helpful tools for planning the implementation of management measures to achieve water quality goals, but the limitations of models and appropriate interpretation of modeling results should be fully understood before implementation decisions are made. The application of models to estimate pollutant loads is discussed further in Chapter 7.

Adaptive Management

Because many of the decisions made regarding the appropriate type, extent, and location of management measures and practices are based upon estimates and partial information, it is highly likely that changes will be needed. If progress is monitored (see Chapter 6) adequately, managers and landowners will be able to adjust implementation plans, schedules, and models as needed to ensure more cost-efficient achievement of water quality objectives. One of the major findings from the Rural Clean Water Program is that water quality monitoring can provide valuable feedback for defining areas needing priority treatment (Gale et al., 1993).

Preventing Unintended Adverse Environmental Effects

As noted in Chapter 2, this guidance does not address all environmental considerations at a particular site or within a watershed. Resource management systems (RMS) are more broad, yet planners and managers should even go beyond the scope of an RMS to consider whether management measure or practice implementation at the site or watershed scale will have any unintended environmental impacts. For example, methane generation from structures implemented to store runoff and facility wastewater from confined animal facilities may be problematic in certain areas. Alternatives to conventional storage structures might be needed.

Similarly, extensive changes to water management could impact baseflows in streams. Different configurations and design specifications for diversions and storage devices might be able to provide needed water quality improvement without causing negative impacts to baseflow patterns. Whole-farm planning approaches such as those specified in Chapter 2 (e.g., Idaho One Plan) can go a long way toward preventing these types of unintended environmental impacts at the farm level, but potential watershed-wide or landscape-scale impacts need to considered from a more global perspective.

Estimating the Effectiveness of Management Measures and Management Practice Systems

It is very difficult to estimate the effectiveness of management practice systems. Some researchers have proposed that the effectiveness of management practice systems should be calculated by adding the average relative effectiveness of individual practices. As an example of this approach, assume a system to control sediment is composed of surface drainage, terraces, and conservation tillage. Based upon data in the literature (Foster et al., 1996), the average sediment load reductions achieved by these practices are 36% for surface drainage, 91% for terraces, and 69% for conservation tillage. Under this approach, the average pollutant load reduction for surface drainage is subtracted from the total load of 100% (100% – 36% = 64%). Thus, 64% of the sediment remains after surface drainage is accounted for. If terraces reduce sediment loads by 91%, then the remaining pollutant load after surface drainage and terraces is about 6% (.64 x [1.00 – .91] = .058 = 5.8%). The remaining practice in the system, conservation tillage, reduces sediment loads by 69%, resulting in a final sediment delivery of approximately 2% (.058 x [1.00 – .69] = .018 = 1.8%).

The Idaho RCWP project, however, demonstrated that the effectiveness of individual practices in a system of practices **are not additive**. The effectiveness of some of the BMPs used in the project was measured by the USDA–Agricultural Research Service, and the results are given in Table 5-1 (Maret et al., 1991).

Table 5-1. Sediment removal effectiveness of selected individual BMPs used in the Snake River RCWP Project (Idaho).		
Individual BMP	Mean % Effectiveness	% Effectiveness Range
Sediment Basins	87	75-95
Mini-basin	86	0-95
Buried Pipe Systems	83	75-95
Vegetative Filters	50	35-70
Straw Mulch	50	40-80

Sediment loads in the Idaho RCWP project were reduced by 75%. Even though the effectiveness of only five of the nineteen BMPs used in the project was measured (Table 5-1), it can be seen that the overall reduction of 75% would not have been estimated accurately by using the above approach in which average effectiveness of practices was considered to be additive. Using the additive approach, the sediment delivery would have been reduced to essentially zero if the mean effectiveness values for the five practices in Table 5-1 were used in the analysis.

In summary, the aggregate effectiveness of any system of management practices is a function of not only the mean effectiveness of individual practices, but also the interactions between the individual practices within the range of site-specific conditions experienced.

Monitoring and Tracking Techniques

Knowledge of land management activities and water quality conditions is important in many ways to efforts involving implementation of management measures and practices. As discussed in Chapter 5, the watershed planning process includes an understanding of the hydrologic resources, an assessment of environmental problems, goal setting, and priority setting. The development of action plans and implementation follow, with evaluation of effectiveness and revisions of plans as needed. Good water quality data are essential to problem identification and characterization, goal setting, priority setting, development of implementation plans, and evaluation. In order to have an understanding of what goals have to be met, a baseline must be established. Without good data regarding land management activities, including the control of point sources, accurate interpretation of the causes of water quality problems and improvements is not possible.

Water Quality Monitoring

Since the relationship between public health and water quality began to influence legislation in the early 1900s, water quality management and its related information needs have evolved considerably. Today, the Intergovernmental Task Force on Monitoring Water Quality (ITFM, 1995) defines water quality monitoring as an integrated activity for evaluating the physical, chemical, and biological character of water in relation to human health, ecological conditions, and designated water uses. Water quality monitoring for nonpoint sources (NPS) of pollution facilitates the important element of relating the physical, chemical, and biological characteristics of receiving waters to land use characteristics. Without current information on water quality conditions and pollutant sources, effects of land-based activities on water quality cannot be assessed, effective management and remediation programs cannot be implemented, and program success cannot be evaluated.

The most fundamental step in the development of a monitoring plan is to define the goals and objectives, or purpose, of the monitoring program. In general, monitoring goals are broad statements such as "to measure improvements in Hojnacki Creek" or "to verify nutrient load reductions into Stumpe Lake." In the past, numerous monitoring programs did not document this aspect of the design process and the resulting data collection efforts led to little useful information for decision making (GAO, 1986; MacDonald et al., 1991; National Research Council, 1986; Ward et al., 1990). As a result, the identification of monitoring goals is the first component of the design framework outlined by the ITFM (1995). Figure 6-1 presents one approach for developing a monitoring plan.

Monitoring programs can be grouped according to the following general purposes or expectations (ITFM, 1995; MacDonald et al., 1991):

❏ Describing and ranking existing and emerging problems

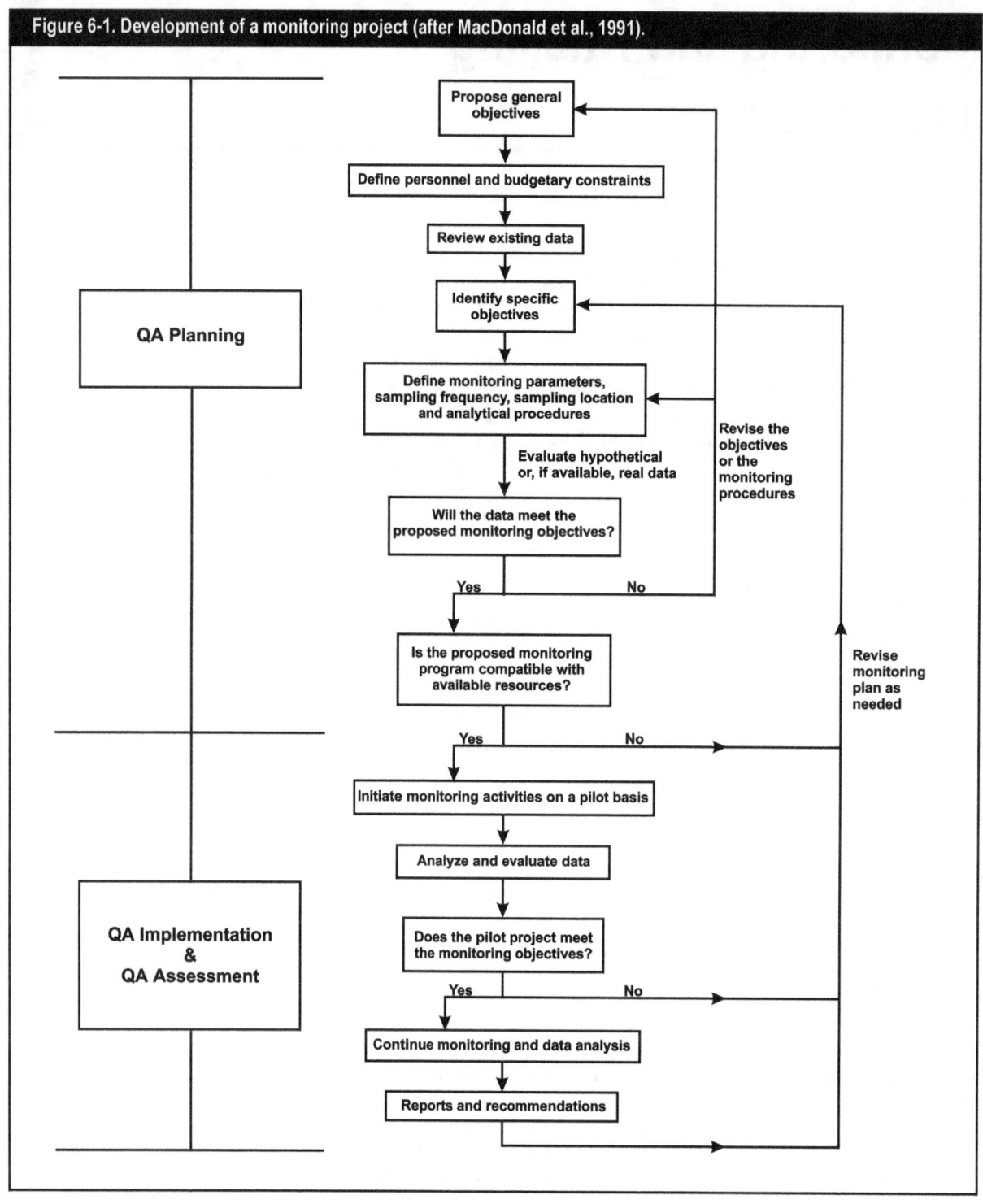

Figure 6-1. Development of a monitoring project (after MacDonald et al., 1991).

❏ Describing status and trends

❏ Designing management and regulatory programs

❏ Evaluating program effectiveness

❏ Responding to emergencies

❏ Describing the implementation of best management practices

❏ Validating a proposed water quality model

❏ Performing research

The importance of problem identification can not be underestimated. The water quality impairment (e.g., algal growth, sediment deposition, turbidity) must first be documented. Second, the pollutant(s) causing the impairments should be identified (e.g., nitrogen, phosphorus, soil erosion or streambank instability). This information can be used to facilitate the identification of pollutant sources. Water quality assessments and land use information are useful in identification of pollutant sources.

Unlike monitoring goals, monitoring objectives are more specific statements that can be used to complete the monitoring design process including scale, variable selection, methods, and sample size (Plafkin et al., 1989; USDA-NRCS, 1996b). Monitoring program objectives must be detailed enough to allow the designer to define precisely what data will be gathered and how the resulting information will be used. An example objective which would facilitate quantitative evaluations is "To detect a decrease in total phosphorus loading to Stumpe Lake via Hajnacki Creek by 50% over the next 6 years." Vague or inaccurate statements of objectives lead to program designs that provide too little or too much data, thereby failing to meet management needs or costing too much.

The remainder of the design framework outlined by the ITFM (1995) includes coordination and collaboration, design, implementation, interpretation, evaluation of the monitoring program, and communication. Numerous guidance documents have been developed, or are in development, to assist resource managers in developing and implementing monitoring programs that address all aspects of the ITFM's design framework. Appendix A in *Monitoring Guidance for Determining the Effectiveness of Nonpoint Source Controls* (EPA, 1997a) presents a review of more than 40 monitoring guidances for both point and NPS pollution. These guidances discuss virtually every aspect of NPS pollution monitoring, including monitoring program design and objectives, sample types and sampling methods, chemical and physical water quality variables, biological monitoring, data analysis and management, and quality assurance and quality control.

Appendix A in Monitoring Guidance for Determining the Effectiveness of Nonpoint Source Controls (EPA, 1997a) presents a review of more than 40 monitoring guidances for both point and NPS pollution.

Once the monitoring goals and objectives have been established, existing data and constraints should be considered. A thorough review of literature pertaining to water quality studies previously conducted in the geographic region of interest should be completed before starting a new study. The review should help determine whether existing data provide sufficient information to address the monitoring goals and what data gaps exist.

Identification of project constraints should address financial, staffing, and temporal elements. Clear and detailed information should be obtained in the time frame within which management decisions need to be made, the amounts and types of data that must be collected, the level of effort required to collect the necessary data, and equipment and personnel needed to conduct the monitoring. From this information it can be determined whether available personnel and budget are sufficient to implement or expand the monitoring program.

As with monitoring program design, the level of monitoring that will be conducted is largely determined when goals and objectives are set for a monitoring program, although there is some flexibility for achieving most monitoring objectives. Table 6-1 provides a summary of general characteristics of various types of monitoring.

The overall scale of a monitoring program has two components—a temporal scale and a geographic scale. The temporal scale is the amount of time required to accomplish the program objectives. It can vary from an afternoon to many years. The geographic scale can also vary from quite small, such as plots along a single stream reach, to very large, such as an entire river basin. The temporal and geographic scales, like a program's design and monitoring level, are primarily determined by the program's objectives.

If the main objective is to determine the current biological condition of a stream, sampling at a few stations in a stream reach over 1 or 2 days might suffice. Similarly, if the monitoring objective is to determine the presence or absence of a NPS impact, a synoptic survey might be conducted in a few select locations. If the objective is to determine the effectiveness of a nutrient management program for reducing nutrient inputs to a downstream lake, however, monitoring a subwatershed for 5 years or longer might be necessary. Collection of baseline information prior to implementation of improved management practices is important so that an improvement can be quantified. If the objective is to calibrate or verify a model, more intensive sampling might be necessary.

Depending on the objectives of the monitoring program, it might be necessary to monitor only the waterbody with the water quality problem or it might be

Table 6-1. General characteristics of monitoring types (MacDonald et al., 1991).

Type of Monitoring	Number and Type of Water Quality Parameters	Frequency of Measurements	Duration of Monitoring	Intensity of Data Analysis
Trend	Usually water column	Low	Long	Low to moderate
Baseline	Variable	Low	Short to medium	Low to moderate
Implementation	None	Variable	Duration of project	Low
Effectiveness	Near activity	Medium to high	Usually short to medium	Medium
Project	Variable	Medium to high	Greater than project duration	Medium
Validation	Few	High	Usually medium to long	High
Compliance	Few	Variable	Dependent on project	Moderate to high

necessary to include areas that have contributed to the problem in the past, areas containing suspected sources of the problem, or a combination of these areas. A monitoring program conducted on a watershed scale must include a decision about a watershed's size. The effective size of a watershed is influenced by drainage patterns, stream order, stream permanence, climate, number of land-owners in the area, homogeneity of land uses, watershed geology, and geomorphology. Each factor is important because each has an influence on stream characteristics.

There is no formula for determining appropriate geographic and temporal scales for any particular monitoring program. Rather, once the objectives of the monitoring program have been determined, a combined analysis of them and any background information on the water quality problem being addressed should make it clear what overall monitoring scale is necessary to reach the objectives.

Other factors that should be considered to determine appropriate temporal and geographic scales include the type of water resource being monitored and the complexity of the NPS problem. Some of the constraints mentioned earlier, such as the availability of resources (staff and money) and the time frame within which managers require monitoring information, will also contribute to determination of the scales of the monitoring program.

For additional details regarding NPS monitoring techniques, including chemical and biological monitoring, the reader is referred to *Monitoring Guidance for Determining the Effectiveness of Nonpoint Source Controls* (EPA, 1997a). This technical document focuses on monitoring to evaluate the effectiveness of management practices, but also includes approximately 300 references and summaries of more than 40 other monitoring guides. In addition, Chapter 8 of EPA's management measures guidance for Section 6217 contains a detailed discussion of monitoring with emphasis on coastal areas (EPA, 1993a). Another useful reference for monitoring design is the National Handbook of Water Quality Monitoring (USDA-NRCS, 1996b).

Tracking Implementation of Management Measures

The implementation of management measures may be tracked to determine the extent to which management measures are implemented in a watershed, recharge area, or other geographic area.

Implementation and trend monitoring can be used to address the following goals:

- ❏ Determine the extent to which management measures and practices are implemented in accordance with relevant standards and specifications.

- ❏ Determine whether there has been a change in the extent to which management measures and practices are being implemented.

- ❏ Establish a baseline from which decisions can be made regarding the need for additional incentives for implementation of management measures,

- ❏ Measure the extent of voluntary implementation efforts,

See EPA's *Monitoring Guidance for Determining Effectiveness of Nonpoint Source Controls* for details on NPS monitoring techniques.

❏ Support work-load and costing analyses for assistance or regulatory programs,

❏ Determine the relative adoption rates of various management measures across different geographic areas,

❏ Determine the extent to which management measures are properly maintained and operated.

Methods to assess the implementation of management measures are a key focus of technical assistance provided by EPA and NOAA.

Implementation assessments can be performed on several scales. Site-specific assessments can be used to assess individual management measures or practices, and watershed assessments can be used to look at the cumulative effects of implementing multiple management measures. With regard to "site-specific" assessments, individual practices must be assessed at the appropriate scale for the practice of interest. For example, to assess the implementation of management measures and practices for animal waste handling and disposal on a farm, only the structures, areas, and practices implemented specifically for animal waste management (e.g., dikes, diversions, storage ponds, composting facility, and manure application records) would need to be inspected. In this instance, the animal waste storage facility would be the appropriate scale and "site." To assess erosion control, the proper scale might be fields over 10 acres and the site could be 100-meter transect measurements of crop residue. For nutrient management, the scale and site might be an entire farm. Site-specific measurements can then be used to extrapolate to a watershed or statewide assessment. It is recognized that some studies might require a complete inventory of management measures and practice implementation across an entire watershed or other geographic area.

Sampling design, approaches to conducting the evaluation, data analysis techniques, and ways to present evaluation results are described in EPA's *Techniques for Tracking, Evaluating, and Reporting the Implementation of Nonpoint Source Control Measures – Agriculture* (EPA, 1997b). Chapter 8 of EPA's management measures guidance for Section 6217 contains a detailed discussion of techniques and procedures to assess implementation, operation, and maintenance of management measures (EPA, 1993a).

Determining Effectiveness of Implemented Management Measures

By tracking management measures and water quality simultaneously, analysts will be in a position to evaluate the performance of those management measures implemented. Management measure tracking will provide the necessary information to determine whether pollution controls have been implemented, operated, and maintained adequately. Without this information, analysts will not be able to fully interpret their water quality monitoring data. For example, analysts cannot determine whether the management measures have been effective unless they know the extent to which these controls were implemented, maintained, and operated.

A major challenge in attempting to relate implementation of management measures to water quality changes is determining the appropriate land management attributes to track. For example, a "bean count" of the number of management measures implemented in a watershed has little chance of being useful in statistical analyses that relate water quality to land treatment since the count will be only remotely related (i.e., a mechanism is lacking) to the measured water quality parameter (e.g., phosphorus concentration). Land treatment and land use monitoring should relate directly to the pollutants or impacts monitored at the water quality station (Coffey and Smolen, 1990). For example, the tons of animal waste managed may be a much more useful parameter to track than the number of confined animal facilities constructed. Since the impact of management measures on water quality may not be immediate or implementation may not be sustained, information on other relevant watershed activities (e.g., urbanization, growth in animal numbers) will be essential for the final analysis.

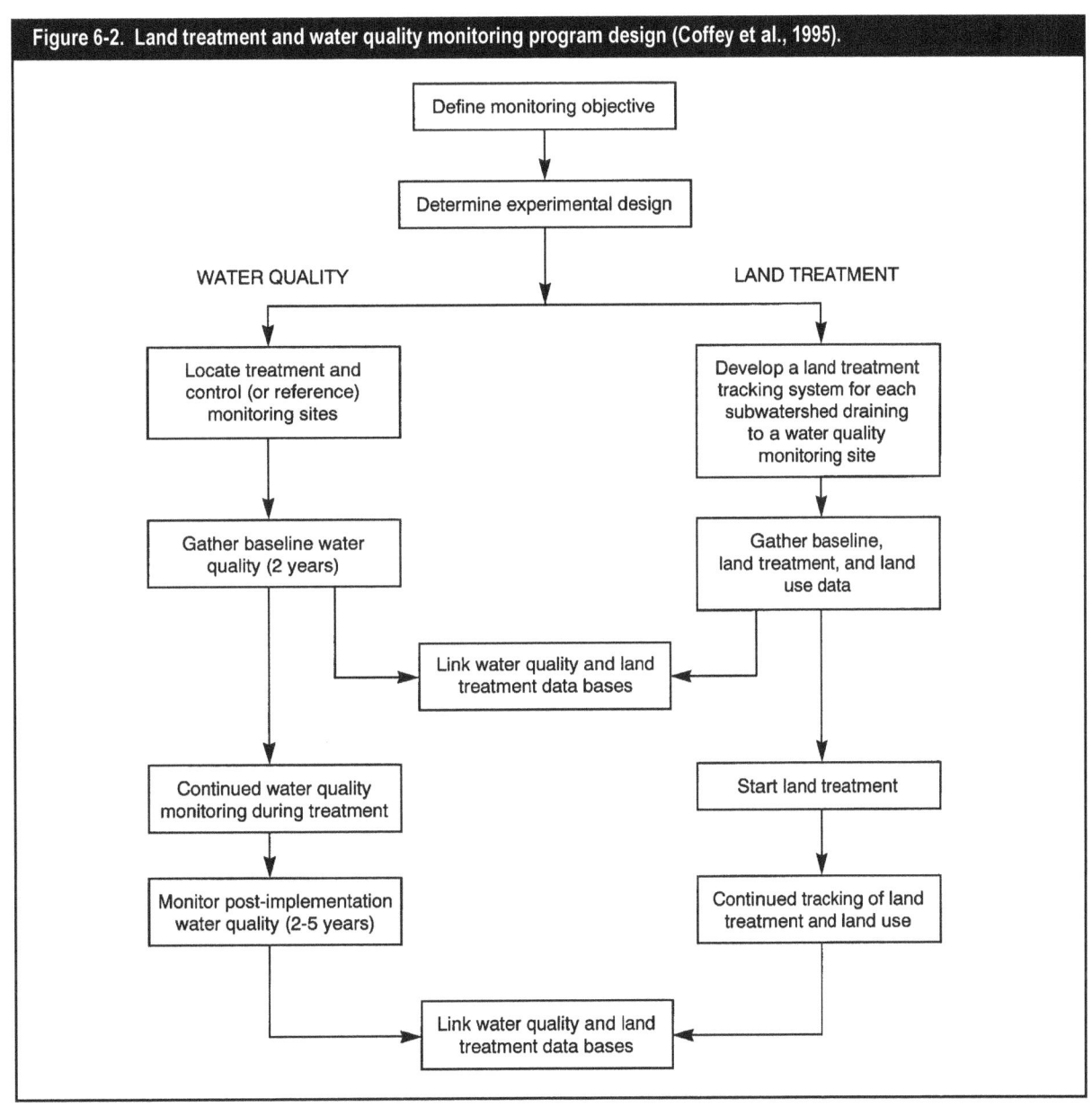

Figure 6-2. Land treatment and water quality monitoring program design (Coffey et al., 1995).

Water quality and land treatment monitoring must be coordinated to maximize the chance of meaningful results. In order to provide the manager with a sense of the nature of the coordination needed, an overview of monitoring program design is provided in Figure 6-2.

Monitoring program design, as shown in Figure 6-2, begins by defining the monitoring objective. Once the objective is defined, the experimental design (e.g., upstream/downstream, pre- and post-BMP, and paired watershed) is determined. Based on the experimental design, separate but coordinated parallel water quality and land treatment activities are specified.

Appropriately collected water quality information can be evaluated with trend analysis to determine whether pollutant loads have been reduced or whether water quality has improved. Valid statistical associations drawn between implementation and water quality data can be used to indicate:

(1) Whether management measures have been successful in improving water quality in a watershed or recharge area, and

(2) The need for additional management measures to meet water quality objectives in the watershed or recharge area.

Greater detail regarding methods to evaluate the effectiveness of land treatment efforts can be found in EPA's NPS monitoring guidance (EPA, 1997a) and management measures guidance for section 6217 (EPA, 1993a).

Quality Assurance and Quality Control

Introduction

Quality assurance (QA) and quality control (QC) are commonly thought of as procedures used in the laboratory to ensure that all analytical measurements made are accurate. Yet QA and QC extend beyond the laboratory and are essential components of all phases and all activities within each phase of a NPS monitoring project. This section defines QA and QC, discusses their value in NPS monitoring programs, and explains EPA's policy on these topics. The following sections provide detailed information and recent references for planning and ensuring quality data and deliverables that can be used to support specific decisions involving NPS pollution.

Definitions of Quality Assurance and Quality Control

Quality assurance is

> an integrated system of management procedures and activities used to verify that the quality control system is operating within acceptable limits and to evaluate the quality of data (Taylor, 1993; EPA, 1994a).

Quality control is

> a system of technical procedures and activities developed and implemented to produce measurements of requisite quality (Taylor, 1993; EPA, 1994a).

Quality control procedures include proper collection, handling, and storage of samples; analysis of blank, duplicate, and spiked samples; and use of standard reference materials to ensure the integrity of analyses. QC procedures also include regular inspection of equipment to ensure proper operation. Quality assurance activities are more managerial in nature and include assignment of roles and responsibilities to project staff, staff training, development of data quality objectives, data validation, and laboratory audits. Table 6-2 lists some common activities that fall under the headings of QA and QC. Such procedures and activities are planned and executed by diverse organizations through carefully designed quality management programs that reflect the importance of the work and the degree of confidence needed in the quality of the results.

Table 6-2. Common quality management activities (adapted from Drouse et al., 1986, and Erickson et al., 1991).

Quality Assurance

- Organization of project into component parts
- Assignment of roles and responsibilities to project staff
- Use of statistics to determine the number of samples and sampling sites needed to obtain data of a required confidence level
- Tracking of sample custody from field collection through final analysis
- Development and use of data quality objectives to guide data collection efforts
- Audits of field and laboratory operations
- Maintenance of accurate and complete records of all project activities
- Personnel training to ensure consistency of sample collection techniques and equipment use

Quality Control

- Collection of duplicate samples for analysis
- Analysis of blank and spike samples
- Replicate sample analysis
- Regular inspection and calibration of analytical equipment
- Regular inspection of reagents and water for contamination
- Regular inspection of refrigerators, ovens, etc. for proper operation

Importance of Quality Management Programs

Although the value of a quality management program might seem questionable while a project is under way, its value should be quite clear after a project is completed. If the objectives of the project were used to design an appropriate data collection and analysis plan, all procedures were followed for all project activities, and accurate and complete records were kept throughout the project, the data and information collected from the project will be adequate to support a choice from among alternative courses of action. In addition, the course of action chosen will be defensible based on the data and information collected. Development and implementation of a quality management program can require up to 10 to 20% of project resources (Cross-Smiecinski and Stetzenback, 1994), but this cost can be recaptured in lower overall costs due to the project's being well planned and executed. Likely problems are anticipated and accounted for before they arise, eliminating the need to spend countless hours and dollars resampling, reanalyzing data, or mentally reconstructing portions of the project to determine where an error was introduced. QA procedures and QC activities are cost-effective measures used to determine how to allocate project energies and resources toward improving the quality of research and the usefulness of project results (Erickson et al., 1991).

EPA Quality Policy

EPA has established a quality policy that requires the implementation of a quality system by EPA and by non-EPA organizations receiving financial assistance from EPA to ensure that data used in research and monitoring are of known and documented quality to satisfy project objectives. A quality system is developed by an organization and documented in writing. The system provides the policies, objectives, responsibilities, and procedures to be followed to ensure the quality of work processes, services, or products. A quality system is typically documented in a quality management plan (QMP). When conducting monitoring or tracking the implementation of management measures by collecting environmental data, site-specific written plans are needed to describe the quality objectives (acceptance or performance criteria) to be met so that the data can be used to support the particular decision(s) for which the data are being collected. Such site-specific plans are known as quality assurance project plans (QAPPs). The use of different methodologies, lack of data comparability, unknown data quality, and poor coordination of sampling and analysis efforts can delay the progress of a project or render the data and information collected from it insufficient for decision making. Whether or not EPA funding is involved, quality practices should be used as an integral part of the development, design, and implementation of an NPS monitoring project to minimize or eliminate these problems (Erickson et al., 1991; Pritt and Raese, 1992; EPA, 1997a).

Additional information on developing quality programs can be found in EPA publications (e.g., EPA, 2000; 2001a, b;), available on the Internet at http://www.epa.gov/quality/qa_tools.html.

Load Estimation Techniques

7

A pollutant *load* is the mass or weight of pollutant transported in a specified unit of time from pollutant sources to a waterbody. The loading rate, or *flux*, is the instantaneous rate at which the load is passing a point of reference on a river, such as a sampling station, and has units of mass/time such as grams/second or tons/day (Richards, 1997). Mathematically, the load is the integral over time of the flux.

Pollutant load estimation is a fundamental element in the development of many watershed management plans. Reliable estimates of the quantity of pollutants delivered from various sources within a watershed are needed to develop a watershed plan that will address the identified water quality problems or issues. Establishing the link between an identified water quality problem and the sources causing the problem often entails a mass balance analysis, a quantitative accounting of the sources and sinks of the pollutants of interest.

There are many reasons for developing management plans, including the development and implementation of a total maximum daily load (TMDL) pursuant to the requirements of section 303(d) of the Clean Water Act (see Highlight). For those waters either not supporting or not projected to support designated uses even after the implementation of point source or other required pollution controls, a TMDL is needed. The components of TMDL development are:

1. Problem Identification

2. Identification of Water Quality Indicators and Target Values

3. Source Assessment

4. Linkage Between Water Quality Targets and Sources

5. Allocation

6. Follow-up Monitoring and Evaluation Plan

7. Assembling the TMDL

It is important to note that TMDL development is a very site-specific process. Therefore, these components are not necessarily sequential steps but can be conducted concurrently or iteratively depending upon the situation (EPA, 1999b).

In source analysis for a TMDL, the relative contributions of different sources are assessed. An estimate of pollutant loads from both point sources and nonpoint sources is essential to this analysis, as is the ability to determine if the load reduction needed to meet water quality standards can be achieved under different management scenarios (e.g., implementation of the management measures). The load allocation for nonpoint sources (and the wasteload allocation for point sources) is determined from an analysis that links the desired endpoints (e.g., achievement of a water quality standard) to various management alternatives that could be applied to the identified sources.

Clean Water Act
Total Maximum Daily Load (TMDL) Program

Section 303(d) of the Clean Water Act and EPA's implementing regulations at 40 CFR Section 130.7 require States to develop TMDLs for their waterbodies that do not or are not expected to meet applicable water quality standards after the application of technology-based point source or other required pollution controls. EPA's regulations at 40 CFR Section 130.2 define some of the elements of the TMDL programs. These include:

- **Loading capacity** – The greatest amount of loading that a water can receive without violating water quality standards.

- **Load allocation** – The portion of a receiving water's loading capacity that is attributed either to one of its existing or future nonpoint sources of pollution or to natural background sources.

- **Wasteload allocation** – The portion of a receiving water's loading capacity that is allocated to one of its existing or future point sources of pollution.

- **Total maximum daily load (TMDL)** – The sum of the individual wasteload allocations for point sources, load allocations for nonpoint sources, natural background, and a margin of safety. TMDLs can be expressed in terms of either mass per time, toxicity, or other appropriate measure that relate to a State's water quality standard. A margin of safety is required as part of each TMDL to account for the uncertainty about the relationship between the pollutant loads and the quality of the receiving waterbody.

- **Water quality-limited segments** – Those water segments that do not or are not expected to meet applicable water quality standards by the next listing even after the application of technology-based effluent limitations for point sources as required by sections 301(b) and 306 of the Clean Water Act. Technology-based controls include, but are not limited to, best practicable control technology currently available and secondary treatment.

- **Margin of Safety** – Element of a TMDL that accounts for uncertainty and lack of knowledge. A margin of safety may be expressed as unallocated assimilative capacity or conservative analytical assumptions used in establishing the TMDL and its maximum allowable pollutant load.

EPA Protocols for TMDL Development

Protocol for Developing Pathogen TMDLs: First Edition, January 2001, EPA 841-R-00-0002.
www.epa.gov/owow/tmdl/pathogen_all.pdf

Protocol for Developing Nutrient TMDLs: First Edition, November 1999, EPA 841-B-99-007.
www.epa.gov/owow/tmdl/nutrient/pdf/nutrient.pdf

Protocol for Developing Sediment TMDLs: First Edition, October 1999, EPA 841-B-99-004.
www.epa.gov/owow/tmdl/sediment/pdf/sediment.pdf

The following sections present some basic information regarding monitoring and modeling to estimate pollutant loads. References to more detailed treatments of the topics are included as well. Additional information on TMDL is available at *www.epa.gov/owow/tmdl/*.

Estimating Pollutant Loads Through Monitoring

Every monitoring effort should have clearly stated objectives. The estimation of pollutant loads is a general objective that should be refined to clarify the monitoring needs. The specific reasons why the pollutant loads are to be estimated could affect decisions regarding the required precision and the conditions under which monitoring should be conducted. For example, if the pollutant is bacteria and the watershed management concerns are associated with the instantaneous value and the 30-day geometric mean (of 5 or more samples), then the sampling protocol should consider multiple samples at a sufficient frequency to calculate the geometric mean as well as evaluate the various conditions under which loading occurs (wet and dry weather). On the other hand, if nutrients are causing accelerated eutrophication in a reservoir then it may only be important to estimate seasonal loads. The time scales and frequency of monitoring needed will be a function of the critical conditions and the receiving water response to the loading of the pollutant of concern.

The averaging period for loading estimates may be hourly, daily, monthly, or longer depending upon site-specific conditions and needs. The variability of loads within the average period of interest and the certainty with which water quality standards violations need to be documented will drive decisions regarding sampling design and frequency. The importance of clearly stated objectives is described more fully in existing monitoring guides (EPA, 1997a; EPA, 1991c; USDA-NRCS, 1996b). Due to the importance of statistical considerations, those designing monitoring plans are strongly encouraged to seek assistance from a trained statistician with experience in water monitoring.

Components of a Load

To estimate pollutant loading, it is necessary to sum the flux, which is commonly expressed as mass per unit time, over the period of interest. Since the flux varies with time, this summing process can be expressed in integral form as shown in the first equation of the following text box. Since flux cannot be measures directly, flux is often expressed as the product of concentration and flow (see second equation of the text box). Thus the three basic steps for estimating pollutant load are:

- ❏ measuring water discharge (e.g., cubic meters per second),
- ❏ measuring pollutant concentration (e.g., milligrams per liter), and
- ❏ calculating pollutant loads (multiplying discharge times concentration over the time frame of interest).

Since concentration and flow vary with time, the key challenge in measuring loads is to determine when to sample to obtain the best estimate at least cost. Richards (1997) points out that it is not uncommon for 80 to 90% or more of the annual load to be delivered during the 10% of the time which corresponds with high fluxes. Depending on the constituent being evaluated, fluxes during snowmelt and storm events are often many times greater than those during periods of low flow (i.e., dry weather conditions). Thus, monitoring programs must be

Load and Flux

The pollutant load is the integral over time of the flux:

$$Load = k \int_t flux(t)\ dt$$

where k is a constant for converting units, and t is time.

Since we cannot measure flux directly, we measure it as the product of concentration and discharge.

$$Load = k \int_t c(t)q(t)dt$$

where c(t) is the concentration at time=t, and q(t) is the water discharge at time=t.

designed with full consideration given to both periods of pollutant flux. The following equations present the mathematical relationship between load, flux, and time.

Measuring Water Discharge

The major options for monitoring stream discharge are flumes, weirs, natural channels, and existing structures (USDA-NRCS, 1996b; Brakensiek et al., 1979). Device selection for *stream discharge* is a function of site-specific conditions such as slope, sediment load, and stream size. Selection of a device for *runoff* measurement depends on peak runoff rate, runoff variability, the extent to which trash and debris are carried in the runoff, icing conditions, and other factors (Brakensiek et al., 1979). Discharge monitoring approaches, and the selection, implementation, and use of various devices are described by Brakensiek et al. (1979) and USDA-NRCS (1996b).

For established gaging stations, flow measurements are relatively inexpensive to make, and are available almost on a continuous basis (Richards, 1997). It is, however, likely that gaps in the flow record will still occur as a result of equipment failure, operational errors, or extreme flow events. Methods to fill gaps in flow records are described by Brakensiek et al. (1979) and USGS (Rantz et al., 1982).

Measuring Pollutant Concentration

Periodic measurements of pollutant levels in water are used in load estimation. The frequency of the measurements required to adequately characterize pollutant concentrations over time is often difficult to determine. Pollutants such as nitrate-nitrogen often do not vary greatly over weekly or monthly intervals while pollutants such as fecal coliform can vary by several orders of magnitude during a week depending on hydrologic and other conditions. The vast majority of nonpoint source load estimations will require storm event sampling. The choice of sampling frequency for load estimation is a complex function of watershed hydrology, pollutant(s) of interest, land use/management, the duration of monitoring and the water resource type. Periodic measurements in the field (in situ or sample analysis with a field kit) or laboratory measurements performed on collected water samples are typically used to provide the pollutant concentration values that will be used in load estimation.

Water sampling approaches have been categorized in several ways, some based more upon the equipment used, and others based more upon the statistical design employed (USDA-NRCS, 1996b; EPA, 1979; EPA, 1991c). Grab, point, composite, integrated, continuous, random, systematic, and stratified sampling are frequently described in the literature. In practice, sampling involves a decision regarding the population and population units to be sampled (e.g., instantaneous concentration at single point or integrated over depth, average concentration at single point or integrated over depth for a specified time interval or flow interval), a determination of the statistical approach to be used (e.g., simple random sampling, stratified random sampling, systematic sampling), and a choice of sampling equipment and configuration (e.g., grab sample taken manually or automatically with a mechanical sampler, time-weighted or flow-weighted sampling with a programmed mechanical sampler).

For any given watershed, the best approach for estimating loads will be determined based upon the needs and characteristics of the watershed. Still, some general rules-of-thumb should be considered (USDA-NRCS, 1996b; Richards, 1997).

❑ **Accuracy and precision increase with increased frequency of sampling.**

❑ **Grab, Point, or Instantaneous Samples** — may be insufficient to determine loads unless concentrations are correlated to discharge which is measured continuously.

❑ **Depth-Integrated and Width-Integrated Grab Samples** — can account for stratification in concentration with depth or horizontally across a stream, but still depends upon correlation to discharge for suitability in load estimation.

❑ **Time-Weighted Composite Samples** — not generally sufficient for load estimation since they may not adequately reflect changes in discharge and concentration during the period over which samples are composited.

❑ **Flow-Weighted Composite Samples** — well-suited to load estimation, but difficult to collect since stage-discharge relationship is needed and a "smart sampler" is needed to trigger sampling as a function of flow rate. Projecting sample size and number of bottles needed is difficult.

❑ **Systematic Sampling** — as efficient as, or more efficient than, simple random sampling if the sampling interval is not equal to a multiple of any strong period of fluctuation in the sampled population (e.g., sampling weekly on the day when a particular pollutant is always at its peak level due to scheduling by a discharger).

❏ **Stratified Random Sampling** — with most samples taken during periods of high flow, can be of great importance in providing increased precision for a given number of samples.

Types of Water Samples

Grab Sample — A single sample taken at one place a single time.

Composite Sample — A series of grab samples, usually collected in the same location but at different times, combined to form one sample for analysis. Composite samples are usually:

Flow-Weighted – Sample is taken after a specified quantity of water has passed the monitoring station (e.g., draw 10 ml sample every 750,000 liters of flow); or

Time-Weighted – A pre-determined sample volume is taken at a predetermined time interval (e.g., draw 10 ml sample every 15 minutes).

Integrated Sample — Subsamples are taken at various depths or distances from the stream bank, and integrated into a single sample.

Continuous Sample — Probes are used to continuously record contaminant concentration in stream. Not widely applicable to nonpoint source programs.

For many TMDLs, the daily pollutant load may be the population unit of greatest importance. In these cases, sampling should emphasize obtaining accurate estimates of daily loads for the pollutant of interest. Since TMDLs establish maximum wasteload and load allocations that can be discharged without violating water quality standards, the monitoring effort should provide the data necessary for determining whether or not quality standards are met. For example, if water quality standards are more likely violated under low-flow (dry weather) conditions, then the monitoring should provide reliable data regarding low-flow loads. Conversely, in cases where water quality standards are violated during high-flows (wet weather or snowmelt) or as a result of loads from high flows, the monitoring should emphasize high-flow monitoring. In other cases, such as those in which annual or seasonal loads are critical, high quality estimates of low-flow and high-flow loads may be equally important.

Sampling location should be determined based upon the monitoring objectives, water resource characteristics, and source characteristics. For example, it may be appropriate to sample at the outlets of tributaries to a lake, or above and below a farm or set of farms, depending upon whether the objective is to estimate lake loading from tributary watersheds or stream loading from an individual farm or farms. Additional information regarding sampling location can be found in existing guides (EPA, 1997a; USDA-NRCS, 1996b; Ponce, 1980).

Detailed discussions of statistical sampling approaches (e.g., random sampling) can be found in several sources (EPA, 1997a; Richards, 1997; USDA-NRCS, 1996b; Gilbert, 1987). Older sampling equipment is described by Brakensiek, et al. (1979), while USDA-NRCS (1996b) provides an overview of more current devices, including a helpful list of references regarding sampling equipment.

Calculating Pollutant Loads

The pollutant load is the integral of flux over time, but flux cannot be measured directly (Richards, 1997). In Figure 7-1 the flux is calculated as the product of concentration and discharge, with appropriate conversion units. Each calculated flux is a discrete value that is assumed to apply across the sampling interval, which is 24 hours in this hypothetical example (daily composites). The cumulative load in Figure 7-1 is determined by adding the calculated fluxes over all sampling intervals.

Because there will be more discharge data than concentration data in almost all chemical monitoring efforts, there will be a need to make estimates of concentration, and therefore pollutant flux, for periods between water quality observations (Richards, 1997). Figure 7-2 illustrates how missing values can greatly affect the calculated load estimates. Load A is the same load as shown in Figure 7-1, whereas Load B was calculated after deleting every other concentration value used to calculate Load A.

Data gaps can be filled by estimating missing concentration values for pairing with the flow data, or by adjusting the load estimate made from the observations where both flow and concentration were measured (Richards, 1997). Flow data typically form the basis for making flux estimates for periods during which water quality (concentration) data are lacking.

Some of the methods for estimating pollutant loads include numeric integration, the worked record procedure, averaging approaches, the flow interval technique, ratio estimators, regression approaches, and flow-proportional sampling (Richards, 1997). A review of evaluative studies of loading approaches has resulted in the following points of consensus (Richards, 1997):

❑ **Averaging methods** (e.g., for monthly or quarterly loads) are generally biased, and the bias increases as the size of the averaging window increases and/or the number of samples decreases. For example, an annual load determined by adding four quarterly loads will generally be

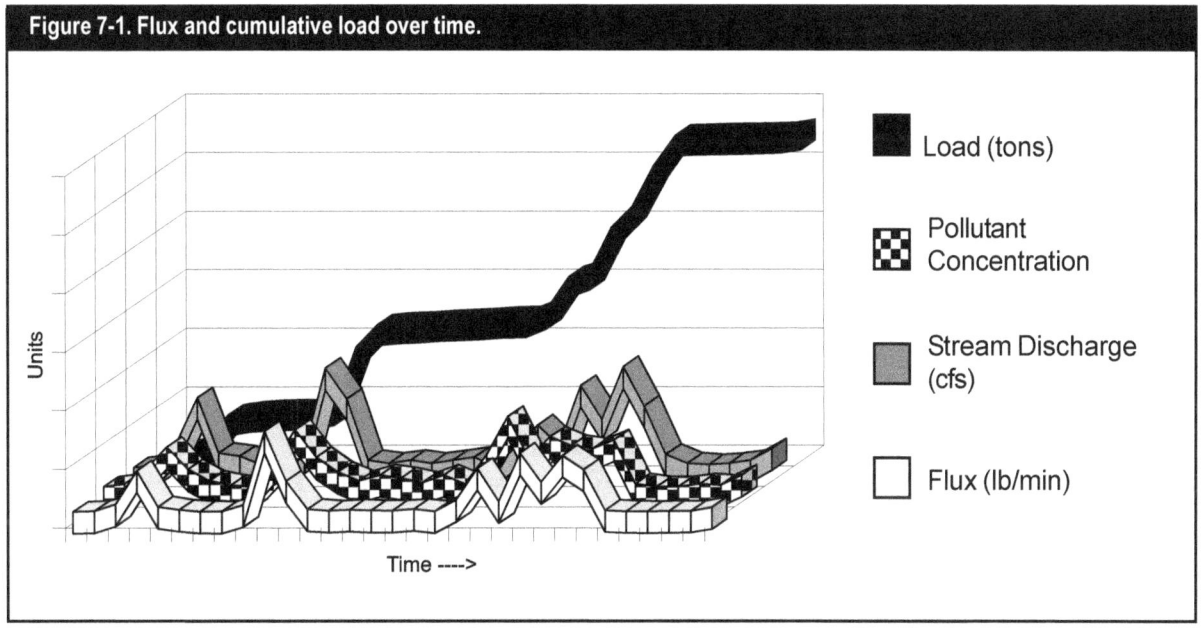

Figure 7-1. Flux and cumulative load over time.

Load (tons)

Pollutant Concentration

Stream Discharge (cfs)

Flux (lb/min)

Units

Time ---->

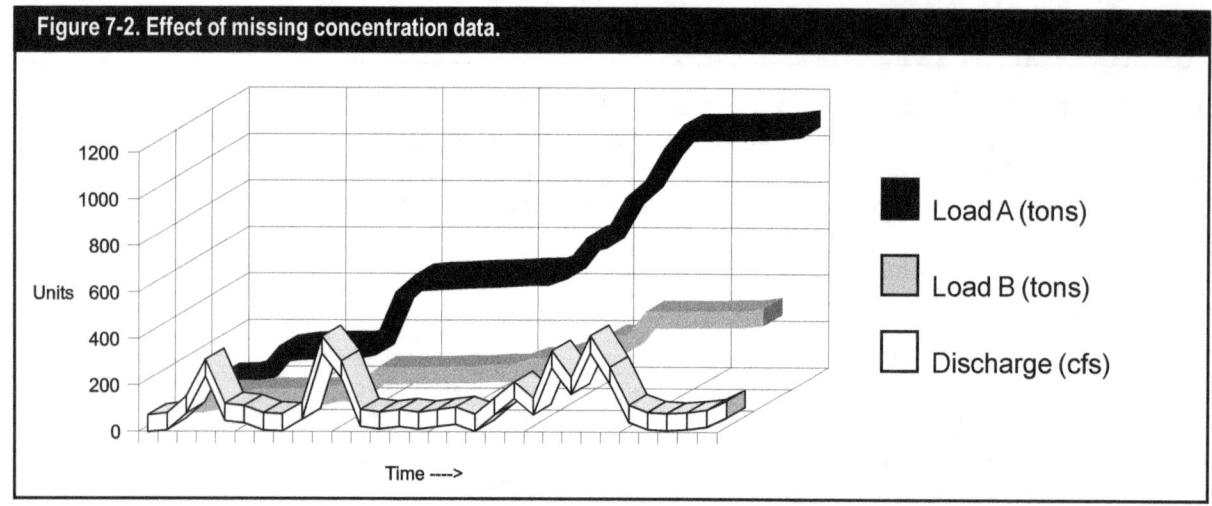

Figure 7-2. Effect of missing concentration data.

more biased than an annual load determined by adding 12 monthly loads.

❏ In most studies, **ratio approaches** performed better than **regression approaches**, and both performed better than **averaging approaches**.

❏ **Regression approaches** can perform well if the relationship between flow and concentration is well-defined, linear throughout the range of flows, and constant throughout the year.

Greater detail and illustrative examples regarding averaging approaches, regression approaches, ratio estimators, and sampling approaches can be found in Richards (1997).

Estimating Pollutant Loads Through Modeling

Types of Models Available

Loading models include techniques primarily designed to predict pollutant movement from the land surface to waterbodies (EPA, 1997d). *Watershed loading models* range from simple loading rate assessments in which loads are a function of land use type only, to complex simulation techniques that more explicitly describe the processes of rainfall, runoff, sediment detachment, and transport to receiving waters. Some loading models operate on a watershed scale, integrating all loads within a watershed, and some allow for the subdivision of the watershed into contributing subbasins.

Field-scale models, which have traditionally specialized in agricultural systems, are loading models that are designed to operate on a smaller, more localized scale. Field-scale models have often been employed to aid in the selection of management measures and practices. For example, a dynamic simulation model was used to predict the long-term patterns of phosphorus export from fields under a variety of management scenarios (Cassell and Clausen, 1993). The process model simulated the annual inputs and outputs of phosphorus, and was determined by the authors to be useful for simulating long-term patterns. Process models such as this one, however, are dependent upon local export coefficients and a thorough understanding of pollutant transport processes.

Water runoff, sediment delivery, and nutrient loading can be estimated using watershed models. Match modeling objectives, staff expertise, data requirements, and available budget for proper model selection.

Methods for Estimating Pollutant Loads (Richards, 1997)

Numeric Integration — Total load is calculated as the sum of the individual loads calculated for each sample.

Worked Record Procedure — Chemical observations are plotted onto a detailed hydrograph, and smooth curves are drawn through chemical data points based upon analyst's experience with the relationship of concentration and flow.

Averaging Approaches — Calculation that uses averaging of concentration and/or flow to estimate loads. For example, analyst might multiply average weekly suspended solids concentration by daily flow to estimate daily loads for the week.

Flow Interval Technique — Semi-graphical technique that calculates "interval loads" as the product of average flux for a range of daily flow values times the number of days in which flows were within the particular flow range.

Ratio Estimators — Total loads are estimated using a known relationship between the less-frequently sampled parameter of interest and a more-frequently sampled parameter (e.g., discharge) to fill gaps in the data record for the parameter of interest.

Regression Approaches — Relationship is established between concentration and flow based on samples taken, and then applied to estimate concentration for days not sampled.

Flow-Proportional Sampling — Mechanical approach in which representative samples are taken to determine concentration for a known discharge. Pollutant load is calculated as the sum of the sample concentrations multiplied by the measured discharge.

Other types of models include *receiving-water models*, which emphasize the response of a waterbody to pollutant loadings, flows, and ambient conditions, and *ecological models* that simulate biological communities and their response to stressors such as toxics and habitat modification (EPA, 1997d). *Integrated modeling systems* link models, data, and a user interface within a single system. The advent of geographic information systems (GIS) has facilitated the development of and expanded the capabilities of integrated modeling systems.

The emphasis of this section will be on watershed loading models. The reader is encouraged to seek additional information regarding field-scale, ecological, and integrated models in existing documents (EPA, 1997d; EPA, 1992b). The reader can also consult Chapter 5 of this manual for information on *field-* and *water-shed-scale* models.

Watershed Loading Models

Watershed loading models are configured and characterized in several ways (see Modeling Jargon), but they can be grouped into three general categories: *simple methods, mid-range models*, and *detailed models* (EPA, 1997d). The defining characteristics of models are the degree to which processes (and complexities of systems) are simplified and the time scale that is used for analysis and display of output information.

Simple methods are generally used to provide quick and easy identification of critical pollutant sources in the watershed. Detailed watershed models represent the other extreme, featuring costly and time-consuming efforts to provide quantitative estimates of pollutant loads from a range of management alternatives. Richards (1997) cautions that modeling of agricultural settings is often inadequate to evaluate the success of management practices in reducing loads because there are mixed land uses that change annually and these land uses have different loading rates. An additional concern is that most models fail to adequately address stream channel and bank dynamics, including the impact of management practices on these factors. Some detailed models such as GLEAMS, however, attempt to capture the variability associated with cropping practices and rotations in the agricultural setting.

Mid-range watershed models are generally midway between the cost, complexity, and accuracy of simple methods and detailed watershed models. Mid-range models provide qualitative estimates of management alternatives (EPA, 1997d).

Figure 7-3 shows examples of models and integrated modeling systems for load estimation. EPA's *Compendium of Tools for Watershed Assessment and TMDL Development* has additional details regarding the capabilities, limitations, and data requirements for these and other models (EPA, 1997d).

Simple Watershed Methods

Uses

- ❏ Support assessment of relative significance of sources
- ❏ Guide decisions for management plans
- ❏ Focus continuing monitoring efforts

Features

- ❏ Typically derived from empirical relationships between physiographic characteristics of the watershed and pollutant export
- ❏ Often applied using a spreadsheet or hand-held calculator

Pros

- ❏ Rapid
- ❏ Minimal data requirements (large-scale aggregation; low resolution)
- ❏ Minimal effort

Cons

- ❏ Output is typically mean annual values or storm loads
- ❏ Rough estimates of loadings
- ❏ Very limited predictive capability
- ❏ Low transferability to other regions due to empirical basis
- ❏ Do not consider degradation and transformation processes
- ❏ Few incorporate detailed representation of pollutant transport within and from watershed
- ❏ Cannot adequately account for most management practices

Figure 7-3. Load estimation models.

Simple Methods

- EPA Screening
- Simple Method
- Regression Method
- SLOSS-PHOSPH
- Federal Highway Administration Model
- Watershed Mangement Model

Mid-Range Models

- SITEMAP
- GWLF
- Urban Catchment Model
- Automated Q-ILLUDAS
- AnnAGNPS
- SLAMM

Detailed Models

- STORM
- DR3M-QUAL
- SWRRBWQ
- SWMM
- HSPF

Field-Scale Loading Models

- CREAM/GLEAMS
- Opus
- WEPP

Integrated Modeling Systems

- PC-VIRGIS
- WSTT
- LWMM
- GISPLM
- BASINS

Mid-Range Watershed Models

Uses

❏ Assist in defining target areas for pollution mitigation programs on watershed basis

❏ Support relative comparisons of management alternatives

Features

❏ Compromise between empiricism of simple methods and complexity of detailed mechanistic models

- Use simplified relationships for the generation and transport of pollutants
- Greater reliance on site-specific data than for simple methods
- Can address land use patterns and landscape configurations in watersheds

❏ Typically require some calibration with additional data sets

❏ Often tailored to site-specific applications (e.g., agriculture only)

Pros

❏ Can assess seasonal or inter-annual variability of loadings, and long-term water quality trends

❏ Those with continuous simulation can compare storm-driven loads over a range of storm events or conditions

❏ Those with GIS interface facilitate parameter estimation

❏ Relatively broad range of regional applicability

❏ Usually include detailed input-output features to simplify processing

❏ Often have built-in graphical and statistical capabilities

Cons

❏ Use of simplifying assumptions can limit accuracy of predictions

❏ Most do not consider degradation and transformation processes

❏ Few incorporate detailed representation of pollutant transport within and from watershed

❏ Can not account for most management practices

Detailed Watershed Models

Uses

❑ If properly applied, can provide accurate estimates of pollutant loads and impacts on water

❑ Identify causes of problems rather than simply describing overall conditions

Features

❑ Use storm event or continuous simulation to predict flow and pollutant concentrations for a range of flow conditions (small calculation time steps)

❑ Algorithms more closely simulate the physical processes of infiltration, runoff, pollutant accumulation, instream effects, and ground/surface water interaction

Pros

❑ Input/output have greater spatial and temporal resolution than simple and mid-range models

❑ Detailed hydrologic simulations can be used to design potential control actions

❑ Linkage to biological modeling is possible

❑ Those with new interfaces and GIS linkages facilitate use of models

❑ Provide relatively accurate predictions of variable flows and water quality at any point in a watershed if properly applied and calibrated

Cons

❑ Considerable time and expenditure required for data collection and model application

❑ Complex — not easily utilized by untrained staff

❑ Require rate parameters for flow velocities, settling, decay, and other processes

❑ Input data file preparation and calibration require professional training and adequate resources

---

Planning and Selection of Models

Setting modeling objectives should be the first step in developing a modeling approach. In some cases, the objectives may be achievable using a simple model, but in other cases it may be necessary to perform complex modeling involving more than one model. Criteria that apply in selecting a model may include the value of the resource under consideration, data needs, hardware needs, cost, accuracy required, type of pollutants/stressors, management considerations such as long-term commitment to the modeling effort, availability of trained personnel, user experience with the model, and acceptance of the model (EPA, 1997d). It is also important in many cases to involve stakeholders from the outset of modeling exercises to increase the potential for broad acceptance of modeling results.

The following steps can be used to define the modeling approach (EPA, 1997e):

1. Use available information to develop a good understanding of watershed characteristics, watershed problems, and watershed hydrology.

2. Consult with program and project managers to develop a clear understanding of project needs and modeling objectives.

3. Select a model or models that best meet the project needs and modeling objectives.

---

4. Choose the processes to be simulated and the level of complexity, and focus on
    the processes that govern the problems of concern.

5. Segment the watershed to the desired degree of complexity including the number of subwatersheds, reaches, and land use categories.

6. Choose a simulation process such as single-event or continuous simulation based upon the specified modeling objectives and the system being modeled.

7. Select the time step and imulation time frame necessary to meet the modeling
    objectives.

Modeling Jargon
Terms You Should Know When Communicating With Modelers

Deterministic models — Mathematical relationships based on physical or mechanistic processes are represented in the model. For example, runoff output is produced in response to precipitation input.

Empirical models — Mathematical relationships in the model (i.e., coefficients for parameters) are based upon measured data rather than theoretical relationships. Must be calibrated.

Steady-state models — Mathematical model of fate and transport that uses constant values of input variables to predict constant values (e.g., receiving water quality concentrations).

Dynamic models — Mathematical model describing the physical behavior of a system or process and its temporal variability.

Hydrodynamic models — Mathematical model that describes circulation, transport, and deposition processes in receiving waters.

Physical models — The building of a scale model of the system and testing it.

Distributed parameter models — Incorporate the influences of the spatially variable, controlling parameters (e.g., topography, soils, land use) in a manner internal to its computational algorithms (EPA, 1982b). Allows simultaneous simulation of conditions at all points within the watershed. Also facilitates incorporation of equations that represent unique processes that occur at only specific points in the watershed.

Lumped parameter models — Use average values for characterizing the influence of specific, non-uniform distributions of each parameter (e.g., soil type, cover, slope steepness).

Calibrated models — Require calibration with measured data for each site-specific application.

"Uncalibrated" or measured-parameter models — Can be used without calibration. Use measured or estimated parameters.

Event-based simulation — Modeling of individual storms. Does not simulate, or account for, periods between storms.

Annualized — Modeling of a longer time series than individual storms. Event-based model outputs can be annualized.

---

8. Design a model calibration and validation process, including data requirements.

9. Evaluate the assumptions and limitations of the modeling approach.

10. Develop a post-processing data analysis and data interpretation plan.

For applications to nonpoint source problems, the key features of nonpoint sources of pollution need to be fully considered, including but not limited to the following:

1. Hydrology (i.e., rainfall, snowmelt, and sometimes irrigation) drives the process.

2. Pollutant sources are land-based and distributed, with pollutant loads often highly variable in both space and time.

3. Land use types range from highly urbanized to undisturbed forest.

4. Management measures and practices vary from non-structural (e.g., nutrient management) to structural (e.g., waste storage ponds).

5. Land management and land cover change over time, including seasonal fertilization, tillage, crop growth, road maintenance, and off-season inactivity.

Additional considerations and details regarding modeling approach, model selection, and data requirements can be found in existing guidance documents (EPA, 1997d; EPA, 1985).

Model Calibration and Validation

The analyst must evaluate how the model will be used to address management or future conditions. The adequacy of the calibration and validation can be evaluated based on consideration of the type of changes expected to occur, the types of management expected, and the loading and assimilation processes that dominate the system. In some cases, changes in land use distribution can be modeled well by a calibrated system. In other cases, a new land use, such as a new crop, may require that supplemental calibration be performed to account for its unique features. Detailed discussions of model calibration and validation steps and procedures can be found in existing documents (EPA, 1997d; EPA, 1993b; EPA, 1989b; EPA, 1985; ASCE, 1993; Haan et al., 1995; Donigian, 1983).

A very important consideration in estimating nonpoint source loads is the quality and representativeness of the water quality data used in model *calibration*. A water quality data set that does not include a representative sample of high-flow events is unlikely to yield a calibration that is relevant to the concern addressed in the modeling effort. For example, if the goal is to determine the extent to which phosphorus loads are reduced through the implementation of management measures in a watershed dominated by agricultural nonpoint source impacts, it is important that runoff conditions are represented adequately in the calibration.

It is also important that the water quality data used in model calibration cover the same range of wet and dry conditions that are to be used in model validation and prediction. For example, measured loads to New York's Owasco Lake were greater than estimates generated by a simple unit-area loading method due largely to the fact that the measured loads were based on sampling during wet years (Heidtke and Auer, 1993). The simple model used in this example does not explicitly represent rainfall runoff processes, and is therefore very sensitive to the conditions under which it is developed. An adjustment of loading coefficients based upon data from the wet years would likely result in over-prediction of long-term average annual loads.

Successful model *validation* should not be blindly interpreted to prove that a model has predictive capabilities. In some cases, the calibration and validation data sets may come from the same period prior to implementation of control measures and practices. For example, if a data set from a period prior to implementation of measures or practices is arbitrarily split in half, with half of the data used for calibration and the other half used for validation, then validation merely confirms that the model can represent conditions prior to implementation of controls. If the measures and practices are intended to change pollutant loads through source reduction, delivery reduction, and/or runoff attenuation, then post-implementation water quality and flow may (and are expected to) respond very differently to precipitation events as compared to pre-implementation conditions. Thus, the model has not really been proven as a predictive tool because the ability to forecast a change in water quality and flow has not been tested with a data set that reflects the changed response to precipitation. Even if the calibration and validation data sets are determined to be independent through statistical analyses, the predictive capabilities are not proven through successful validation unless the validation data set is derived from or reflects conditions of the modeled "future" condition. This is not to say, however, that validation is not important. Successful validation will increase the credibility of modeling results, but the results must be interpreted with care.

> **Calibration** — process of adjusting model input parameters to cause model output values to more closely agree with corresponding observed values.
>
> **Validation** — comparison of model results with an independent data set (without further adjustment).
>
> **Verification** — examination of the numerical technique in the computer code to ascertain that it truly represents the conceptual model and that there are not inherent numerical problems.

Model Calibration and Validation
A good calibration using bad data is a bad calibration.

❏ Ensure that the water quality data used in the calibration and validation process are representative of the true distribution of water quality conditions in the watershed.

- *Don't* use data sets with only low-flow concentrations to simulate high-flow conditions.
- *Do* use data sets with concentration values covering the range of flow and land management conditions in the watershed.

❏ Land use and land management data should be logically linked both to the water quality parameters simulated and to the sources and management measures and practices that will be implemented.

- *Don't* calibrate nutrient concentrations against general land use variables that cannot be logically linked to nutrient management.
- *Do* incorporate to the extent possible data that reflect long-term crop rotations, erosion control, nutrient control, management at other significant sources, and the control of other pollutants that will be managed and simulated in the modeling.

Unit Loads

Several simple methods (see "Simple Watershed Methods" on p. 234) for watershed loading determination use unit loads, or unit-area loads, to represent pollutant contributions from various land uses. Unit loads are expressed as mass per unit area per unit time. One concern associated with unit-load approaches is the availability of good local data regarding the unit loads for watershed-specific physical, chemical, and climatological conditions (Heidtke and Auer, 1993). In the absence of local data, unit loads are approximated using values that may come from nearby studies or studies conducted in distant regions, thus introducing error to the analysis.

Scale should be considered when selecting unit loads, or export coefficients. A study of 210 paired observations of total phosphorus (TP) export taken from 38 studies showed that TP export in agricultural catchments is not a linear function of catchment area, but instead varies as the 0.77 power of drainage basin area (T.-Prairie and Kalff, 1986). This decline in unit-area export was attributable to the TP export from row crops and pasture catchments. However, the study found that the unit-area export of TP from forested catchments did not change as catchment size increased.

Addressing Uncertainty in Modeling Predictions

Because models simplify the real world, the predictions from a model are uncertain, and quantification of the prediction uncertainty should be included in the modeling approach (EPA, 1980). Prediction uncertainty is caused by natural process variability, and bias and error in sampling, measurement, and modeling. Reliably estimated prediction uncertainty can be useful to the planner as a means for judging the value of the prediction and assessing the risk of not achieving management objectives (e.g., meeting the load allocation of a TMDL). Modeling may also result in "unquantified supplemental uncertainty," which is uncertainty introduced through such things as the use of inappropriate export coefficients. This uncertainty, which is unknown to the analyst, is unquantified, and therefore introduces hidden planning risks.

To address the high variability of pesticide loads, a Monte Carlo simulation approach was developed and applied to estimate atrazine and carbofuran loads from hypothetical corn fields in Georgia and Iowa (Haith, 1985). The approach incorporated mathematical models of weather, hydrology, and soil chemistry. One advantage of this approach is the ability to generate a frequency distribution of pollutant loads rather than just a single value, thus allowing an assessment of the probability that any given single value for the pollutant load will occur.

Because of the complexity of quantifying modeling uncertainty, modelers are encouraged to consult with trained statisticians to devise the best approach for their modeling applications. Detailed examples of uncertainty analyses can be found in existing documents (EPA, 1980; EPA, 1989b; Haan, 1989; Beck, 1987).

Model Applications Using GIS Technology

A unit-load approach for estimating phosphorus loads to Owasco Lake in New York used geographic information system (GIS) technology to distribute land-based attributes within the watershed (Heidtke and Auer, 1993). The GIS enabled the modelers to match unit loads with the appropriate areas within the watershed in a distributed manner. GIS technology was also used to facilitate watershed modeling with models such as AGNPS (Agricultural Non-Point Source Pollution) (Line et al., 1997) and SWAT (Soil and Water Assessment Tool) (Engel et al. 1993).

EPA's BASINS (Better Assessment Science Integrating Point and Nonpoint Sources) is an integrated modeling system for performing watershed- and water-quality-based studies (EPA, 2001d). BASINS is intended to facilitate examination of environmental information, support analysis of environmental systems, and provide a framework for examining management alternatives. BASINS includes assessment tools, spatial data, and watershed and water quality modeling components, with GIS providing the integrating framework. An example illustrating the application of BASINS to estimating the impacts of agricultural management measures and practices is given in the BASINS Highlight.

Using BASINS to Develop a TMDL for Fecal Coliform Bacteria

Problem: The Lost River in the state of West Virginia exhibits water quality impairment due to elevated levels of fecal coliform bacteria. Suspected sources of contamination include cattle grazing and feedlots, poultry houses, failing septic systems, geese, wild turkey, and deer, as well as point source dischargers. Section 303(d) of the Clean Water Act and EPA's Water Quality Planning and Management Regulations (40 CFR Part 130) require states to develop Total Maximum Daily Loads (TMDLs) for waterbodies that are not meeting designated uses under technology-based controls. The TMDL process establishes the allowable loadings of pollutants or other quantifiable parameters for a waterbody based on the relationship between pollution sources and instream water quality conditions.

Approach: The U.S. EPA Better Assessment Science Integrating Point and Nonpoint Sources (BASINS) system Version 2.0 (EPA, 1998) and the Nonpoint Source Model (NPSM) were selected to predict the significance of fecal coliform sources and fecal coliform levels in the Lost River watershed. To obtain a spatial variation of the concentration of bacteria along the Lost River, the watershed was subdivided into 11 subwatersheds. This allowed analysts to address the relative contribution of sources within each subwatershed to the different segments of the river. The watershed subdivision was based on a number of factors, including the locations of flow monitoring stations, the locations of stream sampling stations, the locations of feedlots and poultry houses, and land use coverage. To develop a representative linkage between the sources and the instream water quality response in the 11 reaches of the Lost River, model parameters were adjusted to the extent possible for both hydrology and bacteria loading.

Results: Output from NPSM indicates violations of the 200 cfu/100 mL geometric mean standard throughout the Lost River watershed for the existing conditions using the representative time period (October 1990 through September 1991). After applying the load allocations, the NPSM model indicated that all 11 subwatersheds were in compliance with the fecal coliform bacteria standard. The model analysis indicates that water quality standards will be achieved if fecal coliform loads from pastureland are reduced by 38 percent, loads from forestland are reduced 12.8 percent, and loads from cropland are reduced by 37 percent. No change in the point source load was required. The load reductions at the source are expected to be sufficient to meet the 30-day geometric mean, on a daily basis, throughout the year. The margin of safety, an evaluation of the uncertainty in the TMDL, was included implicitly in the model setup and formulation. Conservative assumptions included loads associated with wildlife, septic systems, and existing BMP implementation. Further refinement and corresponding higher accuracy in the analysis could be achieved by more detailed source characterization (actual daily or monthly manure application rates), further evaluation of the viability and dieoff of fecal coliform in the various types of manure, and continued data collection and calibration.

Attainment of the load reductions is expected through implementation of manure storage and application guidelines, crop and pasture management, and wildlife management. No explicit modeling of the BMP effectiveness was performed. Follow-up monitoring is expected to track water quality improvements.

Glossary

10-year, 24-hour storm — A rainfall event of 24-hour duration and 10-year frequency that is used to calculate the runoff volume and peak discharge rate to a BMP.

25-year, 24-hour storm — A rainfall event of 24-hour duration and 25-year frequency that is used to calculate the runoff volume and peak discharge rate to a BMP.

ACP — Agricultural Conservation Program (the ACP is no longer an active USDA program; it was replaced by EQIP).

Adsorption — The adhesion of one substance to the surface of another.

Allelopathy — The inhibition of growth in one species of plants by chemicals produced in another species.

Animal unit (au) — A unit of measurement for any animal feeding operation calculated by adding the following numbers: the number of slaughter and feeder cattle multiplied by 1.0, plus the number of mature dairy cattle multiplied by 1.4, plus the number of swine weighing over 25 kilograms (approximately 55 pounds) multiplied by 0.4, plus the number of sheep multiplied by 0.1, plus the number of horses multiplied by 2.0.

Aquifer — A saturated, permeable geologic unit of sediment or rock that can transmit significant quantities of water under hydraulic gradients.

ASCS — Agricultural Stabilization and Conservation Service of USDA (now called Farm Service Agency).

AUM — Animal unit month. A measure of average monthly stocking rate that is the tenure of one animal unit for a period of 1 month. With respect to the literature reviewed for the grazing management measure, an animal unit is a mature, 1,000-pound cow or the equivalent based on average daily forage consumption of 26 pounds of dry matter per day (Platts, 1990). Alternatively, an AUM is the amount of forage that is required to maintain a mature, 1,000-pound cow or the equivalent for a one-month period. See animal unit for the NPDES definition.

Best management practice (BMP) — A practice or combination of practices that are determined to be the most effective and practicable (including technological, economic, and institutional considerations) means of controlling point and nonpoint pollutants at levels compatible with economic and environmental quality goals.

BMP system — A combination of two or more individual BMPs into a "system" that functions to reduce the same pollutant.

Biochemical oxygen demand (BOD) — A quantitative measure of the strength of contamination by organic carbon materials.

Chemigation — The addition of one or more chemicals to the irrigation water.

Conservation management system (CMS) — a generic term used by the NRCS that includes any combination of conservation practices and management that achieves a level of treatment of the five natural resources that satisfies criteria contained in the USDA–Natural Resources Conservation Service *National Handbook of Conservation Practices*, such as a resource management system or an acceptable management system.

Critical area — An area identified in a watershed or project area as having a significant impact on the impaired use of the receiving waters.

Conservation Reserve Enhancement Program (CREP) — A new initiative of CRP which uses financial incentives to encourage farmers and ranchers to voluntarily protect soil, water, and wildlife resources.

Conservation Reserve Program (CRP) — A volunteer program offering annual rental payments, incentive payments, and cost-share assistance for establishing long-term, resource-conserving cover crops on highly erodible land.

CZARA — Coastal Zone Act Reauthorization Amendments of 1990.

Denitrification — The chemical or biochemical reduction of nitrate or nitrite to gaseous nitrogen, either as molecular nitrogen or as an oxide of nitrogen.

Deposition — The accumulation of material left in a new position by a natural transporting agent such as water, wind, ice, or gravity, or by the activity of man.

Designated use — A beneficial use type established by a State for each water resource and specified in water quality standards, whether or not it is being attained.

Drainage area — Watershed; an area of land that drains to one point.

Ecosystem — A network of interactions between biological communities and the associated physical environment.

EPA — United States Environmental Protection Agency

Environmental Quality Incentives Program (EQIP) — A voluntary conservation program for farmers and ranchers, offering financial, technical, and educational help to install or implement practices to conserve soil and other natural resources.

Erosion — Wearing away of the land surface by running water, glaciers, winds, and waves. The term erosion is usually preceded by a definitive term denoting the type or source of erosion such as gully erosion, sheet erosion, or bank erosion.

Eutrophication — The natural process whereby a lake or other body of water evolves from low productivity and low nutrient concentrations to high productivity and high nutrient levels that is greatly accelerated by nutrient enrichment from human activities. Results of eutrophication can include algal blooms, low dissolved oxygen, and changes in community composition.

Fertigation — Application of plant nutrients in irrigation water.

FOTG — USDA-NRCS's Field Office Technical Guide.

FSA — Farm Service Agency, part of the U.S. Department of Agriculture.

Integrated Pest Management (IPM) — A pest population management system that anticipates and prevents pests from reaching damaging levels by using all suitable tactics including natural enemies, pest-resistant plants, cultural management, and the judicious use of pesticides, leading to an economically sound and environmentally safe agriculture.

Lateral — Secondary or side channel, ditch, or conduit.

Leachate — Liquids that have percolated through a soil and that contain substances in solution or suspension.

Management measures — As defined in section 6217(g)(5) of CZARA; "economically achievable measures for the control of the addition of pollutants from existing and new categories and classes of nonpoint sources of pollution, which reflect the greatest degree of pollutant reduction achievable through the application of the best available nonpoint source control practices, technologies, processes, siting criteria, operating methods, and other alternatives."

MCL — Maximum contaminant level. The enforceable standard or number against which a system's treated water samples are judged for compliance with U.S. Environmental Protection Agency regulations.

Micronutrient — A plant nutrient found in relatively small amounts (<100 mg kg^{-1}) in plants. These are usually B, Cl, Cu, Fe, Mn, Mo, Ni, Co, and Zn.

Natural Resources Conservation Service (NRCS) — An agency of the U.S. Department of Agriculture.

Nitrogen (N) — An element occurring in manure and chemical fertilizer that is essential to the growth and development of plants, but which, in excess, can cause water to become polluted and threaten aquatic animals.

NPS pollution — Nonpoint source pollution; pollution originating from diffuse areas (land surface or atmosphere) having no well-defined source.

Nutrients — Elements or compounds essential as raw materials for organism growth and development, such as carbon, nitrogen, phosphorus, etc.

Pasture — Those improved lands that are primarily used for the production of adapted domesticated forage plants for livestock.

Phosphorus (P) — An element occurring in manure and chemical fertilizer that is essential to the growth and development of plants, but which, in excess, can cause water to become polluted and threaten aquatic animals.

Range — Those lands on which the native or introduced vegetation (climax or natural potential plant community) is predominantly grasses, grasslike plants, forbs, or shrubs suitable for grazing or browsing use. Range includes natural grassland, savannas, many wetlands, some deserts, tundra, and certain forb and shrub communities.

Return flow — That portion or the water diverted from a stream that finds its way back to the stream channel either as surface or underground flow.

Resource management system (RMS) — A term used by NRCS defined as a combination of NRCS conservation practices and management identified by land or water uses that, when installed, will prevent resource degradation and permit sustained use by meeting criteria established in the FOTG for treatment of soil, water, air, plant, and animal resources.

Riparian areas — Vegetated ecosystems along a water body through which energy, materials, and water pass. Riparian areas characteristically have a high water table and are subject to periodic flooding and influence from the adjacent water body.

Runoff — The portion of rainfall or snow melt that drains off the land into ditches and streams by overland flow.

Rural Clean Water Program (RCWP) — A 15-year federally sponsored nonpoint source pollution control program initiated in 1980 as an experimental effort to address agricultural nonpoint source pollution problems in watersheds throughout the United States. The program concluded in 1995.

Sediment — The solid material, both mineral and organic, that is in suspension, is being transported, or has been moved from its site or origin by air, water, gravity, or ice.

Sedimentation — The process of sediment deposition.

Tailwater — Irrigation water that reaches the lower end of a field.

Tillage — The mechanical manipulation of the soil profile for any purpose; but in agriculture, it is usually restricted to modifying soil conditions, managing crop residues and/or weeds, or incorporating chemicals for crop production.

Total Maximum Daily Load (TMDL) — The maximum amount of pollution that a water body can receive without violating water quality standards. Total Maximum Daily Loads are the sum of point and nonpoint source loads.

Watershed — A geographic area in which water, sediments, and dissolved materials drain to a common outlet- a point on a larger stream, a lake, an underlying aquifer, an estuary, or an ocean. This area is also called the drainage basin of the receiving water body.

Watershed approach — A coordinating framework for environmental management that focuses public and private sector efforts to address the highest priority problems within hydrologically defined geographic areas, taking into consideration both ground and surface water.

References

Ackerman, E.O. and A.G. Taylor. 1995. Stream impacts due to feedlot runoff, In: *Animal Waste and the Land-Water Interface*, Kenneth Steele, Ed., Lewis Publishers, Boca Raton, FL, 589 pp.

Adam, Real, et al. 1986. *Evaluation of Beef Feedlot Runoff Treatment by a Vegetative Filter Strip*. ASAE North Atlantic Regional Meeting. Paper No. NAR 86-208.

Ahl, T. 1988. Background yields of phosphorus from drainage area and atmosphere:an empirical approach. Hydrobiologia 170:35-44.

Alabama Soil Conservation Service. 1990. Soil and Water Conservation Practices: Special ACP Water Quality Project, Sand Mountain/Lake Guntersville. In: *USDA Technical Guide, Section V*. US Department of Agriculture.

Alabama Soil and Water Conservation Committee, NCASI, and Gulf of Mexico Program Nutrient Enrichment Committee. 1997. *Constructed Wetlands and Wastewater Management for Confined Animal Feeding Operations*. CH2MHill, Gainesville, FL.

Alesii, B. 1998. New technology, more no-till acres. *No-Till Notebook*, 4 (1), Monsanto Company, Clayton, MO.

Allen, R.G. 1988, 1991. *IRRISKED irrigation scheduling computer model user's manual.* Dept. of Biological and Irrigation Engineering, Utah State University, 189 pp.

Allen, R.G. 1989, 1991. *REF-ET Standard Reference Evapotranspiration computer model user's manual.* Dept. of Biological and Irrigation Engineering, Utah State University, 40 pp.

Annandale, J.G. and D.J. Mulla. 1995. Nitrate Leaching Losses from Hills and Furrows in Irrigated Potatoes. In: *Clean Water - Clean Environment - 21st Century: Team Agriculture - Working to Protect Water Resources - Volume II: Nutrients* Conference Proceedings, ASAE, March 5-8, Kansas City, Missouri.

Armour, C. 1991. *Guidance for evaluating and recommending temperature regimes to protect fish.* Fish and Wildlife Service, Biological Report 90(22) (Instream Flow Information Paper 27).

Arnold, J.G., J.R. Williams, A.D. Nicks, and N.B. Sammons. 1990. *SWRRB: A Basin Scale Simulation Model for Soil and Water Resources Management.* Texas A & M University Press, College Station, TX.

Arnold, J.G. and J.R. Williams. 1987. Validation of SWRRB – Simulation for Water Resources in Rural Basins. *J. Water Resources Planning and Management* 13:243-256.

Arora, K., S.K. Mickelson, J.L. Baker, D.P. Tierney, and C.J. Peter. 1996. Herbicide retention by vegetative buffer strips from runoff under natural rainfall. *Trans of the ASAE* 30(6):2155-2164.

ASAE. 1989. *Standards, Engineering Practices and Data Developed and Adopted by the American Society of Agricultural Engineers.* Standard EP409. American Society of Agricultural Engineers, St. Joseph, MI.

ASCE. 1993. *Criteria for evaluation of watershed models.* American Society of Civil Engineers Task Committee of the Watershed Management Committee.

Asmussen, L. E., A.W. White, Jr., E. W. Hauser, and J. M. Sheridan. 1977. Reduction of 2,4-D load in surface runoff down a grassed waterway. *J. Environ. Qual.* 6:159-162.

Ator, S.W. and M.J. Ferrari. 1997. Nitrate and Selected Pesticides in Ground Water of the Mid-Atlantic Region. Water Resources Investigations Report 97-4139, U.S. Geological Survey, Baltimore, MD.

Baker, D.B. 1993. The Lake Erie agroecosystem program: water quality assessments, *Agriculture, Ecosystems and Environment*, 46:197-215.

Baker, D.B. and R.P. Richards. 1991. Herbicide concentrations in Ohio's drinking water supplies: a quantitative exposure assessment, In: *Pesticides in the Next Decade: The Challenges Ahead*, Proceedings of the Third National Research Conference on Pesticides, November 8-9, 1990. Diana L. Weigmann, Ed., Virginia Water Resources Research Center, Virginia Polytechnique Institute and State University, Blacksburg, VA, 832 pages.

Baker, J.L., H.P. Johnson, M.A. Borcherding, and W.R. Payne. 1978. Nutrient and pesticide movement from field to stream: a field study. Pages 213-245 in Best Management Practices for Agriculture and Silviculture, R.C. Loehr, D.A. Haith, M.F. Walter, and C.S. Martin, eds. Proc. 1978 Cornell Agricultural Waste Conference, Ann Arbor Science, Ann Arbor, MI.

Baker, J.L., and H.P. Johnson. 1979. The effect of tillage system on pesticides in runoff from small watersheds. *Trans. Am. Soc. Agric. Eng.* 22:554-559.

Baker, J.L., and J.M. Laflen. 1979. Runoff losses of surface-applied herbicides as affected by wheel tracks and incorporation. *J. Environ. Qual.* 8:602-607.

Barbarika, A. Jr. 1987. Costs of soil conservation practices. In: *Optimum Erosion Control at Least Cost: Proceedings of the National Symposium on Conservation Systems.* American Society of Agricultural Engineers St. Joseph, MI, pp. 187-195.

Barbash, J.E. and E.A. Resek. 1996. Pesticides in Ground Water: Current Understanding of Distribution, Trends, and Governing Factors. Vol. 2 in the Series: Pesticides in the Hydrologic System. CRC Press, Boca Raton, FL.

Barbash, J.E., G.P. Thelin, D.W. Kolpin, R.J. Gilliom. 2001. Major herbicides in ground water: results from the National Water Quality Assessment. *J. Environ. Qual.* 30:831-845.

Barthalow, J.M. 1989. *Stream temperature investigations: field and analytical methods.* U.S. Fish and Wildlife Service, Biological Report 89(17) (Instream Flow Paper 13).

Barvenik, F.W., R.E. Sojka, R.D. Lentz, F.F. Andrawes, and L.S. Messner. 1996. Fate of acrylamide monomer following application of polyacrylamide to cropland. Managing Irrigation-Induced Erosion and Infiltration with Polyacrylamide. University of Idaho, Miscellaneous Publication No. 101-96.

Baschita. 1997. Riparian Shade and Stream Temperature: An Alternative Perpective. *Rangelands* 19:25-28.

Bauder, J.W., K.N. Sinclair, and R.E. Lund. 1993. Physiographic and land use characteristics associated with Nitrate-Nitrogen in Montana groundwater. J. Env. Qual. 22(2):255-262.

Baxter-Potter, W.R. and M.W. Gilliland. 1988. Bacterial Pollution in Runoff from Agricultural Lands. *J. Environ. Qual.*, Vol. 17(1):27-34.

Beasley, D.B., L.F. Huggins, and E.J. Monke. 1980. ANSWERS: A Model for Watershed Planning. *Trans. of the ASAE* 23:938-944.

Beauchemin, S., R.R. Simard, and D. Cluis. 1998. Forms and concentration of phosphorus in drainage water of twenty-seven tile-drained soils. *J. Environ. Qual.* 27:721-728.

Beck, M.B. 1987. Water Quality Modeling: A Review of the Analysis of Uncertainty. *Water Resources Research* 23(8):1393-1442.

Berry, J.T., and N. Hargett. 1984. *Fertilizer Summary Data.* Tennessee Valley Authority, National Fertilizer Development Center, Mussel Shoals, AL.

Bingner, R.L., C.E. Murphree, and C.K. Mutchler. 1987. *Comparison of Sediment Yield Models on Various Watersheds in Mississippi.* ASAE paper 87-2008, American Society of Agricultural Engineers, St. Joseph, MI.

Bohn, C.C. and J.C. Buckhouse. 1985. Coliforms as an indicator of water quality in wildland streams. *J. Soil and Water Conserv.* 40(1):95-98.

Bosch, D.D., R.L. Bingner, F.D. Theurer, G. Felton, and I. Chaubey. Evalution of the AnnAGNPS Water Quality Model. Presented at the 1998 ASAE Annual International Meeting. July 12-16, 1998, Orlando, Florida, Paper No. 982195, ASAE, 2950 Niles Rd., St. Joseph, MI 49085-9659.

Bouldin, D., W. Reid, and D. Lathwell. 1971. Fertilizer Practices Which Minimize Nutrient Loss. In: *Proceedings of Cornell University Conference on Agricultural Waste Management, Agricultural Wastes: Principles and Guidelines for Practical Solutions*, Syracuse, NY.

Boyd, P.M., L.W. Wulf, J.L. Baker, and S.K. Mickelson. 1999. Pesticide transport over and through the soil profile of a vegetative filter strip. American Society of Agricultural Engineers. ASAE Paper no. 992077.

Boyle Engineering Corp. 1986. *Evaluation of On-Farm Management Alternatives.* Prepared for the San Joaquin Valley Drainage Program, Sacramento, CA.

Brakensiek, D.L., H.B. Osborn, and W.J. Rawls. 1979. *Field Manual for Research in Agricultural Hydrology.* Agriculture Handbook No. 224. U.S. Department of Agriculture, Science and Education Administration, Beltsville, MD.

Breeuwsma, A., J.G.A. Reijerink, and O.F. Schoumans. 1995. Impact of manure on accumulation and leaching of phosphate in areas of intensive livestock farming. Pages 239-249 in *Animal Waste and the Land-Water Interface,* K. Steele, ed., Lewis Publishers, Boca Raton, FL.

Brence, L. and R. Sheley. 1997. *Determining Forage Production and Stocking Rates: A Clipping Procedure for Rangelands.* MONTGUIDE. MT 9704, A-3. Montana State University Extension Service. Bozeman, Montana.

Breve, M.A., R.W. Skaggs, J.E. Parsons, and J.W. Gilliam. 1997. DRAINMOD-N, A Nitrogen Model for Artificially Drained Soils. *Trans. of the ASAE* Vol 40(4):1067-1075.

Brown, M.P., P. Longabucco, and M.R. Rafferty. 1986. *Nonpoint source control of phosphorus: a watershed evaluation – volume 5, the eutrophication of the Cannonsville Reservoir,* Bureau of Technical Services & Research, New York State Department of Environmental Conservation, Albany, NY, 77 pp.

Buckman, H.O. and N.C. Brady. 1969. *The Nature and Properties of Soils, 7th Ed.* London: Macmillan. Reprinted with permission from Macmillan Inc.

Buckhouse, J.C. 1993. *Grazing Effects on Riparian Areas. Rangeland Watershed Program Fact Sheet 14.* U.C. Extension Service.

Burkhart, M.R. and D.W. Kolpin. 1993. Hydrologic and land-use factors associated with herbicide and nitrate in near-surface waters. *J. Environ. Qual.* 22:646-656.

Burkholder, J.M. 1996. Facts About Pfiesteria. Unpublished.

Burkholder, J.M., H.B. Glasgow Jr., C.W. Hobbs. 1995. Fish kills linked to a toxic ambush-predator dinoflagellate: distribution and environmental conditions. *Mar Ecol Prog Ser* 124:43-61.

Burt, C.M. 1995. *The surface irrigation manual: a comprehensive guide to design and operation of surface irrigation systems.* Waterman Industries, Inc., Exeter, CA, 316 pp.

California SWRCB. 1987. *Regulation of Agricultural Drainage to the San Joaquin River: Executive Summary.* California State Water Resources Control Board. Doc. No. WQ-85-1.

California SWRCB. 1992. *Demonstration of Emerging Irrigation Technologies.* Water Conservation Office, Department of Water Resources for Division of Water Quality, Publication No. 91-20-WQ, July.

Camacho, R. 1991. *Financial Cost Effectiveness of Point and Nonpoint Source Nutrient Reduction Technologies in the Chesapeake Bay Basin.* Interstate Commission on the Potomac River Basin, Rockville, Maryland. Unpublished draft.

Cambell Pacific Nuclear. 1998. *Price Quote for 503 DR Hydroprobe.* Martinez, California.

Campbell, C.A., R. DeJong, and R.P. Zentner. 1984. Effect of cropping, summerfallow and fertilizer Nitrogen on Nitrate-Nitrogen lost by leaching on a brown Chernozemic loam. Can. J. Soil Sci. 64(1):61-74.

Cannon, L.E., M. Gamroth, and P.J. Ballerstedt. 1993. *Managing dairy grazing for the most efficient yields.* EM 8412. Oregon State University Extension Service, Corvallis, OR, 4 pp.

Carpenter, A.G. et al. 1994. *Water Quality Control PlanCentral Coast Basin, Region 3.* Sacramento, California: California Regional Water Quality Control Board, Central Coast Region.

Cassell, E.A. and J.C. Clausen. 1993. Dynamic Simulation Modelling for Evaluating Water Quality Response to Agricultural BMP Implementation. *Water Science & Technology.* 28(3-5): 635-648.

Cassell, E.A., J.M. Dorioz, R.L. Kort, J.P. Hoffmann, D.W. Meals, D. Kerschtel, and D.C. Braun. 1998. Modeling phosphorus dynamics in ecosystems: mass balance and dynamic simulation approaches. *J. Environ. Qual.* 27:293-298.

CDC. n.d. Parasitic Disease Information - Fact Sheet - Giardiasis. United States Department of Health and Human Services, Centers for Disease Control and Prevention, Division of Parasitic Diseases, Atlanta, GA. http://www.cdc.gov/ ncidod/dpd/parasites/giardiasis/default.htm. Accessed on March 26, 2001.

Chaney, E., W. Elmore, and W.S. Platts. 1990. *Livestock grazing on western riparian areas.* Northwest Resource Information Center, Inc., Eagle, ID, 45 pp.

Chaney, E., W. Elmore, and W.S. Platts. 1993. *Managing Change: Livestock grazing on western riparian areas.* Northwest Resource Information Center, Inc., Eagle, ID, 31 pp.

Chardon, W. J., R. G. Menon, and S. H. Chien. 1996. Iron oxide impregnated filter paper (Pi test): A review of its development and methodological research. *Nutrient Cycle Agroecosystems* 46: 41-51.

Clary, W.P. and W.C. Leininger. 2000. Stubble Height as a Tool for Management of Riparian Areas. *Journal of Range Management* 53(6):562-673.

Coale, F.J. 2000. Phosphorus dynamics in soils of the Chesapeake Bay: a primer. Pages 44-55 in *Agriculture and Phosphorus Management*, A.N. Sharpley, ed., Lewis Publishers, Boca Raton, FL.

Cochran, P. 1998. Personal communication. Cochran Agronomics, Paris, IL.

Coffey, S.W. and M.D. Smolen. 1990. Results of the Experimental Rural Clean Water Program: Methodology for On-Site Evaluation. National Water Quality Evaluation Project, NCSU Water Quality Group, Biological and Agricultural Engineering Department, North Carolina State University, Raleigh, North Carolina.

Coffey, S.W., J. Spooner, and M.D. Smolen. 1995. *The Nonpoint Source Manager's Guide to Water Quality and Land Treatment Monitoring.* North Carolina Agricultural Extension Service, Department of Biological and Agricultural Engineering, North Carolina State University, Raleigh, NC.

Cole, J.T., J.H. Baird, N.T. Basta, R.L. Huhnke, D.E. Storm, G.V. Johnson, M.E. Payton, M.D. Smolen, D.L. Martin, and J.C. Cole. 1997. Influence of buffers on pesticide and nutrient runoff from bermudagrass turf. *J. Environ. Qual.* 26:1589-1598.

Conservation Technology Information Center (CTIC). 1997. *1997 national crop residue management survey,* West Lafayette, IN.

Conservation Technology Information Center. ca 1997 (undated). *Conservation tillage: a checklist for U.S. farmers plus regional considerations,* West Lafayette, IN, 36 pp.

Cornell Cooperative Extension. 1987. *Cornell Field Crops and Soils Handbook.* Resource Center. Cornell University, Ithaca, NY.

Cornell Cooperative Extension. 1997. *1997 Cornell Recommendations for Integrated Field Crop Management.* Resource Center, Cornell University, Ithaca, NY.

Correll, D.L., T.E. Jordan, and D.E. Weller. 1995. Livestock and pasture land effects on the water quality of Chesapeake Bay watershed streams. Pages 107-117 in Animal Waste and the Land-Water Interface, K. Steele, ed., Lewis Publishers, New York, NY.

Craig, P.H. 1998. Personal communication. Nutrient Management Program, Pennsylvania Department of Agriculture, Harrisburg, PA 717-783-9704.

Crane, S.R., J.A. Moore, M.E. Grismer, and J.R. Miner. 1983. Bacterial Pollution from Agricultural Sources: A Review. *Trans. ASAE* 26:858-866.

Cronshey, R.G. and F.D. Theurer. AnnAGNPS — Nonpoint Pollutant Loading Model. In: *Proceedings of First Federal Interagency Hydrologic Modeling Conference*, 19-23 April 1998, Las Vegas, NV.

Cropper, J. 1998. Grazing ruminations. *Pasture Profit*, Vol. 6(1):1-3, Grazing Lands Technology Institute, University Park, PA.

Cross-Smiecinski, A. and L.D. Stetzenback. 1994. *Quality planning for the life science researcher: Meeting quality assurance requirements.* CRC Press, Boca Raton, Florida.

Cumberland County SWCD, Know-Lincoln SWCD, Maine Department of Environmental Protection, Maine Soil and Water Conservation Commission, Portland Water District, Time and Ride RC and D, U.S. Environmental Protection Agency, and U.S. Department of Agriculture – Soil Conservation Service. *Fact Sheet Series (2, 3, 4, 5, 8, 9, 10, 12).*

Daniel, T.C., A.N. Sharpley, and J.L. Lemunyon. 1998. Agricultural phosphorus and eutrophication: a symposium overview. *J. Environ. Qual.* 27:251-257.

Daniel, T.C., A.N. Sharpley, J.L. Lemunyon, and J.T. Sims. 1997. *Agricultural Phosphorus and Eutrophication: An Overview.* Dept. of Agronomy, Univ. of Arkansas, Fayetteville, AR.

Davenport, T.E. and R.P. Clarke. 1984. *Blue Creek watershed project executive summary and recommendations.* Illinois EPA, Springfield, IL, 32 pp.

Dawson, Spofford, Pfeiffer. May 1996. The Physical Effects of Polyacrylamide on Natural Resources, Conference Proceedings - Managing Irrigation-Induced Erosion and Infiltration with Polyacrylamide. Misc. Pub. 101-96, University of Idaho.

Deal, J.C., J.W. Gilliam, R.W. Skaggs, and K.D. Konyha. 1986. Prediction of Nitrogen and Phosphorus Losses as Related to Agricultural Drainage System Design. *Agriculture Ecosystem and Environment* 18:37-51.

Dickey, E.C. 1981. Performance and design of vegetative filters for feedlot runoff treatment. In: *Proceedings of the Fourth International Symposium on Livestock Wastes, Livestock Waste: A Renewable Resource.*

Donigian, A.S. 1983. Model Predictions vs. Field Observations: The Model Validation/Testing Process. In: *Fate of Chemicals in the Environment*, ACS Symposium Series 225, American Chemical Society, Washington, DC. p.151-171.

Doorenbos, J. and W.O. Pruitt. 1975. *Crop Water Requirements.* Irrigation and Drainage Paper 24, Food and Agricultural Organization of the United Nations (FAO). Rome, Italy.

Dosskey, M., D. Schultz, and T. Isenhart. 1997. Riparian Buffers for Agricultural Land. USDA Forest Service, Rocky Mountain Station, USDA-NRCS, *Agroforestry Notes* 1/97:3(4).

DPRA. 1989. *Evaluation of the Cost Effectiveness of Agricultural Best Management Practices and Publicly Owned Treatment Works in Controlling Phosphorus Pollution in the Great Lakes Basin.* Prepared by DPRA Inc. for U.S. Environmental Protection Agency. Under contract no. 68-01-7947, Manhattan, KS.

DPRA. 1992. *Draft Economic Impact Analysis of Coastal Zone Management Measures Affecting Confined Animal Facilities.* Prepared by DPRA Inc. for U.S. Environmental Protection Agency under contract no. 68-C99-0009, Manhattan, KS.

Drake, D.J. and J. Oltjen. 1994. Intensively managed rotational grazing systems for irrigated pasture. In: S. Blank and J. Oltjen, Eds., *California Rancher's Management Guide*, Cooperative Extension, University of California, Berkeley, CA, pp. 33-41.

Drouse, S.K., D.C. Hillman, J.L. Engles, L.W. Creelman and S.J. Simon. 1986. *National surface water survey. National stream survey (Phase 1-Pilot, mid-Atlantic Phase 1 southeast screening, and episodes pilot) quality assurance plan.* EPA/600/4-86/044. NTIS No. PB87-145819. Prepared by Lockheed Engineering and Management Services Co., Inc., Las Vegas, Nevada, for U.S.

Environmental Protection Agency, Office of Research and Development, Environmental Monitoring Systems Laboratory, Las Vegas, Nevada. December.

Duke, H.R. (editor). 1987. *Scheduling irrigations: A guide for improved irrigation water management through proper timing and amount of water application.* USDA-Agricultural Research Service, Ft. Collins, CO.

Eckert, R.E., and J.S. Spencer. 1987. Growth and Reproduction of Grasses Heavily Grazed under Rest-Rotation Management. *J. Range Mgt.* 40(2):156-159.

EduSelf Multimedia Publishers Ltd. 1994. Courseware handbook irrigation systems. In: *Biology & agriculture courseware an interactive multimedia kit*, Holon, Israel, 150 pp.

Edwards, D.R., V.W. Benson, J.R. Williams, T.C. Daniel, J. Lemunyon, and R.G. Gilbert. 1994. Use of the EPIC Model to Predict Runoff Transport of Surface-applied Inorganic Fertilizer and Poultry Manure Constituents. *Trans. of the ASAE* 37(2):403-409.

Edwards, D.R., B.T. Larsen, and T.T. Lim. 2000. Runoff nutrient and fecal coliform content from cattle manure application to fescue plots. *J. AWRA* 36(4):711-721.

Edwards, W.M., L.B. Owens, and R.K. White. 1983. Managing Runoff from a Small, Paved Beef Feedlot. *J. Environ. Qual.* 12(2).

Edwards, W.M., L.B. Owens, R.K. White, and N.R. Fausey. 1986. Managing feedlot runoff with a settling basin plus tiled infiltration bed. *Transactions of the ASAE*, 29(1): 243-247.

El-Swaify, S.A. and K.R. Cooley. 1980. Sediment losses from small agricultural watersheds in Hawaii (1971-77). Agricultural Reviews and Manuals ARM-W-17, Science and Education Administration, Oakland, CA.

Engel, B.A., R. Srinivasan, J. Arnold, C. Rewerts, and S. J. Brown. 1993. *Nonpoint Source (NPS) Pollution Modeling Using Models Integrated with Geographic Information Systems (GIS)*, Water Science & Technology, 28(3-5): 685-690.

EPA. 1979. *Quantitative Techniques for the Assessment of Lake Quality*, EPA-440/5-79-015, Office of Water, Washington, DC, 146 pp.

EPA. 1980. *Modeling Phosphorus Loading and Lake Response Under Uncertainty: A Manual and Compilation of Export Coefficients*, EPA 440/5-80-011, Office of Water, Washington, DC, 214 pp.

EPA. 1982a. *Planning Guide for Evaluating Agricultural Nonpoint Source Water Quality Controls.* U.S. Environmental Protection Agency, Office of Research and Development, Environmental Research Laboratory, Athens, GA. EPA-600/3-82-021.

EPA. 1982b. *ANSWERS - Users Manual*, EPA-905/9-82-001, Great Lakes National Program Office, Chicago, 54 pp. EPA. 1982. *Planning Guide for Evaluating Agricultural Nonpoint Source Water Quality Controls.* U.S. Environmental Protection Agency, Office of Research and Development, Environmental Research Laboratory, Athens, GA. EPA-600/3-82-021.

EPA. 1984. USEPA Manual of Methods for Virology. United States Environmental Protection Agency, National Exposure Research Laboratory, Cincinnati, OH.

EPA. 1985. *Field Agricultural Runoff Monitoring (FARM) Manual*, EPA/600/3-85/043, Environmental Research Laboratory, Athens, GA, 230 pp.

EPA. 1986. *Ambient Water Quality Criteria for Bacteria - 1986.* United States Environmental Protection Agency, Washington, D.C.

EPA. 1989a. *National Primary and Secondary Drinking Water Standards; Proposed Rule.* 40 CFR Parts 141, 142, and 143. U.S. Environmental Protection Agency, Washington, DC.

EPA, 1989b. *Final Report: A Probabilistic Methodology for Analyzing Water Quality Effects of Urban Runoff on Rivers and Streams*, Office of Water, Washington, DC, 84 pp.

EPA. 1990a. *Lessons learned: Rock Creek, Idaho and Tillamook Bay, Oregon Rural Clean Water Programs.* EPA 910/9-90-004, Water Division, Region 10, Seattle, WA. 25 pp.

EPA. 1990b. *Rural Clean Water Program, lessons learned from a voluntary monpoint source control experiment.* EPA 440/4-90-012, Office of Water, Washington, DC, 29 pp.

EPA. 1991a. *1990 Annual Progress Report for the Baywide Nutrient Reduction Strategy.* U.S. Environmental Protection Agency, Chesapeake Bay Program, Annapolis, MD.

EPA. 1991b. *Pesticides and Groundwater Strategy.* U.S. Environmental Protection Agency, Office of Prevention, Pesticides and Toxic Substances, Washington, DC.

EPA. 1991c. *Monitoring Guidelines to Evaluate Effects of Forestry Activities on Streams in the Pacific Northwest and Alaska*, EPA/910/9-91-001, Water Division, Region 10, Seattle, WA, 166 pp.

EPA. 1992a. *Preliminary Economic Achievability Analysis: Agricultural Management Measures.* U.S. Environmental Protection Agency, Office of Policy, Planning and Evaluation, Washington, DC.

EPA. 1992b. *Compendium of Watershed-Scale Models for TMDL Development*, EPA 841-R-92-002, Office of Water, Washington, DC, 86 pp.

EPA. 1993a. *Guidance Specifying Management Measures for Sources of Nonpoint Source Pollution in Coastal Waters.* U.S. Environmental Protection Agency, Office of Water, Washington, DC. EPA-840-B-92-002.

EPA. 1993b. *Post-Audit Verification of the Model AGNPS in Vermont Agricultural Watersheds*, EPA 841-R-93-006, Office of Water, Washington, DC, 104 p.

EPA. 1993c. Evaluation of the Experimental Rural Clean Water Program. EPA 841-R-93-005. May 1993. Office of Wetlands, Oceans, and Watersheds, Nonpoint Source Control Branch, Washington, D.C.

EPA. 1995a. *Guide Manual on NPDES Regulations for Concentrated Animal Feeding Operations*. Final. U.S. Environmental Protection Agency, Office of Water. EPA833-B-95-001.

EPA. 1995b. *Watershed Protection: A Project Focus*. EPA-841-R-95-003, Office of Water, Washington, DC.

EPA. 1996. *Section 319 national monitoring program projects*. EPA-841-S-96-002, Office of Water, Washington, DC. 253 pp.

EPA. 1997a. *Monitoring Guidance for Determining the Effectiveness of Nonpoint Source Controls*. EPA 841-B-96-004, Office of Water, Washington, DC. 236p.

EPA. 1997b. *Techniques for Tracking, Evaluating, and Reporting the Implementation of Nonpoint Source Control Measures – I. Agriculture*. EPA/841-B-97-010, Office of Water, Washington, DC. 70p.

EPA. 1997c. Water on Tap: A Consumers Guide to the Nation's Drinking Water <http://www.epa.gov/ogwdw000/wot/howsafe.html>.

EPA. 1997d. *Compendium of Tools for Watershed Assessment and TMDL Development*, EPA 841-B-97-006, Office of Water, Washington, DC, 118 pp.

EPA. 1997e. *BASINS Training Course*. December 8-12, 1997, Office of Water, Washington, DC.

EPA. 1998. Better Assessment Science Integrating Point and Nonpoint Sources: BASINS Version 2.0. User's Manual. EPA-823-B-98-006, U.S. Environmental Protection Agency, Office of Water, Washington, D.C. 20460.

EPA. 1999a. Uncovered Finished Water Reservoirs Guidance Manual. United States Environmental Protection Agency, Office of Water, Washington, D.C.

EPA. 1999b. *Protocol for Developing Sediment TMDLs*. EPA 841-B-99-004. Office of Water (4503-F), United States Environmental Protection Agency, Washington, D.C. 132 pp.

EPA. 1999c. *Protocol for Developing Nutrient TMDLs: First Edition*. EPA 841-B-99-007. Office of Water (4503-F), U.S. Environmental Protection Agency, Washington, D.C.

EPA. 2000a. *National Water Quality Inventory: 1998 Report to Congress*. EPA 841-R-00-001. U.S. Environmental Protection Agency, Office of Water, Washington, D.C.

EPA. 2000b. *Implementation Guidance for Ambient Water Quality Criteria for Bacteria - 1986*. United States Environmental Protection Agency, Office of Water, Washington, D.C.

EPA. 2000. Guidance for the data quality objectives process (QA/G-4). EPA/600/R-96/055. U.S. Environmental Protection Agency, Office of Environmental Information, Washington, DC.

EPA. 2001a. EPA requirements for quality management plans (QA/R-2). EPA/240/B-01/002. U.S. Environmental Protection Agency, Office of Environmental Information, Washington, DC.

EPA. 2001b. EPA requirements for quality assurance project plans (QA/R-5). EPA/240/B-01/003. U.S. Environmental Protection Agency, Office of Environmental Information, Washington, DC.

EPA. 2001c. *Protocol for Developing Pathogen TMDLs: First Edition*. EPA 841-R-00-0002. Office of Water (4503-F), U.S. Environmental Protection Agency, Washington, D.C.

EPA. 2001 draft. National Management Measures to Protect and Restore Wetlands and Riparian Areas for the Abatement of Nonpoint Source Pollution. <http://www.epa.gov/owow/nps/wetmeasures/>

EPA. 2002. National Water Quality Inventory - 2000 Report to Congress. EPA 841-F-02-003. Office of Water (4503-T). U.S. Environmental Protection Agency, Washington, D.C.

EPA Office of Inspector General. 1997. Animal Waste Disposal Issues #7100142 <http://www.epa.gov/oigearth/hogchp1.htm>.

Erickson, H.E., M. Morrison, J. Kern, L. Hughes, J. Malcolm, and K. Thornton. 1991. *Watershed manipulation project: Quality assurance implementation plan for 1986-1989*. EPA/600/3-91/008. NTIS No. PB91-148395. Prepared by NSI Technology Services Corporation, Corvallis, Oregon for Corvallis Environmental Research Laboratory, Oregon. January.

Evans, R.O. 1992. Personal communication. Biological and Agricultural Engineering Department, North Carolina State University, Raleigh, NC.

Evans, R.O., D.K. Cassel, and R.E. Sneed. 1991a. *Measuring Soil Water for Irrigation Scheduling: Monitoring Methods and Devices*. North Carolina Cooperative Extension Service, Raleigh, NC. AG-452-2.

Evans, R.O., D.K. Cassel, and R.E. Sneed. 1991b. *Soil, Water, and Crop Characteristics Important to Irrigation Scheduling*. North Carolina Cooperative Extension Service, Raleigh, NC. AG-452-1.

Evans, R.O., R.E. Sneed, and D.K. Cassel. 1991c. *Irrigation Scheduling to Improve Water- and Energy-Use Efficiencies*. North Carolina Cooperative Extension Service, Raleigh, NC. AG-452-4.

FAO, 1997. Management of Agricultural Drainage Water Quality. International Commission on Irrigation and Drainage. Madramootoo, C.A., W.R. Johnston, and L.S. Willardson editors.

Faries, F.C., J.M. Sweeten, and J.C. Reagor. 1998. *Water quality: its relationship to livestock*. L-2374, Texas Agricultural Extension Service, Texas A & M University, College Station, TX, 4 pp.

FISRWG. 1998. *Stream Corridor Restoration: Principles, Processes, and Practices*. Federal Interagency Stream Restoration Working Group. GPO Item No. 0120-A; SuDocs No. A57.6/2:EN3/PT.653. (ISBN-0-934213-59-3).

Feller, M.C. 1981. Effects of clearcutting and slashburning on stream temperature in southwestern British Columbia. *Water Resources Bulletin* 17, 863-867.

Flanagan, D.C. and M.A. Nearing. eds. 1995. *USDA-Water Erosion Prediction Project: Hillslope Profile and Watershed Model Documentation.* NSERL Report No. 10. West Lafayette, IN.

Follet, R.F., D.R. Kenney, and R.M. Cruse. 1991. *NLEAP or Nitrogen Leaching and Economic Analysis Package.* Soil Science of America, Inc., Madison, WI.

Foster, M.A., P.D. Robillard, D.W. Lehning, and R. Zhao. 1996. STEWARD, a knowledge-based system for selection, assessment, and design of water quality control practices in agricultural watersheds. *Water Resource Bulletin,* in review.

Franti, T. G., C. J. Peter, D. P. Tierney, R. S. Fawcett, and S. A. Meyers. 1995. *Best Management Practices to Reduce Herbicide Losses from Tile-outlet Terraces.* Paper 952713, American Society of Agricultural Engineering, St. Joseph, Michigan.

Franzen, D.W., T.F. Scherer, and B.D. Seelig. 1996. *Compatibility of North Dakota soils for irrigation.* NDSU Extension Service. Fargo, ND. EB-68.

Fraser, R.H., P.K. Barten, and D.A.K. Pinney. 1998. Predicting stream pathogen loading from livestock using a geographical information system-based delivery model. *J. Environ. Qual.* 21:935-945.

Fresno Field Office and River Basin Planning Staff. 1979. *Comparison of Alternative Management Practices, Molar Flats Pilot Study Area, Fresno County, California.* Mini-Report. U.S. Department of Agriculture, Soil Conservation Service, Davis, CA.

Frost. B., and G. Ruyle. 1993. Range management terms/definitions. In: *Arizona Ranchers' Management Guide*, R. Gum, G. Ruyle, and R. Rice, Eds., Arizona Cooperative Extension, The University of Arizona, Tucson, AZ, pp. 15-22.

GAO. 1986. *The Nation's Water: Key unanswered questions about the quality of rivers and streams.* Report No. GAO/PEMD-86-6. U.S. General Accounting Office, Washington, D.C.

Gale, J.A., D.E. Line, D.L. Osmond, S.W. Coffey, J. Spooner, J.A. Arnold, T.J. Hoban, and R.C. Wimberley. 1993. *Evaluation of the Experimental Rural Clean Water Program.* National Water Quality Evaluation Project, NCSU Water Quality Group, Biological and Agricultural Engineering Department, North Carolina State University, Raleigh, NC (published by U.S. Environmental Protection Agency). EPA-841-R-93-005. 559p.

Garen, D., D. Woodward, and F. Geter. 1999. A user agency's view of hydrologic, soil erosion and water quality modeling. *Catena* 37:277-289.

Gary, H.L., S.R. Johnson, S.L. Ponce. 1983. Cattle grazing impact in a Colorado Front range stream. *J. Soil and Water Cons.* 38(2) 124 128.

Gburek, W.J., A.N. Sharpley, L. Heathwaite, and G.J. Folmar. 2000a. Phosphorus management at the watershed scale: a modification of the phosphorus index. *J. Environ. Qual.* 29:130-144.

Gburek, W.J., A.N. Sharpley, and G.J. Folmar. 2000b. Critical areas of phosphorus export from agricultural watersheds, Pages 83-104 in *Agriculture and Phosphorus Management*, A.N. Sharpley, ed., Lewis Publishers, Boca Raton, FL.

Gentry, L.E., M.B. David, K.M. Smith-Starks, and D.A. Kovacic. (2000). Nitrogen fertilizer and herbicide transport from tile demand fields. *Journal of Environmental Quality,* Vol. 29, No. 1.

Geohring, L.D. and P.E. Wright. 1998. Preferential flow of liquid manure to subsurface drains in Proceedings of the Manure Management Conference, Feb. 10-12, 1998, Ames, Iowa.

Gerrish, J. 1998. Commonly asked water development questions. *Pasture Profit*, Vol. 6(1):19-20, Grazing Lands Technology Institute, University Park, PA.

Gilbert, R.O. 1987. *Statistical Methods for Environmental Pollution Monitoring*, Van Nostrand Reinhold Company, New York, 320 pp.

Gilliam, J.W., D.L. Osmond, and R.O. Evans. 1997. Selected Best Agricultural Best Management Practices to Control Nitrogen in the Neuse River Basin. North Carolina Agricultural Research Service Technical Bulletin 311, North Carolina State University, Raleigh, NC.

Glenn, S. and J. S. Angle. 1987. Atrazine and simazine in runoff from conventional and no-till corn watersheds. *Agric. Ecosyst. and Environ.* 18:273-280.

Goodman, J. 1992. Personal communication. South Dakota Department of Environment and Natural Resources, Pierre, SD.

Goolsby, D.A., E.M. Thurman, D.W. Kolpin, and W.A. Battaglin. 1995. Occurrence of herbicides and metabolites in surface water, ground water, and rainwater in the midwestern United States, In: *Proceedings 1995 Annual Conference — Water Research, June 18-22, 1995, Anaheim, California*, American Water Works Association, pp. 583-591.

Goolsby, D.A., and W.A. Battaglin, 2000. Nitrogen in the Mississippi Basin - Estimating Sources and Predicting Flux to the Gulf of Mexico: U.S. Geological Survey Fact Sheet 135-00, 6p.

Griffin, M.L., D.B. Beasley, J.I. Fletcher, and G.R. Foster. 1988. Estimating Soil Loss on Topographically Nonuniform Field and Farm Units. *J. Soil and Water Conservation* 43:326-331.

Gum, R. and G. Ruyle. 1993. Grazing cell management. In: *Arizona Ranchers' Management Guide*, R. Gum, G. Ruyle, and R. Rice, Eds., Arizona Cooperative Extension, The University of Arizona, Tucson, AZ, pp. 9-14.

Haan, C.T. 1989. Parametric Uncertainty in Hydrologic Modeling. *Trans. of the ASAE* 32(1): 137-146.

Haan, C.T., B. Allred, D.E. Storm, G.J. Sabbagh, and S. Prabhu. 1995. Statistical Procedure for Evaluating Hydrologic/Water Quality Models. *Trans. of the ASAE* 38(3):725-733.

Hafele, R. 1996. *National Monitoring Program Project Description and Preliminary Results for the Upper Grande Ronde River Nonpoint Source Study - Draft,* Oregon Department of Environmental Quality, Portland, OR.

Haith, D.A. 1985. Variability of Pesticide Loads to Surface Waters. *Journal WPCF,* 57(11): 1062-1067.

Hall, D.W. 1992. Effects of nutrient management on nitrate levels in ground water near Ephrata, Pennsylvania. *Ground Water* 30(5):720-730).

Hall, D.W., P.L. Lietman, and E.H. Koerkle. 1997. *Evaluation of agricultural best-management practices in the Conestoga River headwaters, Pennsylvania: effects of nutrient management on quality of surface runoff and ground water at a small carbonate-rock site near Ephrata, Pennsylvania, 1984-1990, water-quality study of the Conestoga River headwaters, Pennsylvania,* U.S. Geological Survey Water-Resources Investigations Report 95-4143, Lemoyne, PA.

Hall, J.K., N.L. Hartwig, and L.K. Hoffman. 1983. Application mode and alternative cropping effects on atrazine losses from a hillside. *J. Environ. Qual.* 12:336-340.

Hall, J.K., R.O. Mumma, and D.W. Watts. 1991. Leaching and runoff losses of herbicides in a tilled and untilled field. *Agric. Ecosyst. and Environ.* 37:303-314.

Hall, J.K., N.L. Hartwig, and L.D. Hoffman. 1984. Cyanazine losses inrunoff from no-tillage corn in "living" and dead mulches vs. unmulched, conventional tillage. *J. Env. Qual.* 13:105-110

Hallberg, G.R., C.R. Contant, C.A. Chase, G.A. Miller, M.D. Duffy, R.J. Killorn, R.D. Voss, A.M. Blackmer, and S.C. Padgitt. 1991. *A Progress Review of Iowa's Agricultural-Energy-Environmental Initiatives: Nitrogen Management in Iowa.* Technical Information Series 22, Iowa Department of Natural Resources, Iowa City, IA.

Hallberg, G.R., R.D. Libra, Z. Liu, R.D. Rowden, and K.D. Rex. 1993. Watershed-scale water-quality response to changes in landuse and nitrogen management. In: *Proceedings of Agricultural Research to Protect Water Quality,* February 21-24, 1993, Minneapolis, MN, Soil and Water Conservation Society, Ankeny, IA.

Hallock, B.G. 1996. *Morro Bay Watershed Annual Report.* RWQCB paired watershed study. Soil Science Dept., Cal Poly. SLO, CA.

Harper, L.A. and R.R. Sharpe. 1997. Climate and Water Effects on Gaseous Ammonia Emission from a Swine Lagoon. In: *Southeastern Sustainable Animal Waste Management Workshop Proceedings.* February 11-13, 1997, Tifton, Georgia. University of Georgia Publication Number ENG97-001. pp223-230.

Hatfield, J.L., D.B. Jaynes, and M.R. Burkart. 1995. A watershed study to evaluate farming practices on water quality, In: *Proceedings - National Agricultural Ecosystem Management Conference, December 13-15, 1995, New*

Orleans, Louisiana, Conservation Technology Information Center, West Lafayette, IN.

Hauck. R.D. 1995. Perspective on alternative waste utilization strategies. P.463-492 *In*: Animal Waste and the Land-Water Interface, Kenneth Steele, Ed., Lewis Publishers, Boca Raton, FL, 589pp.

Haygarth, P.M. and S.C. Jarvis. 1997. Soil derived phosphorus in surface runoff from grazed grassland lysimeters. *Water Research* 31(1):140-148.

Hegman, W., D. Wang, and C. Borer. 1999. Estimation of Lake Champlain Basinwide Nonpoint Source Phosphorus Export. Technical Report No. 31, Lake Champlain Basin Program, Grand Isle, VT.

Heidtke, T.M. and M.T. Auer. 1993. Application of a GIS-Based Nonpoint Source Nutrient Loading Model for Assessment of Land Development Scenarios and Water Quality in Owasco Lake, New York, *Water Science & Technology,* 28(3-5): 595-604.

Hermsmeyer, B. 1991. *Nebraska Long Pine Creek Rural Clean Water Program ten year report 1981-1991.* Brown County Agricultural Stabilization and Conservation Service, USDA, Brown County, Ainsworth, NE, 275 pp.

Hill, R.W. 1991. Irrigation Scheduling. Chapter 21. In: *Modeling Plant and Soil Systems,* pp. 491-509. Monograph #31. John Hanks and J.T Ritchie (Eds.). American Society of Agronomy, Inc; Crop Science Society of America, Inc; Soil Science Society of America, Inc. Publishers. Madison, Wisconsin. 545 pp. Software PCET available from Biological and Irrigation Department, Utah State University, Logan, Utah.

Hill, R.W. 1994. *Consumptive Use of Irrigated Crops in Utah.* Submitted to Utah Division of Natural Resources, Division of Water Resources and Division of Water Rights. Utah Agricultural Experiment Station Research Report No. 145, Utah State University, Logan, Utah.

Hill, R.W. 1997. *CRPSM: Crop Growth and Irrigation Scheduling Model, Software User Manual* (Unpublished). Available from Biological and Irrigation Department, Utah State University, Logan, Utah.

Hirschi, M., R. Frazee, G. Czapar, and D. Peterson. 1997. *60 Ways Farmers Can Protect Surface Water*, North Central Regional Extension Publication 589, Information Technology and Communication Services, College of Agricultural, Consumer and Environmental Sciences, University of Illinois, Urbana-Champaign, IL, 317 pp.

Hirschi, M.C., F.W. Simmons, D. Peterson, E. Giles. 1993. *50 Ways Farmers Can Protect their Groundwater.* Cooperative Extension Service. University of Illinois, Urbana, Illinois.

Hoffman, Dennis W. 1995. Use of contour grass and wheat filter strips to reduce runoff losses of herbicides. Proc. Austin Water Quality Meeting, Texas A&M Univ., Temple, TX.

Holmes, J.A. and H.A. Regier. 1990. Influence of temperature changes on aquatic ecosystems: an interpretation of empirical data. *Transactions of the American Fisheries Society* 119, 374-389.

Horner, R.R., J.J. Skupien, E.H. Livingston, and H.E. Sharer, 1994. *Fundamentals of Urban Runoff Management: Technical and Institutional Issues.* Terrene Institute, Washington, D.C.

Hubert, W.A., R.P. Lanka, T.A. Wesch, and F. Stabler. 1985. Grazing management influences on two brook trout streams in Wyoming. In: *Riparian Ecosystems and Their Management: Reconciling Conflicting Uses.* U.S. Department of Agriculture, Forest Service, Rocky Mountain Forest and Range Experiment Station. General Technical Report RM-120, pp. 290-294.

Hunt, P.G., K.C. Stone, F.J. Humenik, and J.M. Rice. 1995. Impact of animal waste on water quality in an eastern coastal plain watershed. In: *Animal Waste and the Land-Water Interface*, Kenneth Steele, Ed., Lewis Publishers, Boca Raton, FL, 589 pp.

Hydratec. 1998. *Price quote for Irrometer model SR.* Delano, California.

Interagency Technical Team. 1996a. *Sampling vegetation attributes.* BLM/RS/ST-96/002+1730. National Applied Resource Sciences Center, Bureau of Land Management, Denver, CO.

Interagency Technical Team. 1996b. *Utilization studies and residual.* BLM/RS/ST-96/004+1730, National Applied Resource Sciences Center, Bureau of Land Management, Denver, CO.

Iowa State University. 1991a. Ag. programs bring economic, environmental benefits. In: *Extension News.* Extension Communications, Ames, IA.

Iowa State University. 1991b. *Nitrogen Use in Iowa.* Prepared for the nitrogen use press conference Dec. 5, 1991, University Extension, Ames, IA.

Isensee, A. R. and A. M. Sadeghi. 1993. Impact of tillage practice on runoff and pesticide transport. *J. Soil and Water Cons.* 48(6):523-527.

ITFM. 1995. *The strategy for improving water quality monitoring in the United States. Technical appendixes.* Final report of the Intergovernmental Task Force on Monitoring Water Quality. Intergovernmental Task Force on Monitoring Water Quality. February.

Jackson, G., D. Knox, and L. Nevers. undated. *Farm*A*Syst, Protecting Rural America's Water*, USDA-Extension Service, University of Wisconsin, Madison, WI, 16 pp.

Jensen, M.E., R.D. Burman, and R.G. Allen. 1990. *ASCE Manuals and Reports on Engineering Practice No. 70: Evapotranspiration and Irrigation Water Requirements.* ISBN 0-87262-763-2. 332 pp.

Johnson, J.T., R.D. Lee, and R.L. Stewart. 1997. *Pastures in Georgia.* Bulletin 573, Cooperative Extension Service, College of Agricultural & Environmental Sciences, University of Georgia, Athens, GA, 31 pp.

Johnston, W.R., F. Ittihadieh, R. M. Daum, and A.F. Pillsbury, 1965. Nitrogen and phosphorus in tile drainage effluent. *Soil Sci. Soc. Am. Proc.* 29:287-289.

Justic, D., N.N. Rabalais, and R.E. Turner. 1995. Stoichiometric nutrient balance and origin of coastal eutrophication, *Marine Pollution Bulletin*, 30(1): 41-46.

Kamprath, E. J. and M. E. Watson. 1980. Conventional soil and tissue tests for assessing the phosphorus status of soils. In: *The role of phosphorus in agriculture*, ed. F.E. Khasawneh, pp. 433-469. American Society of Agronomy, Madison, WI.

Kansas State University Cooperative Extension System and The National Association of Wheat Growers Foundation. 1994. *Best Management Practices for Wheat.* NAWG Foundation, Washington, D.C.

Kauffman, J.B., W.C. Krueger, and M. Vavra. 1983a. Effects of late season cattle grazing on riparian plant communities. *J. Range Mgt.* 36(6):685-691.

Kauffman, J.B., W.C. Krueger, and M. Vavra. 1983b. Impacts of cattle on stream banks in northeastern Oregon. *J. Range Mgt.* 36(6) 683685.

Kitchen, N.R., K.A. Sudduth, D.F. Hughes, S.T. Drumond, and S.J. Birrell. 1995. On-the-Go- Changes in Fertilizer Rates to Agree with Claypan Soil Productivity. In: *Clean Water - Clean Environment - 21st Century: Team Agriculture - Working to Protect Water Resources Volume II: Nutrients* Conference Proceedings, ASAE, March 5-8, Kansas City, Missouri.

Kladivko, E.J., J. Grochulska, R.F. Turco, G.E. Van Scoyoc, and J.D. Eigel. May/June 1999. Pesticide and nitrate transport into subsurface tile drains of different spacings. *Journal of Environmental Quality,* Vol.28, No.3.

Klausner, S. 1995. *Nutrient Management: Crop Production and Water Quality.* 95CUWFP1, Cornell University, Ithaca, NY.

Knight, R.L., V. Payne, R.E. Borer, R.A. Clarke, and J.H. Pries. 1996. Livestock Wastewater Treatment Wetland Database. P.J. DuBowy, ed. Proceedings of the Second National Workshop on Constructed Wetlands for Animal Waste Management. May 15-18, 1996, Fort Worth, Texas.

Knisel, G.W. and R.A. Leonard. 1989. Irrigation Impact on Groundwater: Model Study in Humid Region. *J. Irrigation and Drainage Engineering* 15(5):823-837.

Knisel, W.G., R.A. Leonard, and F.M. Davis. 1991. Water Balance Components in the Georgia Coastal Plain: A GLEAMS Model Validation and Simulation. *J. of Soil and Water Conservation* 46(6):450-456.

Koerkle, E.H., D.K. Fishel, M.J. Brown, and K.M. Kostelnik. 1996. *Evaluation of agricultural best-management practices in the Conestoga River headwaters, Pennsylvania: effects of nutrient management on water quality in the Little Conestoga Creek headwaters, 1983-1989*, U.S. Geological Survey Water-Resources Investigations Report 95-4046, Lemoyne, PA.

Koerkle, E.H. and L.C. Gustafson-Minnich. 1997. *Surface-water quality changes after 5 years of nutrient management in the Little Conestoga Creek headwaters, 1989-1991*, U.S. Geological Survey Water-Resources Investigations Report 97-4048, Lemoyne, PA.

Kolpin, D.W., E.M. Thurman, D.A. Goolsby. 1996. Occurrence of selected pesticides and their metabolites in near-surface aquifers of the midwestern United States. *Environmental Science and Technology*, 30(1): 335-339.

Kolpin, D.W., J.E. Barbash, and R.J. Gilliom. 2000. Pesticides in ground water of the United States, 1992-1996. *Ground Water* 38(6):858-863.

Kress, M. and G.F. Gifford. 1984. Fecal Coliform release from cattle fecal deposits. *Water Resources Bulletin.* 20(1):61-66.

Kunkle, S.H. 1970. Sources and transport of bacterial indicators in rural streams. Pages 105-133 in Proc. Symp. on Interdisciplinary Aspects of Watershed Management. *Am. Soc. Civil Eng.* New York, NY.

Lander, C.H., D. Moffitt, and K. Alt. 1998. Nutrients available from livestock manure relative to crop growth requirements. Res. Assess. Strat. Planning Pap. 98-1. USDA-NRCS, Washington, D.C.

Langland, M.J. and D.K. Fishel. 1996. *Effects of agricultural best-management practices on the Brush Run Creek headwaters, Adams County, Pennsylvania, prior to and during nutrient management*, U.S. Geological Survey, Water-Resources Investigations Report 95-4195, Lemoyne, Pennsylvania.

Lal, R. 1983. Soil erosion in the humid tropics with particular reference to agricultural land development and soil management. IAHS Publication No. 140. Hydrology of Humid Tropical Regions, International Association of Hydrological Sciences, Washington, D.C.

Larson, S.J., P.D. Capel, D.A. Goolsby, S.D. Zaugg, and M.W. Sandstrom. 1995. Relations between pesticide use and riverine flux in the Mississippi River basin, *Chemosphere,* 31(5): 3305-3321.

Larson, S.J., P.D. Capel, and M.S. Majewski. 1997. Pesticides in Surface Waters: Distribution, Trends, and Governing Factors. Vol. 3 in the series: Pesticides in the Hydrologic System. Ann Arbor Press, Chelsea, MI.

Lassek, P.J. 1997, August 18. Lake Eucha Drowning in Algae. *Tusla World*.

Laycock, W. A. 1996. *Grazing on Public Lands.* Council for Agricultural Science and Technology (CAST). 4420 W. Lincoln Way, Ames, IA, 50014 3347.

Lemunyon, J.L. and R.G. Gilbert. 1993. The concept and need for a phosphorus assessment tool. *Journal of Production Agriculture*, 6(4): 483-486.

Leonard, S., G. Kinch, V. Elsbernd, M. Borman, and S. Swanson. 1997. Riparian Area Management; Grazing Management for Riparian-Wetland Areas. Bureau of Land Management, BLM/RS/ST-97/002+1737, National Applied Resource Sciences Center, Colorado. 80pp.

Letey, J., C. Roberts, M. Penberth, and C. Vasek. 1986. *An Agricultural Dilemma: Drainage Water and Toxics Disposal in the San Joaquin Valley.* Agricultural Experiment Station, University of California, Division of Agriculture and Natural Resources. Special Publication 3319.

Lietman, P.L., L.C. Gustafson-Minnich, and D.W. Hall. 1997. *Evaluation of agricultural best-management practices in the Conestoga River headwaters, Pennsylvania: effects of pipe-outlet terracing on quantity and quality of surface runoff and ground water at a small carbonate-rock basin near Churchtown, Pennsylvania, 1983-1989, water-quality study of the Conestoga River headwaters, Pennsylvania,* U.S. Geological Survey Water Resources Investigations Report 94-4206, Lemoyne, PA.

Line, D.E., S.W. Coffey, and D.L. Osmond. 1997. WATERSHEDSS — GRASS AGNPS modeling tool. *Trans. of the ASAE* 40(4):971-975.

Line, D.E. and J. Spooner. 1995. *Critical Areas in Agricultural Nonpoint Source Pollution Control Projects: The Rural Clean Water Program Experience.* NCSU Water Quality Group, North Carolina State University, Raleigh, NC. 6p.

Lisle, J.T. and J.B. Rose. 1995. *Cryptosporidium* contamination of water in the USA and UK: a mini-review. *J Water SRT-Aqua* 44/3:103-117.

Liu, B.Y., M.A. Nearing, C. Baffault, and J.C. Ascough. 1997. The WEPP Watershed Model: Three Comparisons to Measured Data from Small Watersheds. *Trans. of the ASAE* 40:945-952.

Logan, T.J. 1990. Agricultural best management practices and groundwater protection. *J. Soil and Water Cons.* 45(2):201-206.

Long, R.H.B., S.M. Benson, T.K. Tokunaga, and A. Lee. 1990. Selenium Immobilization in a Pond Sediment at Kesterson Reservoir. *J. Environ. Qual.* 19:302-311.

Lowrance, R.R., S. McIntyre, and C. Lance. 1988. Erosion and deposition in a field/forest system estimated using Cesium-137 activity. *J. Soil and Water Cons.* 43(2):195-199.

Lowrance, R., L.S. Altier, J.D. Newbold, R.R. Schnabel, P.M. Groffman, J.M. Denver, D.L. Correll, J.W. Gilliam, J.L. Robinson, R.S. Brinsfield, K.W. Staver, W. Lucas, and A.H. Todd. 1995. *Water quality functions of riparian forest buffer systems in the Chesapeake Bay Watershed.* EPA 903-R-95-004.

Lugbill, J. 1990. *Potomac River Basin Nutrient Inventory.* Metropolitan Washington Council of Governments, Washington, DC.

Lundstrom, D.R. and E.C. Stegman. 1991. *Irrigation scheduling by the checkbook method.* NDSU Extension Service. Fargo, ND. AE792.

Luthin, J.N. 1973. Drainage Engineering. Robert E. Krieger Publishing Co.

Lyons, R.K., R. Machen, and T.D.A. Forbes. 1995. *Understanding forage intake in range animals.* L-5152. Texas Agricultural Extension Service, Texas A & M University, College Station, TX, 6 pp.

Lyons, R.K., T.D.A. Forbes, and R. Machen. 1996a. *What range herbivores eat – and why.* B-6037. Texas Agricultural Extension Service, Texas A & M University, College Station, TX, 9 pp.

Lyons, R.K., R. Machen, and T.D.A. Forbes. 1996b. *Why range forage quality changes*. B-6036. Texas Agricultural Extension Service, Texas A & M University, College Station, TX, 7 pp.

Maas, R. 1984. *Best Management Practices for Agricultural Nonpoint Sources: IV. Pesticides.* Biological and Agricultural Engineering Dept., North Carolina State University, Raleigh, NC. 83p. NTIS PB85-114247/WEP.

MacDonald, L.H., A.W. Smart, and R.C. Wissmar. 1991. *Monitoring guidelines to evaluate effects of forestry activities on streams in the Pacific Northwest and Alaska.* EPA/910/9-91-001. U.S. Environmental Protection Agency, Region 10, Seattle, Washington.

MacKenzie, A.J. and F.G. Viets, Jr. 1974. Nutrients and other chemicals in agricultural drainage waters. P. 489-508. In Drainage for agriculture (J. van Schilfgaarde, ed.). Agronomy Monograph No 17. Madison, WI: *Am. Soc. of Agronomy.*

Madramootoo, C.A., K.A. Wiyo, and P. Enright, 1992. Nutrient loses through the tile drains from two potato fields. *Applied Engineering in Agriculture* 8(5):639-646.

Magdoff, F.R., D. Ross, and J. Amadon. 1984. A soil test for nitrogen availability to corn. *Soil Science Society of America Journal,* 48:1301-1304.

Maret, R., R. Yanked, S. Potter, J. McLaughlin, D. Carter, C. Rockway, R. Jesser, and B. Olmstead. 1991. *Rock Creek Rural Clean Water Program Ten Year Report.* Cooperators: USDA-ASCS, USDA-SCS, USDA-ARS, Idaho Division of Environmental Quality, Twin Falls and Sanke River Soil Conservation District.

Maryland Department of Agriculture. 1990. *Nutrient Management Program.* Maryland Department of Agriculture, Annapolis, MD.

Maryland Department of Agriculture. 1998. R. Cuizon, personal communication, Maryland Department of Agriculture, Annapolis, MD, 410-841-5863.

McDougald, N.K., W.E. Frost, and D.E. Jones. 1989. *Use of Supplemental Feeding Locations to Manage Cattle Use on Riparian Areas of Harrdwood Rangelands.* U.S. Department of Agriculture Forest Service. General Technical Report PSW-110, pp. 124-126.

McMahon, G. and M.D. Woodside. 1997. Nutrient mass balance for the Albemarle-Pamlico drainage basin, North Carolina and Virginia, 1990. J. Am. Water Res. Assoc. 33(3):573-589.

McGinty, A. 1996. *Reference guide for Texas ranchers.* L-5097. Texas Agricultural Extension Service, Texas A&M University, College Station, TX. (Internet: http://texnat.tamu.edu/ranchref/guide/)

Meals, D.W. 1993. Assessing nonpoint phosphorus control in the LaPlatte River Watershed. *Lake Reservoir Mgt.* 7(2):197-207.

Meals, D.W. 1996. Watershed-scale response to agricultural diffuse pollution control programs in Vermont, USA, *Wat. Sci. Tech.,* 33(4-5): 197-204.

Meals, D.W. 2001. *Lake Champlain Basin Agricultural Watersheds Section 319 National Monitoring Program Project: Final Project Report, May, 1994 - November 2000.* Vermont Department of Environmental Conservation, Waterbury.

Meals, D.W., J.D. Sutton, and R.H. Griggs. 1996. Assessment of Progress of Selected Water Quality Projects of USDA and State Cooperators. USDA-NRCS, Washington, D.C.

Meals, D.W. and L.F. Budd. 1998. Lake Champlain Basin nonpoint source phosphorus assessment. *J. Am. Water Resources Assoc.* 34(2):251-265.

Meek, B.D., D.L. Carter, D.T. Westerman, J.L. Wright, R.E. Peckenpaugh. 1995. Nitrate leaching under furrow irrigation as affected by crop sequence and tillage. *J. Soil Sci. Soc. Amer.*, 59:204-210.

Michels, D. 1998. Personal communication. Nebraska-Kansas Consulting Services, Superior, NE, 402-879-4401.

Mickelson, S.K. and J.L. Baker. 1993. Buffer strips for controlling herbicide runoff losses. Paper 932084, Amer. Soc. Agric. Eng., St. Joseph, MI.

Midwest Plan Service. 1985. *Livestock Waste Facilities Handbook.* Iowa State University, Ames, IA.

Mielke, L.N. and J.R.C. Leavitt. 1981. *Herbicide Loss in Runoff Water and Sediment as Affected by Center Pivot Irrigation and Tillage Treatments.* U.S. Department of the Interior, Office of Water Research and Technology. Report A-062-NEB.

Miner, J.R., J.C. Buckhouse, and J.A. Moore. 1991. Evaluation of Off-Stream Water Source to Reduce Impact of Winter-Fed Range Cattle on Stream Water Quality. In: *Nonpoint Source Pollution: The Unfinished Agenda for the Protection of Our Water Quality*, 20-21 March, 1991, Tacoma, WA. Washington Water Research Center, Report 78, pp.65-75.

Mitsch, W.J. and J.G. Gosselink. 1986. *Wetlands.* Van Nostrand Reinhold, New York.

Misra, A.K. 1994. *Effectiveness of vegetative buffer strips in reducing herbicide transport with surface runoff under simulated rainfall.* Ph.D. Thesis, Iowa State University, Ames, IA.

Misra, A., J.L. Baker, S.K. Mickelson, and H. Shang. 1996. Contributing area and concentration effects on herbicide removal by vegetative buffer systems. *Trans. of the ASAE* 39(6):2105-2111.

Misra, A.K., J.L. Baker, S.K. Mickelson, and H. Shang. 1994. *Effectiveness of vegetative buffer strips in reducing herbicide transport with surface runoff under simulated rainfall.* Paper 942146, Amer. Soc. Agric. Eng., St. Joseph, MI.

Montana Watershed Coordination Council's Grazing Practices WorkGroup. 1999. Best Management Practices for Grazing Montana.

Moore, J.A., J. Smyth, S. Baker, J.R. Miller. 1988. Evaluating coliform concentrations in runoff from various animal waste management systems.

Special Report 817, Agricultural Experiment Stations, Oregon State University, Corvallis, OR.

Mueller, D.K, and D.R. Helsel. 1996. *Nutrients in the nation's waters: too much of a good thing?* U.S. Geological Survey circular 1136, Denver, CO.

Mullens, J.A., R.F. Carsel, J.E. Scarbrough, and I.A. Ivery. 1993. *PRZM-2, A Model for Predicting Pesticide Fate in the Crop Root and Unsaturated Soil Zones: Users Manual for Release 2.0.* Environmental Research Laboratory, Office of Research and Development, U.S. Environmental Protection Agency, Athens, GA. EPA/600/R-93/046.

Munster, C.L., R.W. Skaggs, J.E. Parsons, R.O. Evans, J.W. Gilliam, and E.W. Harmsen,. 1995. Aldicarb transport in drained coastal plain soil. *J. Irrig. Drain. Eng.* 121(6):378-384.

National Alliance of Independent Crop Consultants (NAICC). 1998. 1055 Petersburg Cove, Colliersville, TN 38017, 901-861-0511

National Atmospheric Deposition Program (NRSP-3)/National Trends Network (June 24, 1998). NADP/NTN Coord. Office, Illinois State Water Survey, 2204 Griffith Drive, Champaign, IL 61820.

National Research Council. 1986. *Ecological knowledge and environmental problem-solving. Concepts and case studies.* National Academy Press, Washington, DC.

National Research Council. 1992. *Restoration of Aquatic Ecosystems: Science, Technology, and Public Policy.* 91-43324. National Academy Press. Washington, D.C.

NCAES. 1982. *Best Management Practices for Agricultural Nonpoint Source Control III: Sediment.* North Carolina Agricultural Extension Service, in cooperation with USEPA and USDA. Raleigh, North Carolina.

NCSU. 2001. Water Sheds - Viruses. North Carolina State University, Water Quality Group. http://h2osparc.wq.ncsu.edu/info/viruses.html. Accessed April 3, 2001.

Nelson, D. 1985. Minimizing nitrogen losses in non-irrigated eastern areas. In: *Proceedings of the Plant Nutrient Use and the Environment Symposium, Plant Nutrient Use and the Environment,* October 21-23, 1985, Kansas City, MO. pp. 173-209. The Fertilizer Institute.

Niemi, R.M. and J.S. Niemi. 1991. Bacterial pollution of waters in pristine and agricultural lands. *J. Environ Qual.* 20(3):620-627.

Nolan, B.T. and M.L. Clark. 1997. Selenium in irrigated agricultural areas of the western United States. *J. Environ. Qual.* 26(3):849-857.

Nolan, B.T., B.C. Ruddy, K.J. Hitt, and D.R. Helsel. 1997. Risk of nitrate in groundwaters of the United States - a national perspective. *Environ. Sci. Technol.* 31:2229-2236.

North Carolina Cooperative Extension Service (NCCES). 1994. SoilFacts: Protecting Groundwater in North Carolina, A Pesticide and Soil Ranking

System. North Carolina State University, College of Agriculture and Life Sciences. AG-439-31.

Northup, B.K., D.T. Goerend, D.M. Hays, and R.A. Nicholson. 1989. Low volume spring developments. *Rangelands,* 11(1):39-41.

Norton, G.W. and J. Mullen. 1994. *Economic evaluation of integrated pest management programs: a literature review.* Va. Coop. Ext. Pub. 448-120, Virginia 20(3):62-627Tech, Blacksburg, VA 24061.

Novotny, V., and H. Olem. 1994. Water Quality Prevention, Identification, and Management of Diffuse Pollution. Van Nostrand Reinhold, New York.

Ongerth, J.E., G.D. Hunter, and F.B. DeWalle. 1995. Watershed use and *Giardia* cyst presence. *Water Res.* 29(5):1295-1299.

Onken, A.B., E. Segarra, and W.M. Lyle. 1995. Plant and Soil Analysis, Irrigation Technology, Economics, and N Use Efficiency. In: *Clean Water - Clean Environment - 21st Century: Team Agriculture - Working to Protect Water Resources Volume II: Nutrients* Conference Proceedings, ASAE, March 5-8, Kansas City, Missouri.

Osmond, D.L., S.W. Coffey, D.E. Line, and J. Spooner. *Section 319 National Monitoring Program Projects - 1997 Summary Report*, EPA 841-S-97-004 (printed by EPA Office of Water, Washington, DC), NCSU Water Quality Group, North Carolina State University, Raleigh, NC, 274 pp.

Owens, L.B., W.M. Edwards, and R.W. Van Keuren. 1989. Sediment and nutrient losses from an unimproved, all year grazed watershed. *J Environ. Qual.* 18(2) 232238.

Owens, L.B., R.W. Van Keuren, and W.M. Edwards. 1982. Environmental effects of a medium-fertility 12-month pasture program: II. Nitrogen. *J. Environ. Qual.* 11(2):241-246.

Patty, L., B. Real, and J.J. Gril. 1997. The use of grassed buffer strips to remove pesticides, nitrate and soluble phosphorus compounds from runoff water. *Pesticide Sci.* 49:243-251.

Pell, A.N. 1997. Manure and microbes: public and animal health problem? *J. Dairy Sci.* 80:2673-2681.

Penn State Cooperative Extension. 1997. *Pequea-Mill Creek Information Series.* Smoketown, PA.

Pennsylvania State University Cooperative Extension. 1997. *Sample Nutrient Management Plan for the Nutrient Management Act.* Penn State University, 5/22/97.

Pennsylvania State University. 1992a. College of Agriculture, Merkle Laboratory – Soil and Forage Testing,University Park, PA.

Pennsylvania State University. 1992b. *Nonpoint Source Database*. Pennsylvania State University, Department of Agricultural and Biological Engineering, University Park, PA.

Pennsylvania State University. 1997. *The Penn State Agronomy Guide, 1997-1998.* University Park, PA.

Plafkin, J.L., M.T. Barbour, K.D. Porter, S.K. Gross, and R.M. Hughes. 1989. *Rapid bioassessment protocols for use in streams and rivers: Benthic macroinvertebrates and fish.* EPA/440/4-89-001. U.S. Environmental Protection Agency, Office of Water, Washington, DC.

Platts, W.S. 1990. *Managing Fisheries and Wildlife on Rangelands Grazed by Livestock, A Guidance and Referenced Document for Biologists.* Nevada Department of Wildlife, Reno, NV.

Platts, W.S. and R.L. Nelson. 1989. Characteristics of riparian plant communities and streambanks with respect to grazing in Northeastern Utah. In: *Practical Approaches to Riparian Resource Management — An Educational Workshop,* ed. R.E. Gressell, B.A. Barton, and J.L. Kershner, pp. 73-81. U.S. Department of the Interior, Bureau of Land Management.

Polenske, J. 1998. Personal communication. Polenske Agronomic Consulting, Appleton, WI.

Ponce, 1980. *Water quality monitoring programs.* WSDG-TP-00002, USDA Forest Service, Fort Collins, CO, 66 pp.

Pote, D.H., T.C. Daniel, A.N. Sharpley, P.A. Moore, Jr., D.R. Edwards, and D.J. Nichols. 1996. Relating extractable soil phosphorus to phosphorus losses in runoff. *Soil Science Society of America Journal.* 60:855-859.

Pote, D.H., T.C. Daniel, D.J. Nichols, A.N. Sharpley, P.A. Moore, D.M. Miller, and D.R. Edwards. 1999. Relationship between phosphorus levels in three ultisols and phosphorus concentrations in runoff. *J. Environ. Qual.* 28:170-175.

Prichard, D., H. Barrett, J. Cagney, R. Clark, J. Fogg, K. Gebhardt, P. Hansen, B. Mitchell, and D. Tippy. 1993. *Riparian Area Management; Process for Assessing Proper Functioning Condition* TR1737-9. Bureau of Land Managment, BLM/SC/ST-93-003+1737, Service Center, Colorado. 60pp.

Pritt, J.W. and J.W. Raese, ed. 1992. *Quality assurance/quality control manual.* Open File Report 92-495. U.S. Geological Survey, National Water Quality Laboratory, Reston, Virginia.

Rabalais, N.N., Turner, R.E., Dubravko, J., Dortsch, Q., and Wisman, W.J., Jr. 1999. Characterization of Hypoxia - Topic 1 Report for the Integrated Assessment on Hypoxia in the Gulf of Mexico: Silver Spring, MD., NOAA Coastal Ocean Office, NOAA Coastal Ocean Program Decision Analysis Series No. 17, 167p.

Rankins, A., Jr., D.R. Shaw, M. Boyette, and S.M. Seifert. 1998. Minimizing herbicide and sediment losses in runoff with vegetative filter strip. Abstracts *Weed Sci. Soc. Am.* 38:59.

Rantz, S.E., et al. 1982. *Computation of discharge.* U.S. Geological Survey Water Supply Paper 2175 (2 vol.) 631 p.

Rasmussen, G.A., M.P. O'Neill, and L.R. Schmidt. 1997. *Monitoring Rangelands: Interpreting What You See.* Utah State University Ext. Bull. NR-503 42p. (Video NR-503 V).

Rasmussen, G.A. and K. Voth. 2001. *Repeat Photography-Monitoring Made Easy.* Utah State University Extension, U.S. Environmental Protection Agency.

Rhode, W.A., L.E. Asmussen, E.W. Hauser, R.D. Wauchope, and H.D. Allison. 1980. Trifluralin movement in runoff from a small agricultural watershed. *J. Environ. Qual.* 9:37-42.

Rice, J.M., A.A. Szogi, S.W. Broome, F.J. Humenik and P.G. Hunt. 1998. Constructed wetland systems for swine wastewater treatment: Animal Production Systems and the Environment. In: *Proceedings of an International Conference on Odor, Water Quality, Nutrient Management and Socio-economic Issues*, Iowa State University. July 1998.

Richards, R.P. 1997. *Estimation of Pollutant Loads in Rivers and Streams: A Guidance Document for NPS Programs.* DRAFT. Water Quality Laboratory, Heidelberg College, Tiffin, OH. 80 pp.

Richards, R.P. and D.B. Baker, 1993. Pesticide concentration patterns in agricultural drainage networks in the Lake Erie basin. *Environmental Toxicology and Chemistry*, 12: 13-26.

Risinger, M. and K. Carver. 1987. *Tensiometers – a gauge for measuring soil moisture.* High Plains Underground Water Conservation District No. 1, USDA-Soil Conservation Service, Lubbock, TX. 4 pp.

Ritter, W. F., H. P. Johnson, W. G. Lovely, and M. Molnau. 1974. Atrazine, propachlor and diazinon residues on small agricultural watersheds: runoff losses, persistence, and movement. *Env. Sci. Technol.* 8:38-42.

Ritzema, H.P., R.A.L. Kselik, and . Chanduvi. 1996. Drainage of Irrigated Lands. FAO Irrigation Water Management Training Manual No. 9.

Robillard, P.D. and M.F. Walter. 1986. *Nonpoint Source Control of Phosphorus — A Watershed Evaluation, Volume 2, Development of Manure Spreading Schedules to Decrease Delivery of Phosphorus to Surface Waters.* U.S. Environmental Protection Agency, Robert S. Kerr Environmental Research Laboratory, Ada, OK. Internal report.

Robinson, C.T. and G.W. Minshall. 1995. Effects of open-range livestock grazing on stream communities, In: *Animal Waste and the Land-Water Interface*, Kenneth Steele, Ed., Lewis Publishers, Boca Raton, FL, 589 pp.

Rowden, R.D., R.D. Libra, and G.R. Hallberg. 1995. *Surface water monitoring in the Big Spring basin 1986-1992: a summary review.* Geological Survey Bureau Technical Information Series 33, Iowa Department of Natural Resources, Des Moines, IA.

Russell, J.R. and L.A. Christensen. 1984. *Use and Cost of Soil Conservation and Water Quality Practices in the Southeast.* U.S. Department of Agriculture, Economic Research Service, Washington, DC.

Ruyle, G. 1993. Nutritional value of range forage for livestock. In: *Arizona Ranchers' Management Guide*, R. Gum, G. Ruyle, and R. Rice, Eds., Arizona Cooperative Extension, The University of Arizona, Tucson, AZ, pp. 27-31.

Ruyle, G. and B. Frost. 1993. Monitoring range land browse vegetation. In: *Arizona Ranchers' Management Guide*, R. Gum, G. Ruyle, and R. Rice, Eds., Arizona Cooperative Extension, The University of Arizona, Tucson, AZ, pp. 1-4.

Saffigna, P.G. and D.R. Keeney. 1977. Nitrate and chloride in ground water under irrigated agriculture in central Wisconsin. *Ground Water* 15:170-177.

Sanders, J.H., D. Valentine, E. Schaeffer, D. Greene, and J. McCoy. 1991. *Double Pipe Creek RCWP: Ten Year Report.* U.S. Department of Agriculture, University of Maryland Cooperative Extension Service, Maryland Department of the Environment, and Carroll County Soil Conservation District.

SARE. 1997. SARE program advances grazing systems. In: *SARE 1997 Project Highlights*, Sustainable Agriculture Research and Education Program, U.S. Department of Agriculture, Washington, DC, 8 pp.

Sawyer, C. N. 1947. Fertilization of lakes by agricultural and urban drainage. *Journal of New England Water Works Association.* 61:109-127.

Scarnecchia, D.L. 1999. Viewpoint: The range utilization concept, allocation arrays, and range management science. *Journal of Range Management* 52(2):157-160.

Schepers, J.S. and R.H. Fox. 1989. Estimation of N budgets for crops. p. 221-246 In: Nitrogen Management and Ground Water Protection. Developments in Agricultural and Managed-Forest Ecology 21, R.F. Follett, ed. Amsterdam: Elsevier.

Schepers, J.S. and D.D. Francis. 1982. Chemical water quality of runoff from grazing land in Nebraska: Influence of grazing livestock. *J. Environ. Qual.* 11(3): 351359.

Schepers, J.S., K.D. Frank, and C. Bourg. 1986. Effect of yield goal and residual soil nitrogen considerations on nitrogen fertilizer recommendations for irrigated maize in Nebraska. *J. Fertilizer Issues* 3:133-139.

Scherer, T. F. 1994. *Selecting a sprinkler irrigation system.* NDSU Extension Service. Fargo, ND. AE-91.

Scherer, T.F., B.D. Seelig, and D.W. Franzen. 1996. *Soil, water and plant characteristics important to irrigation.* NDSU Extension Service. Fargo, ND. EB-66.

Scherer, T.F. and J. Weigel. 1993. *Planning to irrigate: a checklist.* NDSU Extension Service. Fargo, ND. AE-92.

Schoumans, O.F. and P. Groenendijk. 2000. Modeling soil phosphorus levels and phosphorus leaching from agricultural lands in the Netherlands. *J. Environ. Qual.* 29:111-116.

Schwab, G.O., D.D. Fangmeier, W.J. Elliot, and R.K. Frevert. 1993. Soil and Water Conservation Engineering. 4th edition, Wiley and Sons.

Schwennesen, E.P. 1994. Using salt for livestock. In: *Arizona Ranchers' Management Guide*, R. Gum, G. Ruyle, and R. Rice, Eds., Arizona Cooperative Extension, The University of Arizona, Tucson, AZ, pp. 43-45.

Scribner, E.A., D.A. Goolsby, E.M. Thurman, M.T. Meyer, and M.T. Pomes. 1994. *Concentrations of selected herbicides, two triazine metabolites, and nutrients in storm runoff from nine stream basins in the midwestern United States, 1990-1992*, U.S. Geological Survey Open-File Report 94-396, Lawrence, KS.

Scribner, E.A., D.A. Goolsby, E.M. Thurman, M.T. Meyer, and W.A. Battaglin. 1996. *Concentrations of selected herbicides, herbicide metabolites, and nutrients in outflow from selected midwestern reservoirs, April 1992 through September 1993*, U.S. Geological Survey Open-File Report 96-393, Lawrence, KS.

Sedivec, K. 1992. *Water quality: the rangeland component*. R-1028, North Dakota State University Extension Service, North Dakota State University of Agriculture and Applied Science, Fargo, ND, 9 pp.

Seelig, B.D. and J.L. Richardson. 1991. *Salinity and sodicity in North Dakota soils.* NDSU Extension Service. Fargo, ND. EB 57.

Seta, A. K., R. L. Blevins, W. W. Frye, B. J. Barfield. 1993. Reducing soil erosion and agricultural chemical losses with conservation tillage. *J. Environ. Qual.* 22:661-665.

Shaffer, M.J. 1985. Simulation Model for Soil Erosion-Productivity Relationships. *J. Environ. Qual.* 14(1):144-150.

Shaffer, M.J., S.C. Gupta, D.R. Linden, J.A.E. Molina, C.E. Clapp, and W.E. Larson. 1986. Simulation of Nitrogen, Tillage, and Residue Management Effects on Soil Fertility. In: *Analysis of Ecological Systems: State-of-the-art in Ecological Modelling.* W.K. Lauenroth, G.V. Skogerboe, and M. Flug, eds. Developments in Environmental Modelling 5. Elsevier Scientific Publishing Company.

Shaffer, M.J. and W.E. Larson, eds. 1985. *NTRM: A Soil-Crop Simulation Model for Nitrogen, Tillage, and Crop Residue Management.* USDA Agricultural Research Service, St. Paul, MN.

Sharpley, A.N. 1985. Depth of surface soil-runoff interaction as affected by rainfall, soil slope and management. *Soil Sci. Soc. Am. J.* 49:1010-1015.

Sharpley, A.N. 1993. An innovative approach to estimate bioavailable phosphorus in agricultural runoff using iron-oxide impregnated paper. *J. Environ. Qual.* 22:597-601.

Sharpley, A.N. 1995a. Identifying sites vulnerable to phosphorus loss in agricultural runoff. *J. Environ. Qual.* 24(5): 947-951.

Sharpley, A.N. 1995b. Dependence of runoff phosphorus on extractable soil phosphorus. *J. Environ. Qual.* 24:920-926.

Sharpley, A.N. 2000. Practical and innovative measures for the control of agricultural phosphorus losses to water: an overview. *J. Environ. Qual.* 29:1-9.

Sharpley, A.N. and R.G. Menzel. 1987. The impact of soil and fertilizer phosphorus on the environment. Advances in Agronomy 41:297:324.

Sharpley, A.N. and J.R. Williams, eds. 1990. *EPIC–Erosion/Productivity Impact Calculator: 1. Model Documentation.* U.S. Department of Agriculture Bulletin No. 1768. 235 pgs.

Sharpley, A.N., S.J. Smith, O.R. Jones, W.A. Berg, and G.A. Coleman. 1992. The transport of bioavialable phosphorus in agricultural runoff. J. Environ. Qual. 21:30-35.

Sharpley, A.N., T.C. Daniel, and D.R. Edwards. 1993. Phosphorus movement in the landscape. *J. Prod. Agric.* 6(4):492-500.

Sharpley, A.N., S.C. Chapra, R. Wedepohl, J.T. Sims, T.C. Daniel, and K.R. Reddy. 1994. Managing agricultural phosphorus for protection of surface waters: Issues and options. *Journal of Environmental Quality* 23: 437-451. 1994.

Sharpley, A.N., T.C. Daniel, J.T. Sims, and D.H. Pote. 1996. Determining environmentally sound soil phosphorus levels. *Journal of Soil and Water Conservation* 51(2):160-166.

Sherer, B.M., J.R. Miner, J.A. Moore, and J.C. Buckhouse. 1992. Indicator bacterial survival in stream sediments. *J. Environ. Qual.* 21:591-595.

Shipitalo, M.J., W.A. Dick, and W.M. Edwards. 2000. Conservation tillage and macorpore factors that affect water movement and the fate of chemicals. *Soil and Tillage Research* 53(3-4):167-183.

Simard, R.R., S. Beauchemin, and P.M. Haygarth. 2000. Potential for preferential pathways of phosphorus transport. *J. Environ. Qual.* 29:97-105.

Simon, A. and S. Guzman-Rio. 1990. *Sediment discharge from a montane basin, Puerto Rico: implications of erosion processes and rates in the humid tropics.* p. 35-47 In: Research Needs and Applications to Reduce Erosion and Sedimentation in Tropical Steeplands, IAHS Publication No. 192, International Association of Hydrological Sciences, Washington, D.C.

Sims, J.T. 1993. Environmental soil testing for phosphorus. *J. Prod. Agric.* 6(4):501-507.

Sims, J.T. 2000. The role of soil testing in environmental risk assessment for phosphorus. Pages 57-81 in *Agriculture and Phosphorus Management*, A.N. Sharpley, ed., Lewis Publishers, Boca Raton, FL.

Sims, J.T., A.C. Edwards, O.F. Schoumans, and R.R. Simard. 2000. Integrating soil phosphorus testing into environmentally based agricultural management practices. *J. Environ. Qual.* 29:60-71.

Skaggs, R.W. 1980. *A Water Management Model for Artificially Drained Soils.* North Carolina Agricultural Research Service Technical Bulletin 267, North Carolina State University, Raleigh, NC.

Smith, J.A., D. Lyon, A. Dickey, and P. Rickey. 1991. *Emergency Wind Erosion Control.* NebGUIDE G75-282. Cooperative Extension, Institute of Agriculture and Natural Resources, University of Nebraska.

Smith, M. 1992. *CROPWAT: A Computer Program for Irrigation Planning and Management. Irrigation and Drainage.* Paper 46. Food and Agricultural Organization of the United Nations (FAO). Rome, Italy.

Smith, M.C., A.B. Bottcher, K.L. Campbell, and D.L. Thomas. 1991. Field Testing and Comparison of the PRZM and GLEAMS Models. *Trans. of the ASAE* 34(3):838-847.

Smolen, M.D. and F.J. Humenik. 1989. *National Water Quality Evaluation Project 1988 Annual Report: Status of Agricultural Nonpoint Source Projects.* U.S. Environmental Protection Agency and U.S. Department of Agriculture, Washington, DC. EPA-506/9-89/002.

Sneed, R. 1992. Personal communication. Biological and Agricultural Engineering Department, North Carolina State University, Raleigh, NC.

Sojka, Robert. 1999. Personal communication. Northwest Irrigation and Soils Research Laboratory. Kimberly, Idaho. May 24, 1999.

Sojka, R.E., and R.D. Lentz (eds). 1996. A PAM primer: A brief history of PAM and PAM-related issues. University of Idaho, Moscow, ID, Miscellaneous Publication No. 101-96.

Spaulding, R.F., and M.E. Exner. 1993. Occurrence of nitrate in groundwater - a review. *J. Environ. Qual.* 22:392-402.

Springman, R.E. 1992. *Source control: Inside-out planning for milking center wastewater disposal.* Milking Center Design Northeast Regional Agricultural Engineering Service, Ithaca, NY. pp 158-167.

Stanley, P. 1998. Personal communication. Paul Stanley Crop Management Services, E. Fairfield, VT.

Steele, D.D., R.E. Knighton, and E.C. Stegman. 1996. Field-scale water quality under continuous, irrigated corn production in the northern Great Plains. In: *ASAE International Meeting,* July 14-18, 1996, Paper No. 96-2020, ASAE. St. Joseph, MI.

Stevens, J.T., and D.D. Sumner. 1991. Herbicides. *In* W.J. Hayes, Jr. and E.R. Laws, Jr., eds., Handbook of Pesticide Toxicology, Vol. 3, pp. 1317-1408. Academic Press, San Diego, CA.

Stoddard, C.S., M.S. Coyne, and J.H. Grove. Nov/Dec 1998. Fecal bacteria survival and infiltration through shallow agricultural soil: Timing and tillage effects. *Journal of Environmental Quality,* Vol. 27, No.6.

Stoltzfus, J.H. and L.E. Lanyon. 1992. *Pequea Mill Creek Project, Smoketown, PA.* February 1992.

Stolzenburg, B. 1992. Personal communication. University of Nebraska, Cherry County Cooperative Extension Service, Valentine, NE.

Sugiharto, T., T.H. McIntosh, R.C. Uhrig, and J.J. Lavdinois. 1994. Modeling Alternatives to Reduce Dairy Farm and Watershed Nonpoint Source Pollution. *J. Environmental Quality* 23:18-24.

Taylor, J.K. 1993. *Standard reference materials handbook for SRM users.* NIST Special Publication 260-100. National Institute of Standards and Technology, Standards and Reference Materials Program, U.S. Department of Commerce, Technology Administration, Washington, DC. February.

Thelin, R. and G.F. Gifford. 1983. Fecal Coliform release patterns from fecal material of cattle. *J. Environ. Qual.* 12(1):57-63.

Tiedemann, A.R., D.A. Higgins, T.M. Quigley, H.R. Sanderson, and C.C. Bohn. 1988. *Bacterial Water Quality Responses to Four Grazing Strategies — Comparison with Oregon Standards. J. Environmental Quality,* 17(3):492-498.

Tingle, C.H., D.R. Shaw, M. Boyette, and G.P. Murphy. 1998. Metolachlor and metribuzin losses in runoff as affected by width of vegetative filter strips. *Weed Sci.* 46:475-479.

Tiscareno-Lopez, M., V.L. Lopes, J.J. Stone, and L.J. Lane. 1993. Sensitivity Analysis of the WEPP Watershed Model for Rangeland Applications: I. Hillslope Processes. *Trans. of the ASAE* 36:1659-1672.

T.-Prairie, Y., and Kalff, J. 1986. Effect of Catchment Size on Phosphorus Export, *Water Resources Bulletin,* American Water Resources Assocation, 22(3): 465-470.

Troeh, F.R., J.A. Hobbs, and R.L. Donahue. 1980. *Soil and Water Conservation for Productivity and Environmental Protection.* Prentice-Hall, Inc., Englewood Cliffs, NJ.

Turner, J. H., Ed. 1980. *Planning for an irrigation system.* American Association for Vocational Instructional Materials, Athens, GA. 120 pp.

University of California. 1988. *Associated Costs of Drainage Water Reduction.* University of California Committee of Consultants on Drainage Water Reduction.

University of Vermont. 1996. *Agricultural Testing Laboratory – Manure Analysis Averages, 1992-1996.* Dept. of Plant & Soil Science, University of Vermont, Burlington, VT.

University of Wisconsin-Extension and Wisconsin Dept. of Agriculture, Trade, and Consumer Protection. 1989. *Nutrient and Pesticide Best Management Practices for Wisconsin Farms.* WDATCP Technical Bulletin ARM-1, Madison, WI.

USDA. 1983. Sprinkler Irrigation Design. Chapter 11, Section 15, *Irrigation.* National Engineering Handbook.

USDA. 1987. *Farm Drainage in the United States: History, Status and Prospects.* Misc. Pub. No. 1455, Washington D.C.

USDA. 1991. *An Interagency Report: Rock Creek Rural Clean Water Program Final Report 1981-1991.* U.S. Department of Agriculture, Twin Falls, ID.

USDA. 1998. Agricultural Prices—1997 Summary, National Agricultural Statistics Service.

USDA–ARS. 1983. *Volume II — Comprehensive report, ARS/BLM Cooperative Studies, Reynolds Creek Watershed.* USDA Agricultural Watershed Service, Boise, Idaho.

USDA–ARS. 1987. *User Requirements — USDA–Water Erosion Prediction Project (WEPP). Draft 6.3.* U.S. Department of Agriculture–Agricultural Research Service, Beltsville, MD.

USDA–ASCS. 1988. *Moapa Valley, Colorado River Salinity Control Program, Project Implementation Plan (PIP).* U.S. Department of Agriculture, Agricultural Stabilization and Conservation Service, Washington, DC.

USDA–ASCS. 1991a. *Oakwood Lakes-Poinsett Project 20 Rural Clean Water Program Ten Year Report.* U.S. Department of Agriculture, Agricultural Stabilization and Conservation Service, Brookings, SD.

USDA–ASCS. 1991b. *Agricultural Conservation Program — 1990 Fiscal Year Statistical Summary.* U.S. Department of Agriculture, Agricultural Stabilization and Conservation Service, Washington, DC.

USDA–ASCS. 1992a. *Conestoga Headwaters Project Pennsylvania Rural Clean Water Program Ten Year Report 1981-1991.* U.S. Department of Agriculture, Agricultural Stabilization and Conservation Service, Harrisburg, PA.

USDA–ERS. 1997. *Agricultural resources and environmental indicators, 1996-1997.* Agricultural Handbook No. 712. USDA-Economic Research Service, Natural Resources and Environment Division, Washington, DC. 347 pp.

USDA–FSA. 1996. *Agricultural Conservation Program — 1995 Fiscal Year Statistical Summary.* U.S. Department of Agriculture, Farm Service Agency, Washington, DC.

USDA–NRCS. 1977. *National Handbook of Conservation Practices.* Natural Resources Conservation Service (formerly Soil Conservation Service), U.S. Department of Agriculture, Washington, DC.<http://www.ncg.nrcs.usda.gov/index.html>

USDA–NRCS. 1996a. *Irrigation Training Toolbox.* National Employment Development Center, Natural Resources Conservation Service, Fort Worth, TX.

USDA–NRCS. 1996b. *Part 600 - National Water Quality Handbook, National Handbook of Water Quality Monitoring,* U.S. Department of Agriculture, Natural Resources Conservation Service, Washington, DC, 128 pp. USDA-NRCS 450-vi-NHWQM. http://www.wcc.nrcs.usda.gov/water/quality/common/wqm1.pdf.

USDA–NRCS. 1997a. *National Engineering Handbook.* Part 652, Irrigation Guide, USDA Natural Resources Conservation Service, Washington, DC. 710 pp.

USDA–NRCS. 1997b. *National Range and Pasture Handbook.* USDA-NRCS Grazing Lands Technology Institute, Fort Worth, TX, 449 pp.

USDA-NRCS. 1997c. Conservation Practice Physical Effect Worksheet, Practice Code 606, US Department of Agriculture - Natural Resources Conservation Service.

USDA–NRCS. 1998. *Stream Visual Assessment Protocol.* NWCC Technical Note 99-1. National Water and Climate Center, USDA — Natural Resources Conservation Service, Portland, OR.

USDA–SCS. 1970. *Irrigation Water Requirements.* Soil Conservation Service Technical Release 21.

USDA–SCS. 1984. *Engineering Field Manual.* U.S. Department of Agriculture, Soil Conservation Service, Washington, D.C.

USDA–SCS. 1993. Irrigation Water Requirements, Chapter 2, part 623 of the *National Engineering Handbook*, Soil Conservation Service.

USDA-SCS. 1994. *Soil Erosion by Wind.* Agriculture Information Bulletin Number 555.

USDOC and EPA. 1993. *Coastal Nonpoint Pollution Control Program: Program Development and Approval Guidance*, National Oceanic and Atmospheric Administration, U. S. Department of Commerce and Office of Water, Washington, DC, 78 pp.

USDOI–BLM. 1997. *Effective Cattle Management in Riparian Zones: A Field Survey and Literature Review*, Riparian Tech Bulletin #3, U.S. Department of Interior, Montana BLM, November.

USDOI–BLM. 1998. *Successful Strategies for Grazing Cattle in Riparian Zones*, Riparian Tech Bulletin #4, U.S. Department of Interior, Montana BLM, January.

USDOI-BLM, USDA-Forest Service, and USDA-NRCS. 1998. *A User Guide to Assessing Proper Functioning Condition and the Supporting Science for Lotic Areas.*

USDOI-BLM, USDA-Forest Service, and USDA-NRCS. 1999. *A User Guide to Assessing Proper Functioning Condition and the Supporting Science for Lentic Areas.*

USDOI-BLM and USGS, and USDA-NRCS and ARS. 2000. *Interpreting Indicators of Rangeland Health.* Technical Reference 1734-6.

USGS. 1999. *The Quality of our Nation's Waters: Nutrients and Pesticides.* U.S. Geological Survey, Circular 1225. 82pp.

Valiela, I., M. Alber, and M. LaMontagne. 1991. Fecal Coliform loadings and stocks in Buttermilk Bay, Massachusetts, USA, and management implications. *Environmental Management* 15(5):659-674.

Van Poollen, H.W. and J.R. Lacey. 1979. Herbage response to grazing systems and stocking intensities. *J. Range Mgt.* 32(4):250-253.

Vermont Agency of Natural Resources. 1996. *State of Vermont 1996 water quality assessment 305(B) report*, Department of Environmental Conservation, Water Quality Division, Waterbury, VT, 308 pp.

Vermont Department of Agriculture. 1995. *Vermont Agriculture Nonpoint Source Pollution Reduction Program Law and Regulation.* Montpelier, VT. 21 pp.

Vermont Department of Environmental Conservation and New York State Department of Environmental Conservation. 1997. *A Phosphorus Budget, Model, and Load Reduction Strategy for Lake Champlain. Lake Champlain Diagnostic-Feasibility Study Final Report*, VT Dept. Environmental Conservation, Waterbury, VT.

Vollenweider, R.A. 1968. *Scientific Fundamentals of the Eutrophication of Lakes and Flowing Waters, with Particular Reference to Nitrogen and Phosphorus as Factors in Eutrophication.* OECD Report No. DAS/CSI/68.27, OECD, Paris.

Wall, D.B., S.A. McGuire, and J.A. Magner. 1989. *Water Quality Monitoring and Assessment in the Garvin Brook Rural Clean Water Project Area.* Minnesota Pollution Control Agency, St. Paul, MN.

Wang, G., T. Zhao, and M.P. Doyle. 1996. Fate of enterohemorrhagic *Escherichia coli* 0157:H7 in bovine feces. *Appl. Environ. Microbiol.* 62(7):2567-2570.

Wang, P., L.L. Linker, and J. Storrick. 1997. Chesapeake Bay Watershed Model, Application and Calculation of Nutrient and Sediment Loadings, Appendix D. Phase IV Chesapeake Bay Watershed Model Precipitation and Meteorological Data Development and Atmospheric Nutrient Deposition. Chesapeake Bay Program, Annapolis, MD.

Ward, R.C., J.C. Loftis, and G.B. McBride. 1990. *Design of water quality monitoring systems.* Van Nostrand Reinhold Company, New York.

Washington State University Cooperative Extension. 1995. Irrigation Management Practices to Protect Ground Water and Surface Water Quality - State of Washington. WSU Cooperative Extension Publication, EM 4885.

Watson, J., L. Hardy, T. Cordell, S. Cordell, E. Minch, and C. Pachek. 1995. *How water moves through soil: a guide to the video*. Cooperative Extension College of Agriculture, The University of Arizona, Tucson, AZ. 16 pp.

Webster, C.P. and K.W.T. Goulding. 1995. Effect of one year rotational set-aside on immediate and ensuing nitrogen leasing loss. Plant and Soil 177(2):203-209.

Webster, E.P. and D.R. Shaw. 1996. Impact of vegetative filter strips on herbicide loss in runoff from soybean (Glycine max). *Weed Sci.* 44:662-671.

Werner, H. 1992. *Measuring soil moisture for irrigation water management.* SDSU Estension Service. Brookings, SD. FS876.

Westerman, P.W., L.M. Safley, J.C. Barker, and G.M. Chescheir. 1985. Available nutrients in livestock waste. In: *Proceedings of the Fifth International Symposium on Agricultural Wastes, Agricultural Waste Utilization and Management.* American Society of Agricultural Engineers, St. Joseph, MI, pp. 295-307.

White, L.D. 1995. *Do you have enough forage?* L-5141. Texas Agricultural Extension Service, Texas A & M University, College Station, TX, 4 pp.

Winward, A.H. 2000. *Monitoring the Vegetation Resources in Riparian Areas.* USDA-Forest Service. Rocky Mountain Research Station. General Technical Report RMRS-GTR-47.

Wood, C.W., G. L. Mullins, and B. F. Hajek. 1998. *Phosphorus in Agriculture.* Soil Quality Institute Technical Pamphlet No. 2. Soil Quality Institute. Auburn, AL.

Wood, C.W. and J.A. Hattey. 1995. Impacts of long-term manure applications on soil chemical, microbiological, and physical properties. In: *Animal Waste and the Land-Water Interface*, Kenneth Steele, Ed., Lewis Publishers, Boca Raton, FL, 589 pp.

Workman, J.P. and J.F. Hooper. 1968. Preliminary economic evaluation of cattle distribution practices on mountain rangelands. *J. Range Mgt.* 21(3):301-304.

Wright, Peter E. 1996. Prevention, collection and treatment of concentrated pollution sources on farms. In: *Animal Agriculture and the Environment: Nutrients, Pathogens and Community Relations.* Northeast Regional Agricultural Engineering Service, Ithaca, NY. NRAES-96. pp 142-158.

Wylie, B.K., M.J. Shaffer, M.K. Brodohl, D. Dubois, and D.G. Wagner. 1994. Predicting Spatial Distributions of Nitrate Leaching in Northeastern Colorado. *J. Soil and Water Conservation* 49:288-293.

Wylie, B.K., M.J. Shaffer, and M.D. Hall. 1995. Regional Assessment of NLEAP NO3-N Leaching Indices. *Water Resources Bulletin* 31:399-408.

Yankey, R., S. Potter, T. Maret, J. McLaughling, D. Carter, and C. Brockway. 1991. *Rock Creek Rural Clean Water Program ten year report.* USDA-Soil Conservation Service, Twin Falls, ID.

Young, R.A., C.A. Onstad, D.D. Bosch, and W.P. Anderson. 1994. *Agricultural NonPoint Source Pollution Model, Version 4.03 AGNPS User's Guide.* USDA Agricultural Research Service, Morris, MN.

Zacharias, S., C. Heatwole, T. Dillaha, and S. Mostighimi. 1992. *Evaluation of GLEAMS and PRZM for Predicting Pesticide Leaching Under Field Conditions.* ASAE paper 92-2541. American Society of Agricultural Engineers, St. Joseph, MI.

Zebarth, B.J. and E. DeJong. 1989. Water flow in a hummocky landscape in central Saskatchewan, Canada, III. Unsaturated flow in relation to topography and land use. J Hydrology 110(1/2):199-218.

Zeneca Ag Products. 1994. *Conservation farming - a practical handbook for cotton growers,* Zeneca, Inc., USA, 42 pp.

Zucker, L.A. and L.C. Brown (Eds.). 1998. *Agricultural Drainage: Water Quality Impacts and Subsurface Drainage Studies in the Midwest.* Ohio State University Extension Bulletin 871. The Ohio State University.

🔲🔲🔲🔲🔲

🔲🔲🔲🔲🔲🔲 🔲🔲 🔲🔲🔲 🔲🔲🔲🔲🔲 🔲🔲🔲 🔲🔲🔲🔲🔲 🔲🔲🔲🔲🔲🔲🔲🔲🔲 🔲🔲🔲🔲🔲 🔲🔲🔲🔲

Best management practices mentioned in this guidance are listed in alphabetical order below. This is not a complete list of all the management practices for agricultural nonpoint source pollution control; there are others that may be in use or are under development. The NRCS or other code number, if any, is given for each BMP, followed by a short definition. Additional explanatory text about selected BMPs is presented in italicized text below the practice, code, and definition.

Access Road (560): A travelway constructed as part of a conservation plan.

Animal Trails and Walkways (575): A livestock trail or walkway constructed to improve grazing distribution and access to forage and water.

Bedding (310): Plowing, blading, or otherwise elevating the surface of flat land into a series of broad, low ridges separated by shallow, parallel channels.

Brush Management (314): Removal, reduction, or manipulation of non-herbaceous plants.

Improved vegetation quality and the decrease in runoff from the practice will reduce the amount of erosion and sediment yield. Improved vegetative cover acts as a filter strip to trap the movement of dissolved and sediment attached substances, such as nutrients and chemicals from entering downstream water courses. Mechanical brush management may initially increase sediment yields because of soil disturbances and reduced vegetative cover. This is temporary until revegetation occurs.

Channel Vegetation (322): Establishing and maintaining adequate plants on channel banks, berms, spoil, and associated areas.

Chiseling and Subsoiling (324): Loosening the soil, without inverting and with a minimum of mixing of the surface soil, to shatter restrictive layers below normal plow depth that inhibit water movement or root development.

Composting Facility (317): A facility for the biological stabilization of waste organic material.

The purpose is to treat waste organic material biologically by producing a humus-like material that can be recycled as a soil amendment and fertilizer substitute or otherwise utilized in compliance with all laws, rules, and regulations.

Conservation Cover (327): Establishing and maintaining perennial vegetative cover to protect soil and water resources on land retired from agricultural production.

Agricultural chemicals are usually not applied to this cover in large quantities and surface and ground water quality may improve where these material are not used. Ground cover and crop residue will be increased with this practice. Erosion and yields of sediment and sediment related stream pollutants should decrease. Temperatures of the soil, surface runoff and receiving water may be reduced. Effects will vary during the establishment period and include increases in runoff, erosion and sediment yield. Due to the reduction of deep percolation, the leaching of soluble material will be reduced, as will be the potential for causing saline seeps. Long-term effects of the practice would reduce agricultural nonpoint sources of pollution to all water resources.

Conservation Crop Rotation (328): An adapted sequence of crops designed to provide adequate organic residue for maintenance or improvement of soil tilth.

This practice reduces erosion by increasing organic matter, resulting in a reduction of sediment and associated pollutants to surface waters. Crop rotations that improve soil tilth may also disrupt disease, insect and weed reproduction cycles, reducing the need for pesticides. This removes or reduces the availability of some pollutants in the watershed. Deep percolation may carry soluble nutrients and pesticides to the ground water. Underlying soil layers, rock and unconsolidated parent material may block, delay, or enhance the delivery of these pollutants to ground water. The fate of these pollutants will be site specific, depending on the crop management, the soil and geologic conditions.

Constructed Wetland (656): A wetland that has been constructed for the primary purpose of water quality improvement.

This practice is applied to treat waste waters from confined animal operations, sewage, surface runoff, milkhouse wastewater, silage leachate, and mine drainage by the biological, chemical and physical activities of a constructed wetland.

Contour Buffer Strips (332): Narrow strips of permanent, herbaceous vegetative cover established across the slope and alternated down the slope with parallel, wider cropped strips.

Contour Farming (330): Farming sloping land in such a way that preparing land, planting, and cultivating are done on the contour. This includes following established grades of terraces or diversions.

This practice reduces erosion and sediment production. Less sediment and related pollutants may be transported to the receiving waters.

Increased infiltration may increase the transportation potential for soluble substances to the ground water.

Contour Orchard and Other Fruit Area (331): Planting orchards, vineyards, or small fruits so that all cultural operations are done on the contour.

Contour orchards and fruit areas may reduce erosion, sediment yield, and pesticide concentration in the water lost. Where inward sloping benches are used, the sediment and chemicals will be trapped against the slope. With annual events, the bench may provide 100 percent trap efficiency. Outward sloping benches may allow greater sediment and chemical loss.

The amount of retention depends on the slope of the bench and the amount of cover. In addition, outward sloping benches are subject to erosion from runoff from benches immediately above them. Contouring allows better access to rills, permitting maintenance that reduces additional erosion. Immediately after establishment, contour orchards may be subject to erosion and sedimentation in excess of the now contoured orchard. Contour orchards require more fertilization and pesticide application than did the native grasses that frequently covered the slopes before orchards were started. Sediment leaving the site may carry more adsorbed nutrients and pesticides than did the sediment before the benches were established from uncultivated slopes. If contoured orchards replace other crop or intensive land use, the increase or decrease in chemical transport from the site may be determined by examining the types and amounts of chemicals used on the prior land use as compared to the contour orchard condition.

Soluble pesticides and nutrients may be delivered to and possibly through the root zone in an amount proportional to the amount of soluble pesticides applied, the increase in infiltration, the chemistry of the pesticides, organic and clay content of the soil, and amounts of surface residues. Percolating water below the root zone may carry excess solutes or may dissolve potential pollutants as they move. In either case, these solutes could reach ground water supplies and/or surface downslope from the contour orchard area. The amount depends on soil type, surface water quality, and the availability of soluble material (natural or applied).

Contour Stripcropping (585): Growing crops in a systematic arrangement of strips or bands on the contour to reduce water erosion. The crops are arranged so that a strip of grass or close-growing crop is alternated with a strip of clean-tilled crop or fallow or a strip of grass is alternated with a close-growing crop.

This practice may reduce erosion and the amount of sediment and related substances delivered to the surface waters. The practice may increase the amount of water that infiltrates into the root zone, and, at the time there is an overabundance of soil water, this water may percolate and leach soluble substances into the ground water.

Controlled Drainage (335): Control of surface and subsurface water through use of drainage facilities and water control structures.

The purpose is to conserve water and maintain optimum soil moisture to (1) store and manage infiltrated rainfall for more efficient crop production; (2) improve surface water quality by increasing infiltration, thereby reducing runoff, which may carry sediment and undesirable chemicals; (3) reduce nitrates in the drainage water by enhancing conditions for denitrification; (4) reduce subsidence and wind erosion of organic soils; (5) hold water in channels in forest areas to act as ground fire breaks; and (6) provide water for wildlife and a resting and feeding place for waterfowl.

Cover Crop (340): A crop of close-growing grasses, legumes, or small grain grown primarily for seasonal protection and soil improvement. It usually is grown for 1 year or less, except where there is permanent cover as in orchards.

Erosion, sediment and adsorbed chemical yields could be decreased in conventional tillage systems because of the increased period of vegetal cover. Plants will take up available nitrogen and prevent its undesired movement. Organic nutrients may be added to the nutrient budget reducing the need to supply more soluble forms. Overall volume of chemical application may decrease because the vegetation will supply nutrients and there may be allelopathic effects of some of the types of cover vegetation on weeds. Temperatures of ground and surface waters could slightly decrease.

Critical Area Planting (342): Planting vegetation, such as trees, shrubs, vines, grasses, or legumes, on highly erodible or critically eroding areas. (Does not include tree planting mainly for wood products.)

This practice may reduce soil erosion and sediment delivery to surface waters. Plants may take up more of the nutrients in the soil, reducing the amount that can be washed into surface waters or leached into ground water.

During grading, seedbed preparation, seeding, and mulching, large quantities of sediment and associated chemicals may be washed into surface waters prior to plant establishment.

Cross Wind Ridges/StripCropping/Trap Strips (589): Ridges formed by tillage or planting, crops grown in strips, or herbaceous cover aligned perpendicular to the prevailing wind direction.

Dikes (356): An embankment constructed of earth or other suitable materials to protect land against overflow or to regulate water.

Where dikes are used to prevent water from flowing onto the floodplain, the pollution dispersion effect of the temporary wetlands and backwater are decreased. The sediment, sediment-attached, and soluble materials being transported by the water are carried farther downstream. The final fate of these materials must be investigated on site. Where dikes are used to retain runoff on the floodplain or in wetlands, the pollution dispersion effects of these areas may be enhanced. Sediment and related materials may be deposited, and the quality of the water flowing into the stream from this area will be improved.

Dikes are used to prevent wetlands and to form wetlands. The formed areas may be fresh, brackish, or saltwater wetlands. In tidal areas, dikes are used to stop saltwater intrusion, and to increase the hydraulic head of fresh water which will force intruded salt water out of the aquifer. During construction there is a potential of heavy sediment loadings to the surface waters. When pesticides are used to control the brush on the dikes and fertilizers are used for the establishment and maintenance of vegetation, there is the possibility for these materials to be washed into the surface waters.

Diversion (362): A channel constructed across the slope with a supporting ridge on the lower side.

This practice will assist in the stabilization of a watershed, resulting in the reduction of sheet and rill erosion by reducing the length of slope. Sediment may be reduced by the elimination of ephemeral and large gullies. This may reduce the amount of sediment and related pollutants delivered to the surface waters.

Fence (382): A constructed barrier to livestock, wildlife, or people.

Fencing is a facilitating practice to implement a prescribed grazing system which would improve vegetation and reduce erosion, sediment and nutrient delivery.

Fencing is a practice that can be on the contour or up and down slope. Often a fence line has grass and some shrubs in it. When a fence is built across the slope, the grasses and shrubs that may line the fence will slow down runoff and cause deposition of coarser grained materials, reducing the amount of sediment delivered downslope. Fencing may protect riparian areas which act as sediment traps and filters along water channels and impoundments.

Livestock have a tendency to walk along fences in search of forage when the grazing land is poorly managed or has inadequate forage. The paths become bare channels which concentrate and accelerate runoff causing a greater amount of erosion within the path and where the path/channel outlets into another channel. This can deliver more sediment and associated pollutants to surface waters. Fencing can have the effect of concentrating livestock in small areas, causing a concentration of manure which may wash off into the stream, thus causing surface water pollution.

Field Stripcropping (586): Growing crops in a systematic arrangement of strips or bands across the general slope (not on the contour) to reduce water erosion. The crops are arranged so that a strip of grass or a close-growing crop is alternated with a clean-tilled crop or fallow.

This practice may reduce erosion and the delivery of sediment and related substances to the surface waters. The practice may increase infiltration and, when there is sufficient water available, may increase the amount of leachable pollutants moved toward the ground water.

Since this practice is not on the contour there will be areas of concentrated flow, from which detached sediment, adsorbed chemicals and dissolved substances will be delivered more rapidly to the receiving waters. The sod strips will not be efficient filter areas in these areas of concentrated flow.

Field Border (386): A strip of perennial vegetation established at the edge of a field by planting or by converting it from trees to herbaceous vegetation or shrubs.

This practice reduces erosion by having perennial vegetation on an area of the field. Field borders serve as "anchoring points" for contour rows, terraces, diversions, and contour strip cropping. By elimination of the practice of tilling and planting the ends up and down slopes, erosion from concentrated flow in furrows and long rows may be reduced. This use may reduce the quantity of sediment and related pollutants transported to the surface waters.

Field windbreak (392): A strip or belt of trees or shrubs established in or adjacent to a field as a barrier to wind.

Filter Strip (393): A strip or area of vegetation for removing sediment, organic matter, and other pollutants from runoff and wastewater.

Filter strips for sediment and related pollutants meeting minimum requirements may trap the coarser grained sediment. They may not filter out soluble or suspended fine-grained materials. When a storm causes runoff in excess of the design runoff, the filter may be flooded and may cause large loads of pollutants to be released to the surface water. This type of filter requires high maintenance and has a relative short service life and is effective only as long as the flow through the filter is shallow sheet flow.

Filter strips for runoff from concentrated livestock areas may trap organic material, solids, materials which become adsorbed to the vegetation or the soil within the filter. Often they will not filter out soluble materials. This type of filter is often wet and is difficult to maintain.

Filter strips for controlled overland flow treatment of liquid wastes may effectively filter out pollutants. The filter must be properly managed and maintained, including the proper resting time. Filter strips on forest land may trap coarse sediment, timbering debris, and other deleterious material being transported by runoff. This may improve the quality of surface water and has little effect on soluble material in runoff or on the quality of ground water.

All types of filters may reduce erosion in the area on which they are constructed. Filter strips trap solids from the runoff flowing in sheet flow through the filter. Coarse-grained and fibrous materials are filtered more efficiently than fine-grained and soluble substances. Filter strips work for design conditions, but when flooded or overloaded they may release a slug load of pollutants into the surface water.

Floodwater Diversion (400): A graded channel with a supporting embankment or dike on the lower side constructed on lowland subject to flood damage.

Forage Harvest Management (511): The timely cutting and removal of forages from the field as hay, greenchop, or ensillage.

Forest Land Erosion Control System (408): Application of one or more erosion control measures on forest land. Erosion control system includes the use of conservation plants, cultural practices, and erosion control structures on disturbed forest land for the control of sheet and rill erosion, gully formation, and mass soil movement.

Grade Stabilization Structure (410): A structure used to control the grade and head cutting in natural or artificial channels.

Where reduced stream velocities occur upstream and downstream from the structure, streambank and streambed erosion will be reduced. This will decrease the yield of sediment and sediment-attached substances. Structures that trap sediment will improve downstream water quality. The sediment yield change will be a function of the sediment yield to the structure, reservoir trap efficiency and of velocities of released water. Ground water recharge may affect aquifer quality depending on the quality of the recharging water. If the stored water contains only sediment and chemical with low water solubility, the ground water quality should not be affected.

Grassed Waterway (412): A natural or constructed channel that is shaped or graded to required dimensions and established in suitable vegetation for the stable conveyance of runoff.

This practice may reduce the erosion in a concentrated flow area, such as in a gully or in ephemeral gullies. This may result in the reduction of sediment and substances delivered to receiving waters. Vegetation may act as a filter in removing some of the sediment delivered to the waterway, although this is not the primary function of a grassed waterway.

Any chemicals applied to the waterway in the course of treatment of the adjacent cropland may wash directly into the surface waters in the case where there is a runoff event shortly after spraying.

When used as a stable outlet for another practice, waterways may increase the likelihood of dissolved and suspended pollutants being transported to surface waters when these pollutants are delivered to the waterway.

Grazing Land Mechanical Treatment (548): Modifying physical soil and/or plant conditions with mechanical tools by treatments such as; pitting, contour furrowing, and ripping or subsoiling.

Heavy Use Area Protection (561): Protecting heavily used areas by establishing vegetative cover, by surfacing with suitable materials, or by installing needed structures.

Protection may result in a general improvement of surface water quality through the reduction of erosion and the resulting sedimentation. Some increase in erosion may occur during and immediately after construction until the disturbed areas are fully stabilized.

Some increase in chemicals in surface water may occur due to the introduction of fertilizers for vegetated areas and oils and chemicals associated with paved areas. Fertilizers and pesticides used during operation and maintenance may be a source of water pollution.

Paved areas installed for livestock use will increase organic, bacteria, and nutrient loading to surface waters. Changes in ground water quality will be minor. Nitrate nitrogen applied as fertilizer in excess of vegetation needs may move with infiltrating waters. The extent of the problem, if any, may depend on the actual amount of water percolating below the root zone.

Hedgerow Planting (422): Establishing a living fence of shrubs or trees in, across, or around a field.

Herbaceous Wind Bathers (422A): Herbaceous vegetation established in rows or narrow strips across the prevailing wind direction.

Hillside Ditch (423): A channel that has a supporting ridge on the lower side constructed across the slope at definite vertical intervals and gradient, with or without a vegetative barrier.

Irrigation Canal or Lateral (320): A permanent irrigation canal or lateral constructed to convey water from the source of supply to one or more farms.

Irrigation Field Ditch (388): A permanent irrigation ditch constructed to convey water from the source of supply to a field or fields in a farm distribution system.

The standard for this practice applies to open channels and elevated ditches of 25 ft³/second or less capacity formed in and with earth materials.

Irrigation field ditches typically carry irrigation water from the source of supplying to a field or fields. Salinity changes may occur in both the soil and water. This will depend on the irrigation water quality, the level of water management, and the geologic materials of the area. The quality of ground and surface water may be altered depending on environmental conditions. Water lost from the irrigation system to downstream runoff may contain dissolved substances, sediment, and sediment-attached substances that may degrade water quality and increase water temperature. This practice may make water available for wildlife, but may not significantly increase habitat.

Irrigation Land Leveling (464): Reshaping the surface of land to be irrigated to planned grades.

The effects of this practice depend on the level of irrigation water management. If plant root zone soil water is properly managed, then quality decreases of surface and ground water may be avoided. Under poor management, ground and surface water quality may deteriorate. Deep percolation and recharge with poor quality water may lower aquifer quality. Land leveling may minimize erosion and when runoff occurs concurrent sediment yield reduction. Poor management may cause an increase in salinity of soil, ground and surface waters. High efficiency surface irrigation is more probable when earth moving elevations are laser controlled.

Irrigation Pit or Regulating Reservoir, Irrigation Pit (552A): A small storage reservoir constructed to regulate or store a supply of water for irrigation.

Irrigation Pit or Regulating Reservoir, Regulating Reservoir (552B): A small storage reservoir constructed to regulate or store a supply of water for irrigation.

Irrigation Storage Reservoir (436): An irrigation water storage structure made by constructing a dam.

Irrigation System, Microirrigation (441): A planned irrigation system in which all necessary facilities are installed for efficiently applying water directly to the root zone of plants by means of applicators (orifices, emitters, porous tubing, or perforated pipe) operated under low pressure (Figure 2-20). The applicators can be placed on or below the surface of the ground (Figure 2-21).

Surface water quality may not be significantly affected by transported substances because runoff is largely controlled by the system components (practices). Chemical applications may be applied through the system. Reduction of runoff will result in less sediment and chemical losses from the field during irrigation. If excessive, local, deep percolation should occur, a chemical hazard may exist to shallow ground water or to areas where geologic materials provide easy access to the aquifer.

Irrigation System, Sprinkler (422): A planned irrigation system in which all necessary facilities are installed for efficiently applying water by means of perforated pipes or nozzles operated under pressure.

Proper irrigation management controls runoff and prevents downstream surface water deterioration from sediment and sediment attached substances. Over irrigation through poor management can produce impaired water quality in runoff as well as ground water through increased percolation. Chemigation with this system allows the operator the opportunity to mange nutrients, wastewater and pesticides. For example, nutrients applied in several incremental applications based on the plant needs may reduce ground water contamination considerably, compared to one application during planting. Poor management may cause pollution of surface and ground water. Pesticide drift from chemigation may also be hazardous to vegetation, animals, and surface water resources. Appropriate safety equipment, operation and maintenance of the system is needed with chemigation to prevent accidental environmental pollution or backflows to water sources.

Irrigation System, Surface and Subsurface (443): A planned irrigation system in which all necessary water control structures have been installed for efficient distribution of irrigation water by surface means, such as furrows, borders, contour levees, or contour ditches, or by subsurface means.

Operation and management of the irrigation system in a manner which allows little or no runoff may allow small yields of sediment or sediment-attached substances to downstream waters. Pollutants may increase if irrigation water management is not adequate. Ground water quality from mobile, dissolved chemicals may also be a hazard if irrigation water management does not prevent deep percolation. Subsurface irrigation that requires the drainage and removal of excess water from the field may discharge increased amounts of dissolved substances such as nutrients or other salts to surface water. Temperatures of downstream water courses that receive runoff waters may be increased. Temperatures of downstream waters might be decreased with subsurface systems when excess water is being pumped from the field to lower the water table. Downstream temperatures should not be affected by subsurface irrigation during summer months if lowering the water table is not required. Improved aquatic habitat may occur if runoff or seepage occurs from surface systems or from pumping to lower the water table in subsurface systems.

Irrigation System, Tailwater Recovery (447): A facility to collect, store, and transport irrigation tailwater for reuse in the farm irrigation distribution system.

The reservoir will trap sediment and sediment-attached substances from runoff waters. Sediment and chemicals will accumulate in the collection facility by entrapping which would decrease downstream yields of these substances.

Salts, soluble nutrients, and soluble pesticides will be collected with the runoff and will not be released to surface waters. Recovered irrigation water with high salt and/or metal content will ultimately have to be disposed of in an environmentally safe manner and location. Disposal of these waters should be part of the overall management plan. Although some ground water recharge may occur, little if any pollution hazard is usually expected.

Irrigation Water Conveyance, Ditch and Canal Lining, Flexible Membrane (428B): A fixed lining of impervious material installed in an existing or newly constructed irrigation field ditch or irrigation canal or lateral.

Irrigation Water Conveyance, Ditch and Canal Lining, Galvanized Steel (428C): A fixed lining of impervious material installed in an existing or newly constructed irrigation field ditch or irrigation canal or lateral.

Irrigation Water Conveyance, Ditch and Canal Lining, Nonreinforced Concrete (428A): A fixed lining of impervious material installed in an existing or newly constructed irrigation field ditch or irrigation canal or lateral.

Irrigation Water Conveyance, High-Pressure, Underground, Plastic (430DD): A pipeline and appurtenances installed in an irrigation system.

Irrigation Water Conveyance, Low-Pressure, Underground, Plastic (430EE): A pipeline and appurtenances installed in an irrigation system.

Irrigation Water Conveyance, Pipeline, Aluminum Tubing (430AA): A pipeline and appurtenances installed in an irrigation system.

Irrigation Water Conveyance, Pipeline, Asbestos-Cement (430BB): A pipeline and appurtenances installed in an irrigation system.

Irrigation Water Conveyance, Pipeline, Nonreinforced Concrete (430CC): A pipeline and appurtenances installed in an irrigation system.

Irrigation Water Conveyance, Pipeline, Reinforced Plastic Mortar (430GG): A pipeline and appurtenances installed in an irrigation system.

Irrigation Water Conveyance, Pipeline, Rigid Gated Pipeline (430HH): A rigid pipeline, with closely spaced gates, installed as part of a surface irrigation system.

Irrigation Water Conveyance, Pipeline, Steel (430FF): A pipeline and appurtenances installed in an irrigation system.

Irrigation Water Management (449): Determining and controlling the rate, amount, and timing of irrigation water in a planned and efficient manner.

Management of the irrigation system should provide the control needed to minimize losses of water, and yields of sediment and sediment-attached and dissolved substances, such as plant nutrients and herbicides, from the system. Poor management may allow the loss of dissolved substances from the irrigation system to surface or ground water. Good management may reduce saline percolation from geologic origins. Returns to the surface water system would increase downstream water temperature.

The purpose is to effectively use available irrigation water supply in managing and controlling the moisture environment of crops to promote the desired crop response, to minimize soil erosion and loss of plant nutrients, to control undesirable water loss, and to protect water quality.

To achieve this purpose the irrigator must have knowledge of (1) how to determine when irrigation water should be applied, based on the rate of water used by crops and on the stages of plant growth; (2) how to measure or estimate the amount of water required for each irrigation, including the leaching needs; (3) the normal time needed for the soil to absorb the required amount of water and how to detect changes in intake rate; (4) how to adjust water stream size, application rate, or irrigation time to compensate for changes in such factors as intake rate or the amount of irrigation runoff from an area; (5) how to recognize erosion caused by irrigation; (6) how to estimate the amount of irrigation runoff from an area; and (7) how to evaluate the uniformity of water application.

Land Reclamation Landslide Treatment (453): Treating inplace materials, mine spoil, mine waste, or overburden to reduce downslope movement.

Lined Waterway or Outlet (468): A waterway or outlet having an erosion-resistant lining of concrete, stone, or other permanent material.

The lined section extends up the side slopes to a designed depth. The earth above the permanent lining may be vegetated or otherwise protected.

This practice may reduce the erosion in concentrated flow areas resulting in the reduction of sediment and substances delivered to the receiving waters.

When used as a stable outlet for another practice, lined waterways may increase the likelihood of dissolved and suspended substances being transported to surface waters due to high flow velocities. A lined waterway may also prevent recharge of the water table as would occur with a natural water body.

Mole Drain (482): An underground conduit constructed by pulling a bullet-shaped cylinder through the soil.

Mulching (484): Applying plant residues or other suitable materials not produced on the site to the soil surface.

Nutrient Management (590): Managing the amount, source, placement, form and timing of applications of nutrients and soil amendments.

Pasture and Hay Planting (512): Establishing native or introduced forage species.

The long-term effect will be an increase in the quality of the surface water due to reduced erosion and sediment delivery. Increased infiltration and subsequent percolation may cause more soluble substances to be carried to ground water.

Pipeline (516): Pipeline installed for conveying water for livestock or for recreation

Pipelines may decrease sediment, nutrient, organic, and bacteria pollution from livestock. Pipelines may afford the opportunity for alternative water sources other than streams and lakes, possibly keeping the animals away from the stream or impoundment. This will prevent bank destruction with resulting sedimentation, and will reduce animal waste deposition directly in the water. The reduction of concentrated livestock areas will reduce manure solids, nutrients, and bacteria that accompany surface runoff.

Pond (378): A water impoundment made by constructing a dam or an embankment or by excavation of a pit or dugout.

Ponds may trap nutrients and sediment which wash into the basin. This removes these substances from downstream. Chemical concentrations in the pond may be higher during the summer months. By reducing the amount of water that flows in the channel downstream, the frequency of flushing of the stream is reduced and there is a collection of substances held temporarily within the channel. A pond may cause more leachable substances to be carried into the ground water.

Precision Land Forming (462): Reshaping the surface of land to planned grades.

Prescribed Burning (338): Applying controlled fire to predetermined areas.

When the area is burned in accordance with the specifications of this practice the nitrates with the burned vegetation will be released to the atmosphere. The ash will contain phosphorous and potassium which will be in a relatively highly soluble form. If a runoff event occurs soon after the burn there is a probability that these two materials may be transported into the ground water or into the surface water. When in a soluble state the phosphorous and potassium will be more difficult to trap and hold in place. When done on range grasses the growth of the grasses is increased and there will be an increased tie-up of plant nutrients as the grasses' growth is accelerated.

Prescribed Grazing (528A): The controlled harvest of vegetation with grazing or browsing animals, managed with the intent to achieve a specified objective.

Planned grazing systems normally reduce the system time livestock spend in each pasture. This increases quality and quantity of vegetation. As vegetation quality increases, fiber content in manure decreases which speeds manure decomposition and reduces pollution potential. Freeze-thaw, shrink-swell, and other natural soil mechanisms can reduce compacted layers during the absence of grazing animals. This increases infiltration, increases vegetative growth, slows runoff, and improves the nutrient and moisture filtering and trapping ability of the area.

Decreased runoff will reduce the rate of erosion and movement of sediment and dissolved and sediment-attached substances to downstream water courses. No increase in ground water pollution hazard would be anticipated from the use of this practice.

Increased vegetation slows runoff and acts as a sediment filter for sediments and sediment attached substances, uses more nutrients, and reduces raindrop splash. Adverse chemical effects should not be anticipated from the use of this practice.

Pumped Well Drain (532): A well sunk into an aquifer from which water is pumped to lower the prevailing water table.

Range Planting (Seeding) (550): Establishment of adapted perennial vegetation such as grasses, forbs, legumes, shrubs, and trees.

Increased erosion and sediment yield may occur during the establishment of this practice. This is a temporary situation and sediment yields decrease when reseeded area becomes established. If chemicals are used in the reestablishment process, chances of chemical runoff into downstream water courses are reduced if application is applied according to label instructions. After establishment of the grass cover, grass sod slows runoff, acts as a filter to trap sediment, sediment attached substances, increases infiltration, and decreases sediment yields.

Regulating Water in Drainage Systems (554): Controlling the removal of surface or subsurface runoff, primarily through the operation of water-control structures.

Residue Management (329) (NoTill): Any tillage and planting system in which at least 30 percent of the soil surface is covered by plant residue after planting to reduce soil erosion by water; or, where soil erosion by wind is the primary concern, at least 1,000 pounds per acre of flat small grain residue-equivalent are on the surface during the critical erosion period.

This practice reduces soil erosion, detachment and sediment transport by providing soil cover during critical times in the cropping cycle. Surface residues reduce soil compaction from raindrops, preventing soil sealing and increasing infiltration. This action may increase the leaching of agricultural chemicals into the ground water.

In order to maintain the crop residue on the surface it is difficult to incorporate fertilizers and pesticides. This may increase the amount of these chemicals in the runoff and cause more surface water pollution.

The additional organic material on the surface may increase the bacterial action on and near the soil surface. This may tie-up and then breakdown many pesticides which are surface applied, resulting in less pesticide leaving the field. This practice is more effective in humid regions.

With a no-till operation, generally the only soil disturbance is from a leading coulter, followed by the disk openers. Fertilizer may be injected and applied in a separate operation, including side dressing. The surface applied fertilizers and chemicals are not incorporated and often are not in direct contact with the soil surface. This condition may result in a high surface runoff of pollutants (nutrient and pesticides). Macropores develop under a no-till system. They permit deep percolation and the transmittal of pollutants, both soluble and insoluble to be carried into the deeper soil horizons and into the ground water. If rainfall is relatively light and does not cause rapid runoff, surface applied nutrients and herbicides move into the soil and are no longer subject to surface runoff losses.

Reduced tillage systems disrupt or break down the macropores, incidentally incorporate some of the materials applied to the soil surface, and reduce the effects of wheeltrack compaction. The results are less runoff and less pollutants in the runoff.

Riparian Herbaceous Cover (390): Establishing an area of grasses and/or forbs adjacent to and up-gradient from water bodies.

Riparian Forest Buffer (391A): Establishing an area of trees and or shrubs adjacent to and up-gradient from water bodies.

Rock Barrier (555): A rock retaining wall constructed across the slope to form and support a bench terrace that will control the flow of water and check erosion on sloping land.

Roof Runoff Management (558): A facility for controlling and disposing of runoff water from roofs.

This practice may reduce erosion and the delivery of sediment and related substances to surface waters. It will reduce the volume of water polluted by animal wastes. Loadings of organic waste, nutrients, bacteria, and salts to surface water will be reduced as water is prevented from flowing across concentrated waste areas, barnyards, roads and alleys. Pollution and erosion will be reduced. Flooding may be prevented and drainage may improve.

Runoff Management System (570): A system for controlling excess runoff caused by construction operations at development sites, changes in land use, or other land disturbances.

Seasonal Residue Management (344): Using plant residues to protect cultivated fields during critical erosion periods.

When this practice is employed, raindrops are intercepted by the residue, reducing detachment, soil dispersion, and soil compaction. Erosion may be reduced and the delivery of sediment and associated pollutants to surface water may be reduced. Reduced soil sealing, crusting and compaction allows more water to infiltrate, resulting in an increased potential for leaching of dissolved pollutants into the ground water.

Crop residues on the surface increase the microbial and bacterial action on or near the surface. Nitrates and surface-applied pesticides may be tied-up and less available to be delivered to surface and ground water. Residues trap sediment and reduce the amount carried to surface water. Crop residues promote soil aggregation and improve soil tilth.

Sediment Basin (350): A basin constructed to collect and store debris or sediment.

Sediment basins will remove sediment, sediment-associated materials and other debris from the water which is passed on downstream. Due to the detention of the runoff in the basin, there is an increased opportunity for soluble materials to be leached toward the ground water.

Soil and Crop Water Use Data: From soils information the available water-holding capacity of the soil can be determined along with the amount of water that the plant can extract from the soil before additional irrigation is needed.

Water use information for various crops can be obtained from various USDA publications.

The purpose is to allow the water user to estimate the amount of available water remaining in the root zone at any time, thereby indicating when the next irrigation should be scheduled and the amount of water needed. Methods to measure or estimate the soil moisture should be employed, especially for high-value crops or where the water-holding capacity of the soil is low.

Spring Development (574): Improving springs and seeps by excavating, cleaning, capping, or providing collection and storage facilities.

There will be negligible long-term water quality impacts with spring developments. Erosion and sedimentation may occur from any disturbed areas during and immediately after construction, but should be short-lived. These sediments will have minor amounts of adsorbed nutrients from soil organic matter.

Stream Channel Stabilization (584): Stabilizing the channel of a stream with suitable structures.

Stream Corridor Improvement (interim): Restoration of a modified or damaged stream to a more natural state using bioengineering techniques to protect the banks and reestablish the riparian vegetation.

Stream Crossing (interim): A stabilized area to provide access across a stream for livestock and farm machinery.

The purpose is to provide a controlled crossing or watering access point for livestock along with access for farm equipment, in order to control bank and streambed erosion, reduce sediment and enhance water quality, and maintain or improve wildlife habitat.

Streambank and Shoreline Protection (580): Using vegetation or structures to stabilize and protect banks of streams, lakes, estuaries, or excavated channels against scour and erosion.

Stripcropping, Contour (585): Growing crops in a systematic arrangement of strips or bands on the contour to reduce water erosion. The crops are arranged so that a strip of grass or close-growing crop is alternated with a strip of clean-tilled crop or fallow or a strip of grass is alternated with a close-growing crop.

Structure for Water Control (587): A structure in an irrigation, drainage, or other water management systems that conveys water, controls the direction or rate of flow, or maintains a desired water surface elevation.

Subsurface Drain (606): A conduit, such as corrugated plastic tile, or pipe, installed beneath the ground surface to collect and/or convey drainage water.

Soil water outlet to surface water courses by this practice may be low in concentrations of sediment and sediment-adsorbed substances and that may improve stream water quality. Sometimes the drained soil water is high in the concentration of nitrates and other dissolved substances and drinking water standards may be exceeded. If drainage water that is high in dissolved substances is able to recharge ground water, the aquifer quality may become impaired. Stream water temperatures may be reduced by water drainage discharge. Aquatic habitat may be altered or enhanced with the increased cooler water temperatures.

Surface Drainage Field Ditch (607): A graded ditch for collecting excess water in a field.

From erosive fields, this practice may increase the yields of sediment and sediment-attached substances to downstream water courses because of an increase in runoff. In other fields, the location of the ditches may cause a reduction in sheet and rill erosion and ephemeral gully erosion. Drainage of high salinity areas may raise salinity levels temporarily in receiving waters. Areas of soils with high salinity that are drained by the ditches may increase receiving waters. Phosphorus loads resulting from this practice may increase eutrophication problems in ponded receiving waters. Water temperature changes will probably not be significant. Upland wildlife habitat may be improved or increased although the habitat formed by standing water and wet areas may be decreased.

Surface Drainage, Main or Lateral (608): An open drainage ditch constructed to a designed size and grade.

Surface Roughening (609): Roughening the soil surface by ridge or clod-forming tillage.

Terrace (600): An earthen embankment, a channel, or combination ridge and channel constructed across the slope.

This practice reduces the slope length and the amount of surface runoff which passes over the area downslope from an individual terrace. This may reduce the erosion rate and production of sediment within the terrace interval. Terraces trap sediment and reduce the sediment and associated pollutant content in the runoff water which enhance surface water quality. Terraces may intercept and conduct surface runoff at a nonerosive velocity to stable outlets, thus, reducing the occurrence of ephemeral and classic gullies and the resulting sediment. Increases in infiltration can cause a greater amount of soluble nutrients and pesticides to be leached into the soil. Underground outlets may collect highly soluble nutrient and pesticide leachates and convey runoff directly to an outlet. Terraces may increase the delivery of pollutants to surface waters. Terraces increase the opportunity to leach salts below the root zone in the soil. Terraces may have a detrimental effect on water quality if they concentrate and accelerate delivery of dissolved or suspended nutrient, salt, and pesticide pollutants to surface or ground waters.

Tree/Shrub Establishing (612): To establish woody plants by planting or seeding.

Trough or Tank (614): A trough or tank, with needed devices for water control and waste water disposal, installed to provide drinking water for livestock.

By the installation of a trough or tank, livestock may be better distributed over the pasture, grazing can be better controlled, and surface runoff reduced, thus reducing erosion. By itself this practice will have only a minor effect on water quality; however when coupled with other conservation practices, the beneficial effects of the combined practices may be large. Each site and application should be evaluated on its own merits.

Use Exclusion (472): Excluding livestock from an area not intended for grazing.

Livestock exclusion may improve water quality by preventing livestock from being in the water or walking down the banks, and by preventing manure deposition in the stream. The amount of sediment and manure may be reduced in the surface water. This practice prevents compaction of the soil by livestock and prevents losses of vegetation and undergrowth. This may maintain or increase evapotranspiration. Increased permeability may reduce erosion and lower sediment and substance transportation to the surface waters. Shading along streams and channels resulting from the application of this practice may reduce surface water temperature.

Waste Management System (312): A planned system in which all necessary components are installed for managing liquid and solid waste, including runoff from concentrated waste areas, in a manner that does not degrade air, soil, or water resources.

Waste Storage Facility (313): A waste storage impoundment made by constructing an embankment and/or excavating a pit or dugout, or by fabricating a structure.

This practice may reduce the nutrient, pathogen, and organic loading to surface waters. This is accomplished by intercepting and storing the polluted runoff from manure stacking areas, barnyards and feedlots.

Waste Treatment Lagoon (359): An impoundment made by excavation or earth fill for biological treatment of animal or other agricultural wastes.

This practice may reduce polluted surficial runoff and the loading of organics, pathogens, and nutrients into the surface waters. It decreases the nitrogen content of the surface runoff from feedlots by denitrification. Runoff is retained long enough that the solids and insoluble phosphorus settle and form a sludge in the bottom of the lagoon. There may be some seepage through the sidewalls and the bottom of the lagoon. Usually the long-term seepage rate is low enough, so that the concentration of substances transported into the ground water does not reach an unacceptable level.

Waste Utilization (633): Using agricultural wastes or other wastes on land in an environmentally acceptable manner while maintaining or improving soil and plant resources.

Waste utilization helps reduce the transport of sediment and related pollutants to the surface water. Proper site selection, timing of application and rate of application may reduce the potential for degradation of surface and ground water. This practice may increase microbial action in the surface layers of the soil, causing a reaction which assists in controlling pesticides and other pollutants by keeping them in place in the field.

Mortality and other compost, when applied to agricultural land, will be applied in accordance with the nutrient management measure. The composting facility may be subject to State regulations and will have a written operation and management plan if SCS practice 317 (composting facility) is used.

Water and Sediment Control Basin (638): An earthen embankment or a combination ridge and channel generally constructed across the slope and minor watercourses to form a sediment trap and water detention basin.

The practice traps and removes sediment and sediment-attached substances from runoff. Trap control efficiencies for sediment and total phosphorus that are transported by runoff may exceed 90 percent in silt loam soils. Dissolved substances, such as nitrates, may be removed from discharge to downstream areas because of the increased infiltration. Where geologic condition permit, the practice will lead to increased loadings of dissolved substances toward ground water. Water temperatures of surface runoff, released through underground outlets, may increase slightly because of longer exposure to warming during its impoundment.

Water Table Control (641): Water table control through proper use of subsurface drains, water control structures, and water conveyance facilities for the efficient removal of drainage water and distribution of irrigation water.

The water table control practice reduces runoff, therefore downstream sediment and sediment-attached substances yields will be reduced. When drainage is increased, the dissolved substances in the soil water will be discharged to receiving water and the quality of water reduced. Maintaining a high water table, especially during the nongrowing season, will allow denitrification to occur and reduce the nitrate content of surface and ground water by as much as 75 percent. The use of this practice for salinity control can increase the dissolved substance loading of downstream waters while decreasing the salinity of the soil. Installation of this practice may create temporary erosion and sediment yield hazards but the completed practice will lower erosion and sedimentation levels. The effect of the water table control of this practice on downstream wildlife communities may vary with the purpose and management of the water in the system.

Waterspreading (640): Diverting or collecting runoff from natural channels, gullies, or streams with a system of dams, dikes, ditches, or other means, and spreading it over relatively flat areas.

Water Well (642): A well constructed or improved to provide water for irrigation, livestock, wildlife, or recreation.

The location of the well must consider the natural water quality and the hazards of its use in potentially contaminating the environment. Hazards exist during well development and its operation and maintenance. Care must be taken to prevent contamination of the aquifer from back flushing, accident, or flow down the annular spacing between the well casing and the bore hole.

Water-Measuring Device: An irrigation water meter, flume, weir, or other water-measuring device installed in a pipeline or ditch.

The measuring device must be installed between the point of diversion and water distribution system used on the field. The device should provide a means to measure the rate of flow. Total water volume used may then be calculated using rate of flow and time, or read directly, if a totalizing meter is used.

The purpose is to provide the irrigator the rate of flow and/or application of water, and the total amount of water applied to the field with each irrigation.

Wetland Wildlife Habitat Management (644): Creating, maintaining, or enhancing wetland habitat for desired wildlife species.

Wetland Restoration (657): A rehabilitation of a drained or degraded wetland where the soils, hydrology, vegetative community, and biological habitat are returned to the natural condition to the extent practicable.

Wildlife Upland Habitat Management (645): Creating, maintaining, or enhancing upland habitat for desired wildlife species.

Windbreak/Shelterbelt Establishment (380): Linear plantings of single or multiple rows of trees or shrubs established next to farmstead, feedlots, and rural residences as a barrier to wind.

Windbreak/Shelterbelt Renovation (650): Restoration or preservation of an existing windbreak, including widening, replanting, or replacing trees.

The Natural Resources Conservation Service (NRCS) Field Office Technical
Guide (FOTG) (www.nrcs.usda.gov/technical/efotg/) is a compilation of resource
information about soil, water, air, plant, animal, and socio-economic resources in
each local field office area. It also contains other conservation planning aides,
including standards and specifications for conservation practices that are applicable in the local area.

The driving concept behind the FOTG is that effective conservation must
recognize the inherent variability of natural resources across the land. Each
FOTG represents a continuing commitment of NRCS to provide its field office
professionals with science and technologies that are tuned to resources they will
encounter in their work. Because there are many factors to be considered
through the NRCS conservation planning process, regardless of program or
purpose, the FOTG provides the place to go for those considerations.

The FOTG is a key part of the materials needed to carry out NRCS' technical
assistance. The National Planning Procedures Handbook, NRCS' technical
handbooks and manuals, and the FOTG provide the basic framework for doing
high quality conservation planning assistance.

FOTG is a work continually in progress. Because our professional needs change,
our conservation programs change, our information technologies change, and
our knowledge of resources grows, we know that the FOTG is dynamic.

The FOTG and Conservation Planning:

Conservation planning and the FOTG go hand in hand. Conservation planning is
the vehicle we use to deliver technical information then allows clients to sustain
the productive use of the natural resources they manage. At the same time, feedback from conservation planning, application, and evaluation efforts helps expand
the quantity and improve the quality of the technical material found in the FOTG.

Conservation planning is the cornerstone of the technical work NRCS does with
clients, groups, and conservation partners. It is an integrated, systematic way of
utilizing technical information and knowledge to help people address resource
problems and opportunities.

National Conservation Practice Standards Subcommittee:

The National Conservation Practice Standards Subcommittee (NCPSS) is a
function of the National Technical Guide Committee. It exists to coordinate
development and review of national level practice standards; and, it publishes
those national standards in the NRCS National Handbook of Conservation
Practices. NCPSS does not make selection of practice standards for inclusion in
the FOTG. State Conservationists, through their state-level technical guide
committees, direct which national practices are selected for inclusion in FOTGs
in their respective states. Those state-level selections are made with needs of
each field office in mind.

Selection of national practices for inclusion does not end the process. In most, if not all cases, national practice standards are too general for application through NRCS assistance. There are technical processes, procedures from handbooks and manuals, and other details to be added. State laws and local ordinances may impose performance criteria that must be addressed, too. NRCS state-level and other technical specialists (including NRCS field personnel) may be called upon to adapt the national practice standard and to develop the practice specifications.

Since 1996, state practices that are used with highly erodible land or in wetland programs are required to have public review prior to their placement in the FOTG. This is a requirement of the 1996 Farm Bill. This process is undergoing review along with other parts of NRCS' FOTG policy in order to make it more responsive to field needs.

After all these activities and reviews, the practice standard (and its specifications) are ready for inclusion in the field office FOTG. It is a process that ensures that the technical guidance each standard and specification provides is pertinent to field office conditions.

FOTG Contents:

Section I: General Resource References

Section I lists references and other information for use in understanding natural resources of the field office service area or in making decisions about resource use and management systems. The actual references listed are to be filed, to the extent possible, in the same location as the FOTG. Computer-based tools used in resource analysis and modeling will be listed in Section I. References kept in other locations will be cross-referenced. Examples include texts and publications dealing with databases found in Section II (below) as well as other resource issues.

Section II: Natural Resources Information

Section II contains natural resource data, databases, and procedures for interpretation. These may include Ecological Site Descriptions and Forage Suitability Group Descriptions. This section will have a statement indicating exactly what is used as the "official" copy of the Soil Survey. In some cases separate statements may be needed for maps, tables, and data.

Section III: Resource Management Systems and Quality Criteria

Resource Management Systems (RMS) will address all identified resource concerns at or above the level of sustainability, taking into account human-cultural, economic and social concerns relative to the Soil, Water,

Air, Plant, and Animal natural resources. Quality Criteria for treatment required to achieve a RMS will be established by NRCS and filed in this section of the FOTG. Criteria shall be stated in either qualitative or quantitative terms for each resource consideration. Where national criteria have not been established, the State Conservationist will establish criteria. Where State and/or local regulations establish more restrictive criteria, these must be used in developing the RMSs.

Section IV: Practice Standards, Specifications and Supplements

Section IV of the FOTG contains conservation practice standards applicable in that field office. Practice standards contain minimum quality criteria for each practice while the specifications describe requirements necessary to install the practice. Supplements add new information as it becomes available. It may also include specifications guide sheets developed for use with the standards.

Section V: Conservation Effects

Conservation effects provide indicators of the impacts conservation practices and systems have on the natural and cultural resources. They are based primarily on empirical data and field experience with practices and systems of practices. The effects are listed for each individual practice. States may provide hardcopy effects or refer the user to the Conservation Effects data. The effects of systems can be estimated by evaluating the combined effects of practices included in a specific system. When properly planned and applied, systems of conservation practices are generally complimentary and accumulative. When conservation practices are installed, the effects on all natural resources are considered.